Disaster Governance in Urbanising Asia

Michelle Ann Miller • Mike Douglass
Editors

Disaster Governance in Urbanising Asia

Springer

Editors
Michelle Ann Miller
Asia Research Institute, National University of Singapore
Singapore, Singapore

Mike Douglass
Asia Research Institute, National University of Singapore
Singapore, Singapore

ISBN 978-981-10-6121-9 (PB)
ISBN 978-981-287-648-5 (HB)
ISBN 978-981-287-649-2 (eBook)
DOI 10.1007/978-981-287-649-2

Library of Congress Control Number: 2015955133

Springer Singapore Heidelberg New York Dordrecht London

Photograph was taken by 'Afriadi Hikmal' but rights of the photograph belong to the editors.
Cover illustration: Springer Cover Image by Afriadi Hikmal

Printed on acid-free paper

Springer Science+Business Media Singapore Pte Ltd. is part of Springer Science+Business Media (www.springer.com)

Acknowledgements

Two funding sources enabled the production of this book. The Asia Research Institute of the National University of Singapore made possible the *International Conference on Disaster Governance: The Urban Transition in Asia* (07–08 November 2013), which provided the initial impetus for producing this edited collection through the conversations generated by the event as well as through the written contributions of selected presenters. The project also benefited from the financial support of a Singapore Ministry of Education Academic Research Fund Tier 2 grant entitled 'Governing Compound Disasters in Urbanising Asia' (MOE2014-T2-1-017). For their overall administration of this project, we thank Valerie Yeo, Sharon Ong, Jonathan Lee, Henry Kwan and Kristy Won. For their constructive comments on earlier iterations of the chapters in this book, we are grateful to Robin Bush, Helen James, Rita Padawangi, Andrew Rumbach and Jerome Whitington. Thank you also to the anonymous reviewers who helped to improve the quality of this manuscript and to Jayanthie Krishnan, Vishal Daryanomel and the team at Springer Publishing for seeing this volume through to fruition. For their invaluable editorial support, we are indebted to Tharuka Prematillake, Saharah Abubakar and Wajihah Hamid.

Contents

1 **Disaster Governance in an Urbanising World Region**........................ 1
Michelle Ann Miller and Mike Douglass

2 **The Urban Transition of Disaster Governance in Asia**....................... 13
Mike Douglass

3 **Building Cities in a Subduction Zone: Some Indonesian Dangers**..... 45
Anthony Reid

4 **Muddy Resistance: Community Empowerment in Mudflow Disaster Governance in Porong, Sidoarjo, Indonesia**......................... 61
Rita Padawangi

5 **Disaster Governance in War-Torn Societies: Tsunami Recovery in Urbanising Aceh and Sri Lanka**... 85
Malin Åkebo

6 **Disaster Governance in Small Urban Places: Issues, Trends, and Concerns**.. 109
Andrew Rumbach

7 **'Good' Time for Disaster: The Importance of Temporality in Governance Thinking**.. 127
Zuzana Hrdličková

8 **Intergenerational Transmission of Local Knowledge Towards River Flooding Risk Reduction and Adaptation: The Experience of Dagupan City, Philippines**..................................... 145
Fatima Gay J. Molina

9 **Securing the Safety of Informal Settler Families Along Waterways in Metro Manila, Philippines: Government-Civil Society Organisation Partnership**......................... 177
Benigno C. Balgos

10 Bangkok and the Floods of 2011: Urban Governance and the Struggle for Democratisation 195
Alin Chintraruck and John Walsh

11 Is It Possible to Integrate Disaster Governance into Urbanization? Evidence from Chinese Townships Hit by 2008 Wenchuan Earthquake and 2013 Lushan Earthquake 211
Qiang Zhang, Yameng Hu, and Qibin Lu

12 Post-Disaster Reconstruction Models: The Governance of Urban Disasters in China, Iran and Myanmar 237
Helen James

13 Nuclear Risk Governance in Japan and the Fukushima Triple Disaster: Lessons Unlearned 263
Pablo Figueroa

Index 283

About the Contributors

Malin Åkebo is a lecturer at the Department of Political Science, Umeå University. Her research interests are mainly within the field of peace and conflict studies with an empirical focus on developments in the Asian region, in particular in Indonesia's province of Aceh and in Sri Lanka. In 2013 she defended her dissertation 'The Politics of Ceasefires: On Ceasefire Agreements and Peace Processes in Aceh and Sri Lanka' (Umeå University, 2013), where she analysed ceasefire agreements and their relation to peace processes in protracted intrastate conflicts. Her recent publications include 'International Dimensions of Peace Processes in Aceh and Sri Lanka: The Role of Intermediaries in the 2000s' (*The Security-Development Nexus*, Anthem Press, 2012) and 'The Role of External Actors in Managing Peace Processes in Asia: An Overview of Attempts in Aceh, Mindanao and Sri Lanka' (*Conflict Management and Dispute Settlement in East Asia*, Ashgate, 2011). Malin has taught various courses in peace and conflict studies and in political science, including on conflict management, peace-building and democratisation processes in societies recovering from crises.

Benigno C. Balgos heads the Research, Knowledge Exchange and Management (RKEM) unit of the Center for Disaster Preparedness, the Philippines, and teaches at the University of the Philippines in Diliman. He has written book chapters and journal articles on disaster risk reduction, particularly in Indonesia and the Philippines. He has been involved in research projects with development agencies such as the United Nations Economic and Social Commission for Asia and the Pacific (UNESCAP), AusAID, The Asia Foundation, Plan International, the Institute of Development Studies and the Asian Disaster Preparedness Center. He is interested in comparative analysis of the Philippines and Indonesia in relation to environmental disasters, social protection and risk financing.

Alin Chintraruck is a doctoral candidate at the School of Management, Shinawatra University, Thailand. She is also head of the International Relations Unit of the Department of Groundwater Resources, Ministry of Natural Resources and

Environment, and is a planning and policy analyst in the same department. Her research concerns the prospects for integrated water resources management in Thailand.

Mike Douglass is professor and leader of the Asian Urbanisms Cluster at the Asia Research Institute and professor in the Lee Kuan Yew School of Public Policy, the National University of Singapore, where he teaches and does research on cities in Asia. He was previously professor and chair of the Department of Urban and Regional Planning and Director of the Globalization Research Center at the University of Hawaii. He received his PhD in urban planning from UCLA. He previously taught at the Institute of Social Studies (the Netherlands) and at the School of Development Studies, University of East Anglia (UK). He is the recipient of the Bui Xuan Phai Award given in Vietnam to those who are noted for their contributions to the culture of Hanoi. His recent books include *Globalization, the Rise of Civil Society and Civic Spaces in Pacific Asia Cities*; *Connected Cities: Histories, Hinterlands, Hierarchies and Networks*; and *Building Urban Communities: The Politics of Civic Space in Asia*.

Pablo Figueroa is an assistant professor in the Center for International Education at Waseda University, in Tokyo, Japan. He is interested in contemporary issues of Japanese society and culture, including governance, natural and man-made disasters and perceptions of risk. His paper, 'Risk Communication surrounding the Fukushima Nuclear Disaster: An Anthropological Approach' (Springer-Verlag, 2013), discusses how the Japanese government communicated risk during the catastrophic nuclear meltdowns that followed the Great East Japan Earthquake. Pablo's research currently focuses on health governance of the Fukushima nuclear accident.

Helen James is an associate professor in the Department of Anthropology, School of Culture, History and Language, College of Asia Pacific, the Australian National University (ANU). She took her PhD and MA from the University of Pittsburgh and BA (oriental studies) from the ANU. She is a specialist in Southeast Asia, particularly Myanmar and Thailand. She has published eight books and over 60 articles and chapters in books. She is the principal investigator on an Australian Research Council-funded discovery project on 'The Demographic Consequences of Asian Disasters: Family Dynamics, Social Capital and Migration Patterns' which covers disasters in China, Iran, Myanmar, the Philippines, Indonesia and Taiwan. Her most recent book (with Douglas Paton) is *The Consequences of Asian Disasters: Demographic, Planning and Policy Implications* (C. H. Thomas, Publishers, Illinois, 2014).

Zuzana Hrdličková is a postdoctoral researcher in the Department of Sociology of Goldsmiths College, University of London. She is an anthropologist with a background in South Asian studies and majority of her fieldwork hailing from the region. She is currently working on the European Research Council (ERC)-funded project 'Organizing Disaster: Civil Protection and Population'. She is interested in the anthropology of disaster, conflict and gender as well as STS. Her PhD (2009, Charles University in Prague) was on the impact of war in Sri Lanka on women. She has applied experience in disaster relief and war contexts.

Yameng Hu is a research associate at the School of Social Development and Public Policy of Beijing Normal University. Her research interests revolve around disaster politics, state-civil society relationships, rural public service delivery, social innovation and social entrepreneurship. Previously Yameng worked with the Urban China Initiative, a think tank for design policy frameworks to promote sustainable urbanisation.

Qibin Lu is an assistant professor in the School of Social Development and Public Policy of Beijing Normal University. His research interests have focused on public policy and social protection in relation to natural disasters since the 2008 Wenchuan Earthquake. His previous research concentrated on coordination efforts between government agencies and civil society, as well as the dynamics of cooperation networks among non-governmental organisations and social enterprises.

Michelle Ann Miller is a senior research fellow in the Asian Urbanisms Cluster at the Asia Research Institute, National University of Singapore. She previously taught in the Master of International and Community Development programme at Deakin University and on subjects related to participatory approaches to development at Charles Darwin University. Dr Miller has conducted research in Indonesia for 15 years, focusing particularly on Indonesia's westernmost province of Aceh, but more recently in Yogyakarta and Solo. Her current research investigates the role of decentralised governance in preparing for, responding to and recovering from environmental disasters. She has authored, edited or co-edited a number of books including *Rebellion and Reform in Indonesia: Jakarta's Security and Autonomy Policies in Aceh* (Routledge, 2009), *Autonomy and Armed Separatism in South and Southeast Asia* (ISEAS, 2012), *Ethnic and Racial Minorities in Asia: Inclusion or Exclusion?* (Routledge, 2012), and (with Tim Bunnell) *Asian Cities in an Era of Decentralisation* (Routledge, 2014).

Fatima Gay J. Molina is a senior research associate at the Research, Knowledge Exchange and Management Program of the Center for Disaster Preparedness, Philippines, and a member of the National Anti-Poverty Commission-Victims of Disasters and Calamities sector under the Office of the President of the Philippines. She has been engaged in disaster and climate change practice for research, training and advocacy since 2008 with projects implemented in collaboration with the United Nations Educational, Scientific and Cultural Organization, the United Nations Development Program, Oxfam, the Institute of Development Studies in the University of Sussex, Plan International, Save the Children, World Vision, Japan International Cooperation Agency, Asian Disaster Reduction Center, Asian Disaster Preparedness Center, International Recovery Platform and Asia-Pacific Network for Global Change Research. She holds a diploma on children, youth and development from the International Institute of Social Studies of Erasmus University Rotterdam and is currently completing her MA in anthropology at the University of the Philippines where she also earned her BA in anthropology (cum laude) in 2008.

Rita Padawangi is a senior research fellow of the Asian Urbanisms Cluster at the Asia Research Institute, National University of Singapore. She received her PhD in sociology from Loyola University Chicago where she was also a Fulbright Scholar for her MA studies. She was previously a research fellow at the Lee Kuan Yew School of Public Policy and the Global Asia Institute, National University of Singapore; Center for Urban Research and Learning at Loyola University Chicago; and the Centre for Strategic and International Studies in Jakarta, Indonesia. She has taught at the School of Design and Environment at the National University of Singapore and at the Department of Sociology at Loyola University Chicago, with a special focus on urban sociology and the sociology of the built environment. She holds a bachelor of architecture degree from the Parahyangan Catholic University and was a practising architect in Bandung, Indonesia. Dr Padawangi has conducted research in Southeast Asian cities, including Indonesia, the Philippines and Singapore. Her research theme is 'cities by and for the people', with special focus on public space, social movements and environmental sociology.

Anthony Reid is a Southeast Asian historian, based as emeritus professor at the Australian National University, where he also served as professor of Southeast Asian history for many years before 1999. In between he was founding director of the Center for Southeast Asian Studies at UCLA (1999–2002) and of the Asia Research Institute at NUS, Singapore (2002–2007). He is a fellow of the Australian Academy of the Humanities and a corresponding fellow of the British Academy. He was awarded the Fukuoka Asian Culture Prize in 2002 and the Association of Asian Studies' Distinguished Contributions to Asian Studies Award in 2010. His recent books include *Imperial Alchemy: Nationalism and Political Identity in Southeast Asia* (2010), *To Nation by Revolution: Indonesia in the 20th Century* (2011) and *A History of Southeast Asia* (2015) and as (co-)editor *The New Cambridge History of Islam*, vol. 3 (2011) and *Indonesia Rising: The Repositioning of Asia's Third Giant* (2012).

Andrew Rumbach is an assistant professor in the Department of Planning and Design at the University of Colorado, Denver, USA. His research is broadly concerned with the social and institutional dimensions of disasters and climate change and the relationships between urbanisation and environmental risk. His current projects include a study of new town development and emerging geographies of disaster risk in Calcutta, India; an evaluation of India's National Disaster Management framework and its efficacy across states and territories; an examination of the role of indigenous institutions in tsunami recovery in Samoa and American Samoa; and a history of American planning interventions in India. Andrew holds a PhD in city and regional planning from Cornell University and a BA in political science from Reed College. At UCD he teaches in the areas of environmental planning, disaster management and international development.

John Walsh is director of Shinawatra University's (SIU) Research Centre, Thailand, editor of the *SIU Journal of Management* and assistant professor at the School of Management, Shinawatra University, Thailand. His doctorate was received from the University of Oxford for a thesis related to international management in East Asia. His research mainly focuses on the social and economic development of the *Greater Mekong Sub-Region*, which is also the title of his latest book.

Qiang Zhang is associate dean and associate professor of the School of Social Development and Public Policy at Beijing Normal University, China. He was a Harvard Yenching Visiting Scholar from 2011 to 2012, and he directs the Policy Research Center at the Academy of Disaster Reduction and Emergency Management, which is jointly supported by the Ministry of Civil Affairs and the Ministry of Education of China. Dr Zhang serves as the founder and assistant president of the China Chapter of TIEMS (The International Emergency Management Society), and he is director of the Social Entrepreneurship Lab(seLAB) at Beijing Normal University as well as the general secretary of the China Nonprofit Platform for Disaster Risk Management. Dr Zhang most recently served as an independent consultant for UNDP, UNESCO and UNV. His research interests involve disaster reduction and crisis management, public policy and social innovation. He received his PhD in management from Tsinghua University in Beijing, China.

Chapter 1
Disaster Governance in an Urbanising World Region

Michelle Ann Miller and Mike Douglass

Abstract In the early twenty-first century, Asia's accelerated urban transition is both a major source and target of increasingly frequent and costly environmental disasters. Asia is home to more than half of the global urban population, and its share is increasing. Currently, over 1.5 billion people live in Asia's urban settlements. High rates of rural to urban migration coupled with industrialisation and the advent of automobile societies that are drivers of global climate change and related environmental degradation have amplified the exposure, intensity and human as well as material costs of environmental disasters. The dominant pattern of urban expansion along Asia's coastlines and river deltas is also producing extended urban agglomerations that are at or below sea level, raising the vulnerabilities of their growing populations to floods, storm surges, typhoons and the unpredictable impacts of climate change on local ecologies. As Asia's environmental disasters occur at multiple scales and impact upon urban populations in different ways with unintended and often long-term consequences, a multi-sector and multidisciplinary approach is needed to adequately address the multitude of theoretical and practical dimensions of disaster governance in urbanising Asia.

Keywords Governance • Environmental disasters • Urban • Asia

1.1 Introduction

Dealing with disasters is not a new phenomenon, but disaster studies is still an emerging field that must grapple with a rapidly changing and increasingly unpredictable global environment in which scholarship constantly risks being overtaken by events. At the same time, traditional knowledge systems and trans-generational social mechanisms to cope with environmental risks and hazards are also being displaced and supplanted by new realities. Among the changes is the transition from rural to more urban societies, which is implicated in global climate change

M.A. Miller (✉) • M. Douglass
Asia Research Institute, National University of Singapore, Singapore, Singapore
e-mail: arimam@nus.edu.sg; michaeld@nus.edu.sg

M.A. Miller, M. Douglass (eds.), *Disaster Governance in Urbanising Asia*,
DOI 10.1007/978-981-287-649-2_1

and associated unpredictable weather events that surpass both local experiential understandings and scientific forecasts.

These observations are especially relevant in Asia, which is the most disaster prone[1] and populous region of the world and is also the home of 53 % of the global urban population (World Population Statistics 2014; United Nations 2014, p. 1).[2] Although governance regimes across urbanising Asia are intensifying their efforts to cope with increasingly uncertain conditions, responses have not kept pace with either the magnitude or longer-term implications of extreme environmental disasters that are mounting in severity and timing. New capacities for more effective and inclusive disaster preparation, response and recovery programs are urgently needed. They would require capacities to respond to a diversity of situations in flexibly agile and committed courses of action that are able to extend beyond the disaster event to the social and economic recovery processes that require multiple sources of support beyond physical repair.

Prevailing disaster responses tend towards physical infrastructure solutions and population relocation as management problems are increasingly found to be inadequate in their ability to traverse diverse and complex geographies and changing societies in the context of heightening environmental disasters and urbanisation in Asia. By placing applied research on disasters under the umbrella of governance rather than management, the response in the chapters in this book go beyond existing limitations to improve the understanding of how localised power dynamics associated are socially and spatially uneven. They further explore how entering disaster responses into a more inclusive public sphere of decision-making can better reach poorer and marginalised people who are the most vulnerable to disasters. Disasters do not impact upon societies evenly, and ineffective governance regimes exacerbate social inequality to place significant sections of society at higher risk, especially people living in slums and informal settlements, women, children and the elderly (Hardoy et al. 2004; Kovats and Akhtar 2008).

Another key challenge for urban governance in mitigating and adapting to the growing threats and vulnerabilities of disasters is the difficulty of disentangling the dynamics of ecological disruptions made through human transformations of landscapes and those from seemingly natural sources such as weather events, earthquakes, tsunamis, landslides and coastal erosion. In broader terms, the world has entered into the Anthropocene age of human-generated changes to the Earth's life support systems (Knight and Harrison 2014).[3] With human action now as the dominant force of change in Earth systems, a central feature of the Anthropocene is

---

[1] From all environmental disaster-related fatalities incurred globally between 1970 and 2011, 75 % were in Asia (ESCAP/UNISDR 2012, xxii).

[2] Although Asia is still one of the world's least urbanised regions, its urban population is expected to almost double to reach 2.6 billion by 2030, increasing the urban share of total population from the current 48–55 % (Miller and Bunnell 2014, pp. 1, 6; United Nations 2004; pp. 2–3; United Nations 2014, p. 6).

[3] Earth systems can be defined as 'The totality of systems operating at the Earth's surface' that include 'geomorphological, ecological, hydrological and pedological systems' (Knight and Harrison 2014, p. 2).

that normal science of tracking natural variability is no longer appropriate. Instead, in a new era of 'post-normal science', 'non-linear systems feedbacks are increasingly more important' (Knight and Harrison 2014, p. 2). By extension, research on environmental disasters requires greater site-specific contextual knowledge to go beyond trend analyses to be able to better assess compound impacts of human involvement in Earth systems. Research can no longer be kept only in the domain of natural sciences such as geology; it should now include other fields of knowledge to better enable learning processes that can better understand and anticipate anthropogenic impacts related to disasters in local contexts.

In other words, the need for a de-centred, multi-scalar approach that is multidisciplinary and inclusive of many voices and forms of knowledge is another reason for placing environmental disasters into the realm of governance rather than into silos of expert management. This need is the main conceptual concern of this book, which seeks to interrogate the underlying principles, objectives and outcomes of disaster governance regimes in urbanising Asia in the search for more inclusive and effective policy choices and programs. Disasters always occur in political spaces, which in urbanising societies invariably become spaces fraught with heightened contestation, negotiated compromise and cooperation in times of crisis, rupture and displacement. Our theoretical approach is therefore driven by an emphasis on the political as well as social and economic dimensions of governance that extend beyond generalised operating procedures to encompass the range of actors, processes and structures across formal and informal sectors through which knowledge and information are generated and applied. This wide definition serves as a useful entry point for examining whose knowledge is privileged, how vulnerability is defined and by whom and what priorities and resources are harnessed in the development of policies to deal with environmental disasters.

Our intention in framing this edited collection in terms of a broad and inclusive understanding of disaster governance is to depart from two dominant trends in the extant literature and in mainstream policy practices. As previously noted, one of these trends, which is rooted in traditional disaster studies that grew out of case studies of floods and famines in rural areas, is informed by an environmental determinism that ascribes disasters to natural physical causalities that are beyond human control and therefore necessitate physical science-based interventions to restore manageable relations between society and nature (Pelling 2003). This perspective remains prevalent among city governments in Asia, which tend to rely more heavily on technical solutions to environmental disasters that are viewed as problems for experts to resolve through engineering approaches and an emphasis on physical infrastructure, usually at the expense of local knowledge and social adaptation strategies (Chan et al. 2012).

Referring to disasters as 'natural' has its own political expediency in deflecting attention away from government inadequacies in preparing for them. This tendency is especially problematic in cities where poorer urban settlers locating in high-risk areas are blamed for contributing to disasters. The result is typically one of seeking remedies by removing these people from their settlements to allow for canal dredging or other infrastructure improvements. On both counts these physical and technical approaches have repeatedly proven inadequate. Poorer people tend to relocate in similar degraded areas, and site-specific remedies are unable to keep pace with the increasing frequencies and impacts of disasters that emanate from more systemic forces. For better preparedness and more appropriate responses to be taken, there is thus a pressing need for a paradigmatic shift that recognises human complicity in creating conditions for disasters to occur. Similarly, the narrowness of understanding and responses also risk losing generations of accumulated knowledge and acquired coping strategies that are omitted from official discourses and policy programs.

The second trend from which we depart in this volume is the strong public administration orientation that has characterised much of the recent theorising about disaster governance (e.g. Birkland 2007; Herzog and Austin 2007; Sahni and Medury 2003; Shi et al. 2013). In disaster administration theory, environmental disasters are typically treated as a 'policy domain' within which actors and agencies frame agendas, problematise governance programs and build bridges of routinised responses between localities that are prone to similar types of disaster.[4] Such approaches tend to neglect or overlook critical historical, social and cultural dimensions of environmental disasters in the pursuit of a standardised set of 'lessons learned'. We acknowledge that the experiences of one type of disaster can certainly lend critical insights to another similar empirical context. However, we caution against uncritically presented—and largely depoliticised—'best practices', which assume that certain features of disaster environments are reducible to a set of universalised operating procedures that are prone to cliché in disaster studies and development discourses.

Bearing in mind the many scales, causalities and interconnected impacts of environmental disasters in urbanising Asia, our contributors approach the question of disaster governance with a view to linking knowledge to action in the diverse and often divided populations impacted by such disruptive events. Each of the cases in this collection is therefore concerned with a particular subset of questions associated with the differential vulnerabilities, power relations and scales involved in governing environmental disasters in Asia's cities, as well as in extended urban regions and rural areas that are impacted by the spreading reach of urbanisation and its urban-centred demands and transformations.

The chapters in this book concern the dynamics of disaster governance in urban populations and systems in nine Asian countries: India, China, Japan, Iran, Myanmar, Indonesia, Sri Lanka, the Philippines and Thailand. This selection is not intended to be exhaustive or representative of the entire Asian region. Rather, it reflects a range of individual and comparative cases from within and among nation-states and highlights key dimensions of the diversity in governance regimes for dealing with different forms of disasters in varying sociocultural and environmental contexts. The goal of this compilation, then, is to avoid the sorts of simplistic abstractions that are used to generate the standardised models of universalised governance noted above. While we expect readers to be drawn into particular case

[4] Two notable exceptions to this trend are McEntire (2007) and Wisner et al. (2011).

studies in which they have knowledge or expertise, the broader ambition of the chapters in this book is to collectively challenge the parameters of extant thinking about disaster governance while making important contributions in their own right. In other words, apart from their applied policy relevance to particular peoples and places, the chapters raise awareness of the continuing need to expand our understanding of the interconnected causalities, manifold scales and varied stakeholders involved in environmental disasters. This includes the far-reaching consequences of such events that affect the lives, livelihoods and aspirations of individuals and communities well beyond the immediate stressors created by the disaster itself.

The remainder of this introduction sets the scene for the chapters in this book. We situate our collection of essays within the wider literature on disaster governance. In the process, we highlight a set of more general concerns associated with governing disasters in the rapidly changing societies of urbanising Asia. Following our overview of the individual chapters within this body of literature, we then return to the argument of this volume as a whole in terms of the ways in which the case studies, taken together, demonstrate the sutured nature of environmental disasters in urban systems and the necessity of embracing a multifaceted and inclusive approach to governing environmental disasters. Such an approach, we believe, is key to bringing scholarship into deeper conversation with governance priorities in the ongoing search for more comprehensive and effective multi-sector, multidisciplinary and multi-stakeholder platforms in disaster preparation, response and recovery programs.

1.2 Governing Diverse Urbanising Societies: A Chapter Overview

Collectively, the chapters in this book address critical issues in disaster governance in Asia's heterogeneous urbanising populations. They consider correlations between geographies of socioeconomic inequality and localised power relations in governance regimes that better protect some sections of society from environmental disasters at the neglect and expense of others. The sorts of spatially manifested inequalities and relationships that they describe challenge us to think more widely about the implications of urbanisation for effective governance. This includes how Asia's urban settlements are impacted by and contribute to the expanding ecological reach and demands of cities into rural areas, the escalation of (predominantly rural to urban) internal migration and the increasingly compound nature of environmental disasters that create numerous cascading threats and vulnerabilities at the local, national, regional and even global scales.

These complex challenges for the different scales and shifting dynamics of disaster governance in twenty-first-century Asia are the subject of Chap. 2 of this volume by Mike Douglass. Identifying five types of urban transition taking place across Asia, Douglass makes a strong case for the realignment of managerial and expert-driven approaches to disaster governance towards a more participatory and inclusive framework that encourages wider societal participation at the neighbourhood, city region and transborder riparian region geographical scales. He identifies five types of urban transition effects that necessitate this fundamental change. They are (1) the agglomeration and formation of mega-urban regions, (2) the spatial polarisation of urbanisation in high-risk zones such as coastal regions and major river deltas, (3) new forms and magnitudes of vulnerability in urbanising settlements, (4) the compound character of environmental disasters with multiple causalities and cascading effects and (5) the expanding ecological reach and demands of cities into rural areas and across administrative borders. Taken together, these five types of urban transition at the three spatial scales noted above highlight the multiplicity of ways in which our understanding of and approach to environmental disasters needs to be recalibrated in scholarship and practice. The temporal and spatial complexities of environmental disasters also demand a longitudinal perspective to disaster governance that reflexively considers the compound chain of interdependencies that are set into motion well before the rupture of the disaster event itself.

Like Douglass, Anthony Reid (Chap. 3) argues that there is an imperative to extend our understanding of past disasters to assess their implications for the future and that greater cooperation between the physical sciences, social sciences and policy-makers is critical in this pursuit of translating knowledge more effectively into action. In his role as historian, Reid offers a longitudinal lens for examining environmental disasters in the territories that today constitute the Republic of Indonesia to show how the relative quietude of the twentieth century in terms of frequency of disasters was in hindsight misleading. This is because of the transformation of colonial ports to urban agglomerations in usually endangered coastal locations during this period as well as the vast movement of peoples from rural settlements to coastal cities and their growing exposure to the unintended effects of global climate change. Reid shows how seemingly unprecedented disasters that occurred prior to the twentieth century are situated in longer-term patterns and abrupt moments of societal changes, and how these trends have exacerbated the risks of environmental disasters for urban populations in the twenty-first century.

The remaining chapters in this volume draw from various theoretical and conceptual contributions made in the earlier chapters and use case studies of environmental disasters in the first part of the twenty-first century to make wider contributions to our understanding of disaster governance within and beyond urbanising Asia. Chapters 4 and 5 bring together individual and comparative case studies from Indonesia and Sri Lanka. Writing on the protracted Lapindo mudflow disaster on Indonesia's island of Java, Rita Padawangi (Chap. 4) demonstrates the importance of community resilience strategies centred on adaptive livelihoods rather than on compensation schemes or other temporary interventions in the face of largely unresponsive formal government authorities and institutions. Through her emphasis on grassroots activism in the lifeworlds impacted by the Lapindo mudflow disaster, Padawangi shows how enacting disaster spaces as spheres of everyday life for cultural and economic innovations offers a socially empowering alternative to conventional forms of disaster governance. Rendering spaces impacted by the

Lapindo mudflow as ordinary sites of cultural mobilisation and employment has enabled impacted residents to construct new forms of social identity and longer-term adaptive capacities in spite of their generally adversarial relationship with the state and its big business allies.

Malin Åkebo, in her comparative study of the impact of the 2004 Indian Ocean tsunami on urban and rural populations in Sri Lanka and in Indonesia's westernmost province of Aceh (Chap. 5), takes an in-depth look at the impact of this disaster on pre-existing adversarial relationships between the state and armed separatist insurgents. Prior to the 2004 tsunami, Aceh and Sri Lanka were embroiled in two of Asia's most protracted and bloody secessionist conflicts. Whereas in Aceh, the window of political opportunity created by the tsunami for the warring parties to put aside their differences ushered in the beginning of a successful peace process, in Sri Lanka, it was the reverse as the violence there continued virtually unabated. By mapping the distribution of recovery resources to Acehnese and Sri Lankan cities, towns and villages in the aftermath of the tsunami, Åkebo examines how the logic that framed development agendas and priorities contributed to the cessation of hostilities in some areas while aggravating tensions in other places or sowing the seeds for new forms of social conflict along urban–rural cleavages.

Much of the recent literature on disaster governance in cities has included a disproportionate focus on megacities, often overlooking or neglecting the experiences of smaller urban places (see, e.g. Adikari et al. 2010; Hochrainer and Mechler 2011; Pelling and Blackburn 2014). One reason for this is the strong coastal orientation and associated vulnerabilities of Asia's megacities; some 25 of Asia's 35 largest mega-urban regions with populations exceeding ten million people are located in coastal areas and along major river deltas (Jones and Douglass 2008; McGranahan et al. 2007). Another reason is that the human and material costs of environmental disasters in densely populated and heavily industrialised urban agglomerations are far higher than in less remote and rural areas.

Several of the chapters in this volume offer a complimentary corrective to this trend by examining key issues and concerns for disaster governance in small- and medium-sized urban settlements. In addition to the aforementioned chapters by Rita Padawangi and Malin Åkebo, Andrew Rumbach (Chap. 6) directly addresses the distinction between megacities and secondary cities by identifying four differential characteristics that affect the capacities of the latter to prepare for, respond to and recover from disasters. Illustrating his case through evidence from three landslide-prone secondary cities in West Bengal, India, Rumbach argues that unlike larger cities, small urban populations (a) are geographically, politically and culturally more remote from centres of power and influence; (b) lack proportionate growth in infrastructure and governance capacity to deal with disasters, despite experiencing similar increases in risk accumulation as larger cities; (c) experience rates of urbanisation that exceed and outpace environmental learning and memory; and (d) often lack forms of redundancy that are typically associated with more established and resilient urban systems. In rethinking the scales at which disaster governance functions, Rumbach points out that far more work is needed to understand how city size relates to disaster risk, vulnerability and resilience, as well as disaster governance capacities.

Puducherry, as described by Zuzana Hrdličková (Chap. 7), is an example of an Indian Union Territory comprising small urban settlements and rural villages that emerged as a national innovator in disaster governance through its experiences of the 2004 Indian Ocean tsunami and Cyclone Thane in 2011. This is despite the countervailing wisdom that major urban centres are the engines of progressive approaches to environmental conservation and social justice. Big cities are also typically seen to be better equipped than rural or remote areas to deal with environmental disasters because of their dense concentration of human and material resources and command over the building and distribution of physical infrastructure (Miller and Bunnell 2013). Yet, the Puducherry administration was able to capitalise on the massive influx of funding and humanitarian assistance in the aftermath of these major disasters to develop one of the most efficient and inclusive disaster preparedness, response and recovery programs in India.

Instead of suffering from the 'tyranny of distance' from political and financial centres that plagued the small Indian cities examined by Rumbach, Puducherry transformed itself through disaster into a model for emulation and best practice for other cities and urbanising populations. Moreover, like the case of Aceh described in this volume by Malin Åkebo, Puducherry was able to capitalise on the window of political opportunity created by the rupture of the 2004 tsunami to initiate a series of progressive policy reforms. In this sense, the successful case of Puducherry makes a novel contribution to the emerging literature on the role of (usually major) cities as critical players in setting agendas for adapting to environmental disasters and in influencing meaningful policy choices at higher levels of government (see also Lee and Koski 2014; Schreurs 2010).

Dagupan in the Philippines is another secondary city that is becoming a model of innovation in disaster governance, especially in relation to the protection of significant vulnerable and urban poor populations. Fatima Molina (Chap. 8) explains how a responsive regime of decentralised governance in Dagupan facilitated the effective integration of intergenerational local knowledge systems into official urban disaster governance programs to deal with chronic regular flooding episodes. Molina argues that critical to the successful incorporation of community-level stewardship of the water commons into formal governance structures, there is an awareness and understanding of the fluid and dynamic nature of both local knowledge and scientific approaches to disaster governance over time. It is only by recognising that both forms of knowledge and expertise are nonstatic and transformative that Asia's changing urban populations can build resilience in their mitigation of, and adaptation to, environmental disasters.

From a somewhat different perspective, Benigno Balgos (Chap. 9) proposes that political will by state actors is crucial to the development of inclusive and responsive disaster governance. Citing evidence from flood-prone informal settlements along waterways in the Philippines National Capital Region of Metro Manila, Balgos recognises that many initiatives in dealing with floods are generated from within these poor and marginalised communities themselves. Despite this, Balgos argues that the support of reform-minded government authorities and the mobilisation of political will from above have been necessary to cast aside the stigmas typically associated with informal settlers in order to start to address the multiple risks and vulnerabilities they face through recurrent flooding episodes through collaborative partnerships involving state and non-state actors.

In the case of Thailand's capital of Bangkok, Alin Chintraruck and John Walsh (Chap. 10) explore the ongoing struggle for democratic accommodation within the context of a politically divided society in which the various government agencies responsible for regulating and coordinating water policy have competing agendas driven by fundamentally different objectives. As in Metro Manila, the geographies of risk and vulnerability in Bangkok are simultaneously impacted and exacerbated by entrenched urban governance regimes. Chintraruck and Walsh, in their study of the 2011 floods in Bangkok, highlight the limits of democratisation in a situation where political elite opportunists used the 2011 floods to consolidate their personal power bases while exacerbating the wealth divide between heavily flooded low-income suburbs and the protected affluent inner-city area with its financial district.

The rupture of a disaster often exposes underlying vulnerabilities in the processes of urbanisation, especially as many of Asia's cities are expanding at a rate that exceeds their capacity to build resilience to such events. As towns and cities swell and spatially spread to incorporate peri-urban areas, urbanisation is producing new forms of vulnerability through inappropriate planning extending development into coastal and swamp areas, inland floodplains, unstable hillsides and other lands unsafe for settlement (Jha et al. 2012). The transition from rural to more urban societies in Asia has exacerbated these vulnerabilities as rural–urban migration has fuelled the development of densely populated housing settlements in environmentally degraded locations where disaster risks are magnified. With nearly half-a-billion people in Asia living in slums and their numbers increasing (UNESCAP 2010), the search for more secure, safe and socially just forms of urbanisation can at times appear overwhelming.

This question of whether disasters can trigger the initiation of more sustainable forms of urbanisation in rapidly expanding populations is the subject of Chap. 11 in this volume by Qiang Zhang, Yameng Hu and Qibin Lu. Through their comparative study of the 2008 Wenchuan earthquake and the 2013 Lushan earthquake, Zhang et al. examine how massive rural–urban migration flows and accelerated economic development fuelled rapid urbanisation prior to both disasters in these southwestern Chinese townships, which in turn impacted upon the scale and distribution of the devastation as well as on local capacities in the variegated processes of recovery. Citing insufficient societal participation and cross-sector cooperation as leading obstacles to effective disaster governance after the Wenchuan earthquake, the authors observed that the lessons learned from this disaster were not translated into institutional changes or operational procedures to facilitate a more rapid recovery from the nearby Lushan earthquake 3 years later. Drawing from these Chinese experiences, Zhang et al. propose an integrated conceptual framework for dealing with the vexed relationship between disaster governance and urbanisation that incorporates extant knowledge into a more inclusive urban disaster governance system.

Efforts to develop more effective modes of disaster governance in urbanising societies remain partial and are often mired by a lack of information and wide variations in localised contexts. The increasingly interconnected nature of environmental disasters also makes it difficult to isolate impacts of damages. For instance, while insurance companies have begun to evaluate projected losses of storm damage related to climate change, information remains limited for individual cities (Bulkeley 2013). Reliable data is especially difficult to gather when environmental disasters trigger compound disruptions that impact upon lives and livelihoods well beyond the immediate stressors of the disaster itself.

In the preoccupation in the literature with the direct human costs (health and life) and economic losses (infrastructure, public facilities and services), scant attention has been paid to the longer-term legacy of social trauma in disaster governance regimes. Helen James provides some redress for this oversight in her essay on the cultural contributions of three urban populations in the national contexts of China, Myanmar and Iran to adaptation and recovery from trauma in the aftermath of disaster (Chap. 12). Focusing on sociocultural factors that enable disaster-impacted communities to reconstruct their daily lives, James problematises the difficulties encountered in adapting to new realities around an engaged future through the lens of civil society. For James, differences between rural and urban governance systems were far less important in each of these societies than cultural capital or the ability of survivors to reconstruct their social relations and assets around communities and livelihoods.

Cultural capital was conspicuously lacking from the disaster governance model described by Pablo Figueroa (Chap. 13) in dealing with Japan's triple disaster of the 2011 Fukushima earthquake, tsunami and meltdown at the Fukushima Daiichi power plant. The devastation and social trauma caused by these cascading events was compounded by a political culture of opaque and secretive top-down decision-making in the national government in Tokyo in collusion with the nuclear industry. This lack of transparency alienated civil society from official recovery and reconstruction processes and created a crisis in public confidence. A paralysing outcome of this pervasive political culture of secrecy, argues Figueroa, was that the voices of civil society demonstrators were ignored, leaving communities feeling disoriented with corrosive consequences for their longer-term capacities to build resilience to future disasters.

All of the cases in this edited collection highlight that environmental disasters do not have single causes or isolated causalities and impacts. As such, they must be viewed through an interdisciplinary lens that encompasses multiple stakeholders in the ongoing search for more inclusive and effective modes of disaster governance. This book further shows how the uneven patterns of urbanisation create heterogeneous and divided societies that emerge as spaces of heightened political contestation and fragmentation in times of disaster due to the uneven experiences of exposure, risk and vulnerability. Urban sprawl often exacerbates these issues of conflict, coordination and cooperation because compound disasters cannot be neatly contained within jurisdictional boundaries and tend to spill over into neighbouring regions with overlapping areas of authority and competing interests. More broadly, as a contribution to our understanding of the unfolding field of disaster studies, the chapters in this book reveal the varied scales at which disaster governance must be considered when compound disasters cut across rural and urban areas, and when causal connections have global consequences such as global climate change precipitating extreme or unusual weather events and rising sea levels.

Our hope is that researchers and policy-makers can build upon the ideas and themes explored in this book to further understand the complexities of disaster governance within and beyond urbanising Asia in the pursuit of more socially equitable, inclusive and effective processes and outcomes. Individually, the cases presented in the following pages offer a range of insights and perspectives that consciously defy singular or formulaic prescriptions for disaster governance. Collectively, they take up the challenge of pushing the boundaries of disaster studies by promoting a multi-sector, multidisciplinary and multi-stakeholder approach that links existing knowledge more coherently to action in disaster governance programs. Implicit in this approach is the assumption that in twenty-first-century urbanising populations, disasters can no longer be treated as isolated events and must be considered in terms of interconnected causalities that are set into motion well before the moment of the disaster itself and which have cascading impacts long into its aftermath.

References

Adikari, Y., Osti, R., & Noro, T. (2010). Flood-related disaster vulnerability: An impending crisis of megacities in Asia. *Journal of Flood Risk Management,* 3(3), 185–191.

Birkland, T.A. (2007). *Lessons of disaster. Policy change after catastrophic events.* Washington, DC: Georgetown University Press.

Bulkeley, H. (2013). *Cities and climate change.* London/New York: Routledge.

Chan, F.K.S., Mitchell, G., Adekola, O., & McDonald, A. (2012). Flood risk in Asia's urban mega-deltas: Drivers, impacts and response. *Environment and Urbanization Asia*, 3(1), 41–61.

ESCAP/UNISDR. (2012). *Asia-Pacific disaster report 2012. Reducing vulnerability and exposure to disasters*. http://www.unisdr.org/files/29288_apdrexecsummary.pdf. Accessed 7 Aug 2014.

Hardoy, J.E., Mitlin, D., & Satterthwaite, D. (2004). *Environmental problems in an urbanizing world: Finding solutions in cities in Africa, Asia and Latin America*. London: Earthscan Publications Ltd.

Herzog, R.J., & Austin, S.F. (2007). Symposium-naming and framing public administration theory in natural disaster planning and response. *Administration Theory and Praxis*, 29(4), 493–496.

Hochrainer, S., & Mechler, R. (2011). Natural disaster risk in Asian megacities: A case for risk pooling? *Cities*, 28(1), 53–61.

Jha, A.K., Bloch, R., & Lamond, J. (2012). *Cities and flooding: A guide to integrated urban flood risk management for the 21st century*. Washington, DC: World Bank.

Jones, G., & Douglass, M. (2008). *The rise of mega-urban regions in Pacific Asia – Urban dynamics in a global era*. Singapore: National University Press.

Knight, J., & Harrison, S. (2014). Limitations of uniformitarianism in the Anthropocene. *Anthropocene*. http://dx.doi.org/10.1016/j.ancene.2014.06.001. Accessed 4 Oct 2014.

Kovats, S., & Akhtar, R. (2008). Climate, climate change and human health in Asian cities. *Environment and Urbanization*, 20(1), 165–175.

Lee, T., & Koski, C. (2014). Mitigating global warming in global cities: Comparing participation and climate change policies of C40 cities. *Journal of Comparative Policy Analysis*, 16(5):1–18.

McEntire, D. (Ed.). (2007). *Disciplines, disasters and emergency management.* Springfield: Charles. C. Thomas Publisher Ltd.

McGranahan, G., Balk, D., & Anderson, B. (2007). The rising tide: Assessing the risks of climate change and human settlement in low elevation coastal zones. *Environment and Urbanization,* 19(1), 17–37.

Miller, M.A., & Bunnell, T. (2013). Urban–rural connections: Banda Aceh through conflict, tsunami and decentralization. In T. Bunnell, D. Parthasarathy & E.C. Thompson (Eds.), *Cleavage, connection and conflict in rural, urban and contemporary Asia* (pp. 83–98). Dordrecht: Springer.

Miller, M.A., & Bunnell, T. (2014). Introduction: Asian cities in an era of decentralisation. In M.A. Miller & T. Bunnell (Eds.), *Asian cities in an era of decentralisation* (pp. 1–6). London/New York: Routledge.

Pelling, M. (2003). *The vulnerability of cities: Natural disasters and social resilience.* London: Earthscan Publications Ltd.

Pelling, M., & Blackburn, S. (2014). *Megacities and the coast: Risk, resilience and transformation.* London: Earthscan Publications Ltd.

Sahni, P., & Medury, U. (Eds.). (2003). *Governance for development: Issues and strategies.* New Delhi: PhI Learning Private Limited.

Schreurs, M.A. (2010). Multi-level governance and global climate change in East Asia. *Asian Economic Policy Review,* 5(1), 88–105.

Shi, P., Jaeger, C., & Ye, Q. (Eds.) (2013). *Integrated risk governance. Science plan and case studies of large-scale disasters.* Beijing/Heidelberg: Beijing Normal University Press.

UNESCAP. (2010). *The state of Asian cities 2010/2011.* Bangkok: UNESCAP.

United Nations. (2004). *World urbanization prospects. The 2003 revision.* New York: Department of Economic and Social Affairs, Population Division.

United Nations. (2014). *World urbanization prospects. The 2014 revision highlights.* New York: Department of Economic and Social Affairs.

Wisner, B., Gaillard, J.C., & Kelman, I. (Eds.) (2011). *Handbook of hazards and disaster risk reduction.* London/New York: Routledge.

World Population Statistics. (2014, March 14). *Population of Asia 2014.* http://www.worldpopulationstatistics.com/population-of-asia-2014/. Accessed 16 Aug 2014.

Chapter 2
The Urban Transition of Disaster Governance in Asia

Mike Douglass

Abstract The increasing frequency and severity of natural disasters in Asia are highly correlated with the rapid urban transition now taking place in this world region. Five types of urban transition effects are identified to explain how the urbanization of disasters calls for fundamental changes in approaches to disaster prevention, response, adaptation and resilience. The effects include agglomeration and the formation of mega-urban regions, spatial polarization in high-risk zones, new forms and magnitudes of vulnerability, compound disasters and the expanding ecological reach of cities. Taken together, they call for a shift from expert-centred disaster management to participatory disaster governance as the framework for society-wide engagement in all phases of disaster experiences and responses. Three spatial scales of governance – neighbourhood, city region and transborder riparian region – are among the most critical to be included in the search for innovations in disaster governance.

Keywords Urbanization • Disaster governance • Asia • Compound disasters • Multiscalar planning • Urban form

2.1 Urbanizing Disasters

> Exposure to hazards has multiplied as urban centers grow and people and economic activities expand into increasingly exposed and hazard-prone land. Rapid urbanization expands exposure to hazards, and it also increases people's vulnerability, especially among the poor. (UNISDR/UNESCAP 2012, xxii)

The human and economic costs of environmental disasters in Asia are rising (Fig. 2.1).[1] Between 2000 and 2012, they affected an estimated 1.6 billion people in

[1] The United Nations (2013) finds that direct economic losses over the past three decades in middle- and low-income economies alone totalled more than $300 billion. However, actual costs of disasters are much greater than reported figures indicate (UN 2013). In terms of flooding, from 1970 to 2010 the number of people impacted by annual inundations more than doubled from 30

M. Douglass (✉)
Asia Research Institute, National University of Singapore, Singapore, Singapore
e-mail: michaeld@nus.edu.sg

M.A. Miller, M. Douglass (eds.), *Disaster Governance in Urbanising Asia*,
DOI 10.1007/978-981-287-649-2_2

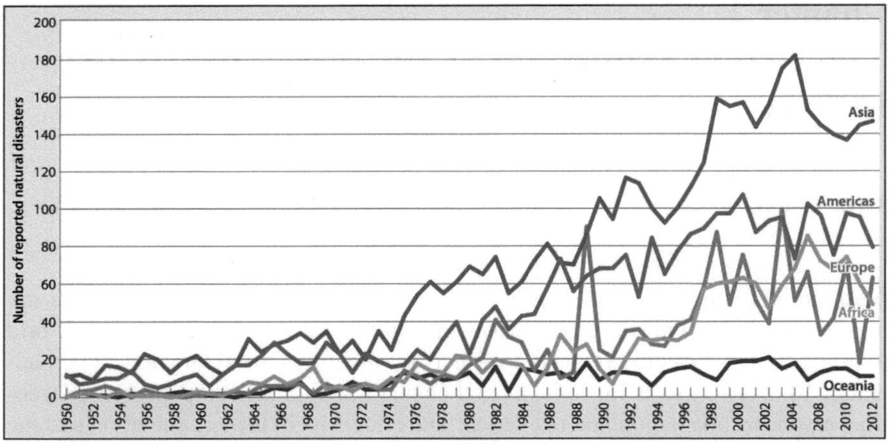

Fig. 2.1 Numbers of disasters by world region 1950–2012 (*Source*: ADB (2013))

Asia (Jha and Stanton-Geddes 2013). While about 40 % of all environmental disasters in the world occur in Asia, 88 % of people affected reside in this region (ADB 2013).

Although the sources of registered increases are manifold, many are increasingly related to Asia's accelerated urban transition that began in earnest in the 1970s. From a level of less than 20 % in 1960, Asia is approaching the 50 % urban mark today, and by the latter half of the twenty-first century, it will have completed a historic transformation from agrarian to urban-based societies (UNISDR/UNESCAP 2012; ADB 2013). From 1980 to 2010, cities in Asia added over one billion people to their populations, and another billion will live in cities by 2040 (UN 2011). Major urban regions in Asia are each annually adding hundreds of thousands of new residents to their populations. These city regions and the distant spaces they bring into their orbit are becoming increasingly vulnerable to disasters (Jha and Stanton-Geddes 2013).

Relationships between risks of environmental disasters and urbanization are well known. Global climate change, for example, is closely linked with urban-industrial growth. In 2012, 98 % of all displacements of people worldwide were from climate change and related extreme weather events (DMC 2013). While highly industrialized countries of Europe and North America have been the principal sources of climate change, Asia's urban-industrial growth is now becoming a major contributor as well. Its major cities also rank among the most environmentally degraded in the world (UNESCAP 2012).

Similarly, disaster-linked transformations of nature in the form of massive deforestation, mega-dams constructed in riparian regions and mining of energy resources

million to 64 million in Asia. In Asia, 21 million people were displaced by natural disasters in 2012 (DMC 2013; UNISDR/UNESCAP 2012).

and construction materials are all associated with the rising demand for natural resources for building cities and fuelling their economies. As communications and transportation spread through expanding urban systems, capacities to organize rural regions in even remote locations to serve urban demands expand, which also transform and can undermine traditional modes of environmental stewardship and local livelihoods.

More recently, cities are being portrayed as "engines of economic growth" substantially replacing agriculture as a national economic base, but are also becoming the major recipients of high impacts of environmental disasters. As mega-urban regions in Asia approach and even surpass the 30 million population threshold, imagining a major disaster hitting any one of them immediately raises the spectre of unparalleled human suffering and economic costs well beyond the affected city region through the compounding of disasters with effects that move through an emerging global system of cities.

In sum, the idea of an isolated environmental disaster that is contained within a relatively small area appears to be passing into history. Instead, we can now say that environmental disasters are occurring within an urban matrix of interrelated impacts or effects. Five effects of Asia's urban transition serve to underscore the profound changes underway in the understanding and the effectiveness of responses to disasters. They include agglomeration and the formation of mega-urban regions, spatial polarization in high-risk zones, new forms and magnitudes of vulnerability, compound disasters and the expanding ecological reach of cities. Together, these effects call for a shift from expert-centred disaster management to participatory disaster governance as the framework for society-wide engagement in all phases of disaster experiences and responses.

2.2 Agglomeration Effects

> By 2025, the number of megacities in Asia is expected to increase to 21 out of a global total of 37. Growth of assets and megacities means that multi-billion-dollar disasters are becoming more widespread in the region. Population density, urbanization, and demographic profiles are context-specific factors that are likely to drive death tolls and victimization.
> (ADB 2013)

> Most Asian cities are poorly equipped to manage the effects of natural disasters, climate change, contaminated or unstable land and health pandemics. Many will need massive investments in infrastructure, public services, institutional capacity and environmental programmes if basic security, health, safety and overall conditions are to improve for the majority of urban residents.
> (UNESCAP 2011, 164)

Up to the 1970s, few cities in Asia were larger than one million in population. Today, and in contrast to the limited "mega-city" measurement of a city demarcated by its administrative boundary, the huge city regions that have emerged in Asia are agglomerations incorporating several municipalities and administrative areas into immense mega-urban regions (MURs) that stretch 100 km or more from core central cities to form *desakota* realms of rural–urban land-use mixes (McGee 1991; Jones and Douglass 2008). Figure 2.2 maps the distribution of MURs in Asia.

Table 2.1 ranks these cities within Asia and globally. It shows that slightly more than half of the largest 66 MURs in the world are in Asia. In 2013, these MURs include 7 that are larger than 20 million in population. The 35 MURs together have a total population of 451 million people. Almost all are in areas of high disaster risk.

The sheer size and human density of these agglomerations take the understanding of environmental disasters into new realms of complexity never experienced before. Responses now require large-scale infrastructure beyond the capacities of disaster victims to mobilize themselves or at the community level. Responses now require cooperation among millions of people and capable, committed governments to be effective. As such, they become part of the politics of governing cities. A flooding disaster, for example, is no longer only a matter of canal dredging or repair, but now includes all forms of urban services, transportation, housing and land-use

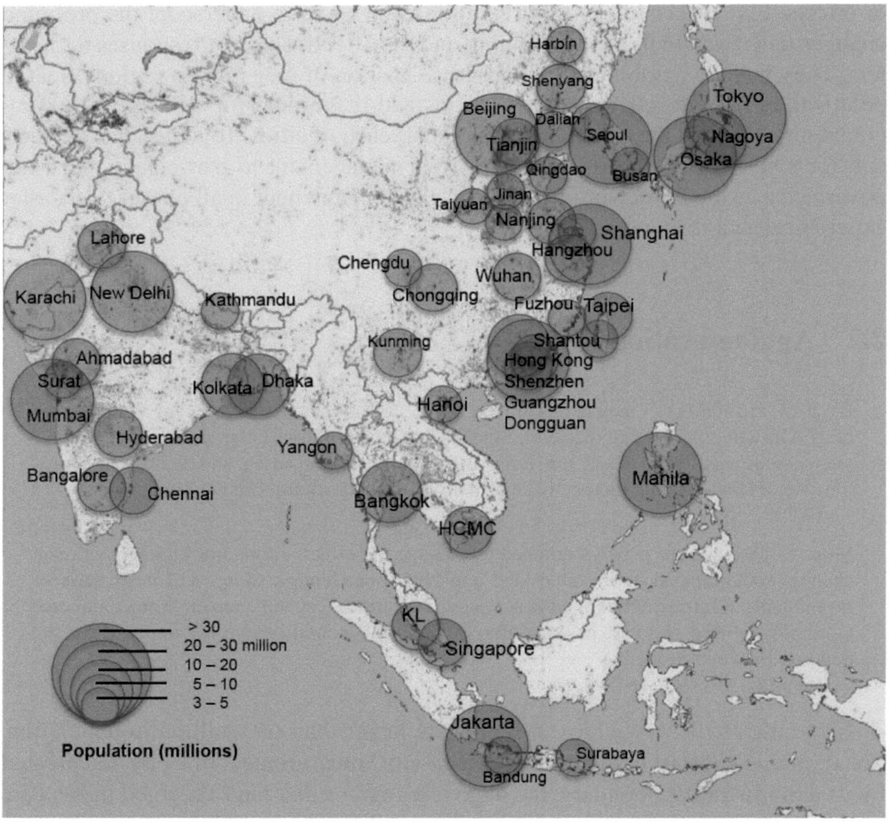

Fig. 2.2 Asia's mega-urban regions (*Source*: Author. Data from Table 2.2)

Table 2.1 Population of Asia's mega-urban regions (2013)

Asia rank	World rank	Country	City	Population (millions)
1	1	Japan	Tokyo-Yokohama	37.2
2	2	Indonesia	Jakarta (Jabotabek)	26.7
3	3	South Korea	Seoul-Incheon	22.9
4	4	India	Delhi, DL-HR-UP	22.8
5	5	China	Shanghai, SHG	21.8
6	6	Philippines	Manila	21.2
7	7	Pakistan	Karachi	20.9
8	11	China	Beijing, BJ	18.2
9	12	China	Guangzhou-Foshan, GD	17.7
10	13	India	Mumbai, MAH	17.3
11	14	Japan	Osaka-Kobe-Kyoto	17.2
12	18	India	Kolkata, WB	14.6
13	19	Thailand	Bangkok	14.5
14	20	Bangladesh	Dhaka	14.4
15	24	China	Shenzhen, GD	12.5
16	28	Japan	Nagoya	10.2
17	32	China	Tianjin, TJ	9.3
18	33	India	Chennai, TN	9.2
19	35	India	Bangalore, KAR	9.0
20	37	Vietnam	Ho Chi Minh City	8.8
21	38	China	Dongguan, GD	8.6
22	39	China	Chengdu, SC	8.4
23	40	China: Taiwan	Taipei	8.4
24	41	India	Hyderabad, AP	8.2
25	42	Pakistan	Lahore	8.0
26	44	China	Wuhan, HUB	7.4
27	46	China	Hong Kong, HK	7.2
28	47	India	Ahmedabad, GUJ	6.7
29	48	China	Chongqing, CQ	6.6
30	49	Malaysia	Kuala Lumpur	6.6
31	50	China	Hangzhou, ZJ	6.6
32	59	China	Nanjing, JS	5.8
33	60	China	Shenyang, LN	5.6
34	65	China	Xi'an, SAA	5.3
35	66	Singapore	Singapore	5.3
Total				451.2

Source: Demographia World Urban Areas (2013)

controls. It also inherently leads to contestations and demands for social justice over land, welfare, livelihoods and urban ecosystems.

Asia's urban transition is a complex of mega-trends that appear in varying combinations in time and space. These trends include shifts from agrarian to urban-centred economies that involve multiple types of increases in the appropriation of natural resources, demographic changes now beginning to move toward chronic rural depopulation and slower-growing and ageing societies, the emergence of new urban classes and the rise of civil society with pushes for political reform, global climate change that is both raising sea levels and melting the Himalayan glaciers from which most continental Asian rivers flow, new forms of communicable diseases that can be quickly diffused through urban systems and enhanced technological capacities and scales of projects that transform cityscapes and countryside alike. When animated in specific settings, the interactions among these trends create new, still rapidly changing and often unstable social, political and economic contexts for responses to disasters.

In the case of the events in Japan in 2011, for example, disaster relief had to cope with an ageing society with heavily depopulated rural regions, towns and villages that were already experiencing economic downturns and declines in basic services. It also had to cope with the compound effects from earthquake and tsunami to nuclear meltdown. With urbanization, disasters can thus be seen as creating vortices of multiple disjunctures in already dynamically changing and turbulent processes. Responses to these dynamics necessarily require flexibility and collaborations across social divides that go beyond both sector approaches and disaster management as professional activities reserved for experts. In involving access to land, scarce resources and, more broadly, the right to the city, they become matters for public consideration.

Somewhat paradoxically, more economically advanced cities experience the highest costs of environmental disasters due to the high value of their infrastructure and assets (Jha and Stanton-Geddes 2013). In Asia, for example, Japan has the highest exposure of produced capital in absolute terms and is the third highest in relative terms in per capita GDP, while smaller economies, including Hong Kong and Macao, have high levels of relative risk. At the same time, however, these cities are shown to be more capable of responding to disasters. Studies show that cities with high levels of poverty and poor infrastructure are least able to effectively respond to disasters, including small repeated disasters, which have cumulatively debilitating impacts. This again turns attention to issues of inequality and their connections to disaster impacts and recovery.

These observations show the complex interrelationships between the economic vitality of a city, the costs of disasters and the capacity for resilience in disaster situations. Beyond specific contexts, all cities have to contend with agglomeration effects that raise the potential for high impacts of disasters, which today involves millions of people in one urban setting. In all cases, too, the real cost of disasters is greatly understated when only measuring near-term mortality, infrastructure costs and business losses incurred in a disaster event. "Invisible risks", longer-term compounding of impacts and social costs that cannot be given monetary value are also in need of accounting, which calls for an open public sphere of disaster governance (Jha and Stanton-Geddes 2013; UNISDR/UNESCAP 2012).

2.3 Spatial Polarization in High-Risk Coastal Zones and Riparian Regions

> The increase in concentrations of people and growth of assets in hazardous areas is the single largest driver of disaster risk and greatest challenge for managing disaster risks.
> (Jha and Stanton-Geddes 2013, 17)

> From 1970 to 2010 the average number of people exposed to yearly flooding in Asia has more than doubled from 29.5 million to 63.8 million and the population resident in cyclone-prone areas has grown from 71.8 million to 120.7 million…Exposure to disaster risk is growing faster than our ability to build resilience.
> (UNISDR/UNESCAP 2012)

As Fig. 2.2 and Table 2.1 suggest, the expansion of MURs is occurring in a larger process of continuing spatial polarization of urbanization in coastal regions and river basins that encounter high risks of environmental disasters. Of the 35 largest cities (MURs) in Asia listed in Table 2.1, 25 are located in coastal areas. Available evidence shows that they continue to accrue larger shares of national populations over time (Jones and Douglass 2008), with most of this growth now in smaller cities located in zones spreading beyond the metropolitan core. The rising risks in these urbanizing regions are not only related to global climate change but also to failures of large-scale attempts to control nature for human uses and the resulting deterioration of ecological conditions in these MURs (UNESCO 2006; Douglass 2010).

The coastal shift of population is most dramatically seen in the case of China, which since the early 1990s has witnessed an unprecedented surge in migration from inland provinces to the major MURs of Beijing, Shanghai and the Pearl River Delta (Fig. 2.2). Elsewhere in the Himalayan regions of Asia, unusual weather events, melting glaciers and large infrastructure projects such as highways and mega-dams are creating high-risk zones of landslides and flooding for cities that are expanding in that region. In the Mekong River Delta, Can Tho, which now has a population of about three million people, potentially faces total inundation from global sea rises in this century. Cyclonic winds such as typhoons pose equally high risks to coastal city regions. Approximately 80 % of the risk from cyclonic wind events is concentrated in Asia (UN 2013). Coastal MURs such as Jakarta and Manila have entered an era of chronic annual flooding for which no sufficient remedy is in sight.

Spatial polarization of Asia's urban transition is propelled by very high levels of rural–urban migration, most of which come from heartland rural regions. In many countries that are reaching mid-point in their urban transitions, MURs are experiencing annual population increases of one-quarter to close to a million new residents annually. In Southeast Asia, the share of migration in urban population growth increased from 45 % in the 1970s to 72 % in the 2010s and is projected to be at 76 % in the 2020s (UNESCAP 2012). South Korea represents the interesting case in that as national fertility rates fall below replacement level, which they have in all the higher-income economies of Asia, negative population growth for

Table 2.2 Urban population living in slums, 1990–2010 (millions)

Population at mid-year by region (1,000 region)	1990	1995	2000	2005	2007	2010
East Asia	160	177	192	195	194	190
South Asia	180	190	194	192	192	191
Southeast Asia	69	76	82	84	84	89
Total Asia	409	443	468	471	470	470

Source: UNESCAP (2011, 260)

the country reaches from rural areas into cities. As a result, migration either from rural areas or among cities becomes the only means by which a city can grow. Thus, in South Korea in 2020 urban growth will be slower than today, but is nonetheless expected to occur with an 86 % contribution from rural–urban migration. This is occurring in a setting in which rural areas have been chronically shrinking in population since the 1980s. Along with spatial polarization, these demographic shifts have resulted in half of the national population of South Korea now living in the Seoul Capital Region.

Where migration fuels rapid urban population growth, even the most well-prepared governments are hard pressed to provide sufficient environmental and other urban services and housing in the face of huge annual increases. In many MURs, the deficits are wide and continue to grow along with widening income inequalities. Such high levels of population growth through large-scale migration have resulted in very large numbers of people settling in slums and informal settlements in sites with exceptionally high risk of disasters, including flooding and industrial accidents. While the share of urban populations living in slums in Asia is decreasing in some countries, the numbers of slum dwellers in Asia as a whole continue to remain very high. In 2010, an official estimate of 470 million people was living in slums in Asia (Table 2.2).

Many slums are locating along canals and in other high disaster risk sites. When disasters such as flooding occur, they are also the first to be evicted to make way for canal widening and flood management needs. This syndrome of channelling poor people to high-risk areas and then evicting them to make way not only for flood control but also for corporate megaprojects along coastal areas has become a chronic feature of the political economy of environmental disaster management in many cities in Asia and elsewhere. These processes add to other new vulnerability effects.

When global climate change and predicted sea rise are fully added to risks involved in the coastal orientation of Asia's urban transition, scenarios can be bleak. Some predict the eventuality of mass evacuations from the huge urban agglomerations appearing in coastal areas within this century (Satterthwaite et al. 2007). Climate change is expected to reduce clean water supplies and productive agricultural areas, such as those in the Mekong Delta. Altered rainfall patterns will affect food production and supplies to cities, with serious implications for food security as desertification also enters the equation in countries such as China and India (Douglas 2009). One-fifth of Asia's global GNP is also concentrated in urbanizing coastal regions.

Although many governments have attempted to redirect urbanization away from growing mega-urban regions, none has had significant success. Agglomeration economies are powerful, and reasons to locate in coastal areas where access to world markets and other linkages are highest are too compelling when contrasted with incentives to relocate business to inland and peripheral regions. As such, approaches to disaster risk reduction will have to focus on these city regions where they are.

2.4 Vulnerability Effects

> Increasing disaster risks in Asia-Pacific are driven by the twin challenge of increasing exposure of its people and economic assets, and the inability of the most vulnerable groups to cope with disasters. (UNISDR/UNESCAP 2012)

The most rapid urbanization is proceeding in some of the least prepared countries that have huge and widening deficits in public urban infrastructure such as drainage, disaster-resilient housing as well as public services that is needed to minimize disaster risks. When combined with the density of urban settlements, vulnerability increases become large and diverse. While disasters are known to disproportionally affect poorer and marginalized people (Neumayer and Plümper 2007), urbanization can compound vulnerability in deeply profound ways that recast the question of who the most vulnerable are when environmental disasters occur.

Reinterpreting earlier work by Sen (1990), vulnerabilities in the city can be seen in part as entitlement failures. Broadly, urbanization represents a shift in entitlements to the means of preparing for, adapting to and recovering from disasters. Whereas in more remote rural regions people might have direct access to collecting or growing food, taking water or directly accessing other resources while also drawing upon reciprocal relationships among kin and community, in the city, residents are increasingly dependent on access to money and government assistance when disasters occur. When jobs are lost due to a disaster and governments or other sources of aid are not sufficient, disasters worsen their plight.

Expressing a similar concern about the ways in which urbanization impacts vulnerability, UNESCAP (2012, 41) states that "vulnerability in urban environments is further heightened by significant structural changes that families undergo in urban settings". These changes include declining extended family structures and support that increase the vulnerability of people, especially children, the elderly and the disabled. As discussed under governance below, among the more promising efforts to improve disaster resilience are those that work with poor communities to cooperatively support their members. In part, these can be seen as efforts to compensate for diminished capacities within urban households.

At a broader urban level, changes in the production and ownership of urban space are also having new vulnerability effects on lower-middle and poorer urban households. From the late 1980s onward, major cities in Asia entered a still continuing era of corporatization of urban spaces. Megaprojects have been particularly intrusive in land development schemes that have pushed low-income residents out of their neighbourhoods in core urban areas and into ever more precarious disaster-prone locations.

In peri-urban areas, gated housing enclaves and vast private cities with no forms of public governance have displaced small farmers, and in the city aggressive privatization of public spaces and the rapid displacement of locally owned enterprises with global franchises have added to a proletarianization process in which wage work for the majority has steadily replaced family and small-scale enterprises. Petty commodity production in the form of street vendors continues to exist, but even these activities require some capital and can be very difficult for poorer members of urban society to enter other than as disguised wage workers or piece workers. For those with very low incomes who are compelled to live in slums in high-risk flooding or other disaster-prone areas, their vulnerability increases due to a lack of access to capital, land and other resources. In many MURs in Asia, these people number in the hundreds of thousands, if not millions.

In these contexts, recovering from a disaster confronts great difficulties as well. This is underscored in research findings showing that in the 2 years after Cyclone Nargis in Myanmar, average debts of labourers and fishermen more than doubled (Jha and Stanton-Geddes 2013). Similarly, in Pakistan while almost all small shops were able to reopen within 6 months of the 2010 flood, the majority found themselves operating at a loss due to a complex interweaving of post-disaster dynamics. The bottlenecks included disruption in supply chains and lifelines, deepening stress on family and community, draining of personal savings needed to restock supplies and loss of trained staff (Asgary et al. 2012). Decline of small business incomes multiplied into declines in employment and the overall vitality of the economy. Businesses that recovered and were able to do well did so through strong familial and social networks in their local areas.

In higher-income economies, vulnerability is taking a new turn toward depopulating cities and towns, rapidly ageing populations and shrinking supplies of labour available for mobilization in disaster situations, as in the case of the Great East Japan earthquake and tsunami in 2011. While younger survivors might simply migrate from the affected areas, the many senior residents became dependent upon government and other social support for an indefinite period of time in settings with depressed economies. The compounding of disasters in this area makes returning to normal life a continuing problem.

One of the complicating factors in assessing who is vulnerable is that answers to this question can vary substantially depending on the position and background of the person making the assessment. As noted by Bankoff and Hilhorst (2009), taking action about disaster risks and impacts involves individual calculations about the prevailing social order and social relations, which are experienced differently by those with little social power compared to those in power. People directly impacted by a disaster can and often do have views that differ from government officials, for example. For these reasons, vulnerability assessments need to be part of a governance process in which different voices can be heard and mechanisms exist to resolve competing claims that arise over this question. Thus, proactive approaches call for the inclusion of many actors in open and transparent public deliberations rather than solely through professional disaster management teams alone.

2.5 Compound Disasters, Disaster Incubation and Network Effects

> The increase in occurrence of multiple large disasters is an inevitable consequence of increases in the population and spatial density in existing urban centers, greater reliance on technological solutions to maintaining growth and development in hazardous environments and the fragility of social, economic, and risk management systems. There is a need to recognize that compound disasters are a result of a series of component disasters in communities that in their aggregate overwhelm existing abilities to respond.
> (ADB 2013, 5)

> The socio-political aspects of risk assessment and risk governance are pronounced in the case of cumulative risks from multiple stressors.
> (Assmuth et al. 2009, 3943)

The combination of agglomeration effects and nodal importance of MURs in an emerging global system of cities in an era of near-instant communications and fast transportation results in profound compounding ripple effects that extend far beyond a specific disaster moment or location. How to include these potential network effects in disaster governance is one of the greatest challenges of today. Compound disasters are multiple sequential disaster events that produce increasingly more serious impacts than do single disasters occurring independently (Kawata 2011).[2] Huge urban agglomerations that have large densely settled populations, economies dependent on global networks of production and consumption and that are dependent upon steady supplies of environmental resources are particularly vulnerable to compound disasters.

Within cities, population growth has put ever greater pressure on both physical and social support infrastructure. Even a moderate technological or environmental hazard can trigger progressive failures in infrastructure, basic services, fuel supplies and housing that cascade into multiple disasters. The 2011 Great East Japan earthquake and tsunami that triggered the Fukushima nuclear accident is a clear case of compound disasters that not only had different forms of immediate impacts but also extended well beyond the particular events themselves.

These disasters also reveal what is being called "disaster incubation" (Mulvihill and Ali 2007). As originally proposed by Turner (1976), this phenomenon refers to the accumulation of disaster risks that are often out of view and unreported in rural and exurban areas but which are crucial for the functioning and well-being of cities,

[2] The Eyjafjallajökull volcanic ash clouds in Iceland in 2010 was one of the most spectacular incidences of an environmental disaster affecting a globalized world as air traffic in most European countries was shut down for six straight days, costing airlines US$1.7 billion in revenues.

including MURs. As city systems create extensive core-periphery structures between MURs and more distant towns and cities, undesirable infrastructure and services, such as toxic waste sites or nuclear power plants, are typically deployed to peripheral regions where capacities to oversee these installations are often weak and not placed under local supervision. Such was the case of the Fukushima nuclear power plant, which involved a decision to construct the facility much closer to sea level than the original plans had specified. In the context of the 2011 disaster that had no historical precedent in Japan, this decision, which was made to lower construction costs, unintentionally incubated a compound disaster generated by the earthquake and tsunami.

Flooding, which has become the most pervasive and frequent form of disaster in Asia, also results from and creates compound disasters that include disaster incubation dimensions. As urbanization reduces impermeable surfaces, drainage is reduced, raising the potential for flooding. Global climate change adds to this through unusually severe weather events of heavy rain and high winds that are on the increase. Land subsidence from overdrawing groundwater, antiquated canal systems, large-scale deforestation of uplands and loss of spaces for natural drainage all add to the incubation of flooding disasters. Nicholls et al. (2007) found that of the 136 port cities worldwide that are exposed to once-in-a-century coastal flooding, 50 are in Asia. Six of the ten major port cities most at risk (in terms of exposed population) of flooding and inundation are Ho Chi Minh City, Guangzhou, Kolkata, Mumbai, Osaka-Kobe and Shanghai (UNESCAP 2011, 182).

Table 2.3 lists the 20 largest cities in the world ranked in terms of size and assessed in terms of preparedness for flooding. It shows that of the 11 in Asia, all, except Tokyo, Osaka and to a lesser extent Seoul, are "critically unprepared" for floods. A major reason for the difficulties in being prepared is the interlocking of disaster dynamics that attend to agglomeration effects and the human involvement in incubating the subsequent compound flooding disasters. East and Southeast Asia account for about 40 % of the total number of floods worldwide over the past 30 years (Jha and Stanton-Geddes 2013). High risks of flooding are expected to reach over 400 million people living in cities in Asia by 2025.

Beyond the scale of individual cities, one of the most prominent themes about urbanization is that it creates networks of cities that include rural as well as urban linkages. From the mid-1980s, for example, urban studies have been fascinated with the appearance of global (or world) cities and global city networks that articulate the global corporate economy. This system can also articulate and accentuate the impacts of disasters. For example, the spread of SARS was city to city on a potentially global scale as it travelled from Hong Kong to Toronto in just a single day.

Spatial network effects of environmental disasters are now endemic among global firms. Toyota lost $1.2 billion in product revenue from the 2011 Japan earthquake and tsunami due to parts shortages that caused 150,000 fewer Toyota automobiles to be manufactured in the USA and reductions in production of 70 % in India and 50 % in China (ADB 2013). Likewise, the Philippines, Thailand and Indonesia also experienced a 10–20 % decline in automobile assembly due to parts shortages from Japan.

Table 2.3 Preparedness for flooding in the largest cities in the world

City	Flood and storm risk
Tokyo	Very well prepared
Seoul	Could be better prepared
Jakarta	Critically unprepared
Delhi	Critically unprepared
Mumbai	Critically unprepared
Mexico City	Could be better prepared
São Paulo	Could be better prepared
New York	Could be better prepared
Osaka	Very well prepared
Shanghai	Critically unprepared
Manila	Critically unprepared
Hong Kong-Shenzhen	Very well prepared
Los Angeles	Could be better prepared
Kolkata	Critically unprepared
London	Could be better prepared
Moscow	No high risks
Cairo	Could be better prepared
Buenos Aires	Could be better prepared
Dhaka	Critically unprepared
Beijing	Critically unprepared
Karachi	Critically unprepared
Rio de Janeiro	Could be better prepared
Paris	Could be better prepared

Source: GreenAsh (2013)

Network distribution systems are now global, linking suppliers in Asia with assembly and sales throughout the world. An environmental disaster that disrupts these supply chains would have global impacts (ADB 2013). In 2010, Indonesia, Malaysia, the Philippines and Thailand taken together were among the most dependent economies on parts, components and industrial materials from Japan (imports 22 % and exports 18 %) (METI 2011). In addition to the disasters in Japan in 2011, the flooding of Bangkok in the same year resulted in impacts beyond the estimated $212 billion in direct costs as they worked through supply chains and markets beyond Thailand. Thai exports in electronics fell by nearly 50 %, which also led to declines in production in Japan.

As a key dimension of globalization, urban networks have the effect of tremendously reducing the time it takes for the effects of an environmental disaster in one location to impact other locations. This annihilation of space with time is one of the most important aspects of the urban matrix of environmental disasters. Global supply chains are based on the belief that no major disruptions in them will occur. At another level, dependence on higher-income countries as markets for exports from Asia was also affected by disasters in those countries. Increases in the frequency of environmental disasters in Asia will undoubtedly increase the vulnerability of supply chains and production networks to sudden economic downturns.

In a related manner, environmental disasters can also lead to a loss of network position that can last many years or even become permanent. In the case of the Great Hanshin-Awaji earthquake in Japan in 1995, Kobe Port, which was already declining in its global position, experienced devastating ripple effects from the destruction of container shipping berths, warehouses, bridges and utilities. Jobs were lost that were never recovered as this event worked to spur the ascendance of Busan and ports in China over Kobe. Prior to 1995, Kobe was the sixth largest port in the world. Even though the port was reconstructed over the next 2 years after the earthquake, by 1997 it had fallen to 17th place, and it fell further to 47th position in 2010 (ADB 2013). Reductions in harbour dues and mounting costs of recovery were like "pouring water into a bamboo basket" by disappearing without filling the port with renewed business.

Compound disasters, disaster incubation and its boomerang effects and local-to-global networks impacted by even a seemingly remote disaster event together present a formidable challenge to disaster preparedness, adaptation and resilience. The expanding environmental resource reach of cities and resulting impacts on local and global ecologies further complicate this challenge.

2.6 Expanding Natural Resource Reach and Ecology Impact Effects

> The unprecedented rates of urban population growth over the past century have occurred on less than 3 % of the global terrestrial surface; yet the impact has been global, with 78 % of carbon emissions, 60 % of residential water use, and 76 % of wood used for industrial purposes attributed to cities. Land change to build cities and to support the demands of urban populations itself drives other types of environmental change. (Grimm et al. 2008, 756)

> Planet Earth can offer a nominal 1.7 global hectares per head (ghph) of habitable land to support the needs of the human race. Now in most Asian cities, the average ecological footprint is in excess of five hectares per head, indicating that current consumption patterns are unsustainable. Although the footprints of Asian cities tend to be smaller than those in developed countries, they are on an upward trend, a phenomenon that is not without consequences for the global environment. (UNESCAP 2011, 168)

Asia's urban transition does not mean that rural or remote areas are not part of the emergent urban matrix of disasters. To the contrary, urbanization entails an expanding appropriation of natural resources to feed, fuel and supply urban demand. One measure of this expansion is the ecological footprint of a city, which indicates, "the area of productive land and aquatic ecosystems required to produce the resources used, and to assimilate the wastes produced, by a defined population at a specified material standard of living" (Rees 1996). Variations among cities are substantial. Singapore, for example, has a reported ecological footprint 7.1 global hectares per head (ghph) compared with Taipei's 4.75, Tokyo's 4.25 and Seoul's 4.20

(UNESCAP 2011, 169). Asia is in general much lower than the USA and Europe, but rapid urban population growth implies steady increases in the ecological footprints of cities accompanying Asia's urban transition.

The ecological footprint measure underestimates the real impacts of cities on local, regional and the world's ecologies (Schneider et al. 2009). For example, the construction of mega-dams that change the course of rivers and transform downstream ecologies are not included in ecological footprint measures; nor does pollution from strip mining or toxic industrial spillage. The tendency of urbanization to occur along rivers and coastlines makes Asia's urban transition a major contributor to disasters related to severe riparian and ocean pollution and eutrophication. In all of these ways, the actual impacts of urbanization on ecologies near and far remain largely unaccounted for, even though they are identified in countless case studies.

Taking all of these urbanization effects together provides sobering scenarios about the future. However, cities are also sites of social, economic, political and technological capabilities to respond to disasters. Taking advantage of these capacities will require far-reaching innovations in governance at local, city region and transborder scales.

2.7 Governance and Disasters

> The exercise of governance greatly influences the nature of socio-economic vulnerabilities and the extent of people's exposure to hazards. (UNISDR/UNESCAP 2012)

> A more resilient city is one with inclusive decision-making processes in the realm of planning, open dialog, accountability, and collaboration. It is one in which people and local stakeholders, including the private sector, various social groups, communities, civil society and grassroots organizations participate…A more resilient city is one with less social inequalities and a fairer distribution of resilience resources. (Jabareen 2013, 223–224)

Substantial evidence has accumulated to show that improving governance capacities is crucial for a locality to gain resilience in facing environmental disasters. Governance can be defined as a process of public decision-making that includes civil society as well as state and business interests (UNDP 2012).[3] Assmuth et al. (2009, 3943) go further to emphasize that the turn toward governance as a focus for disaster risk reduction includes a shift in the balance between state intervention and social autonomy by embracing the change from regulation, with the state acting *in* "multi-actor participation and negotiation and from technical management to legal, institutional, social and economic contexts".

In other terms, governance focuses attention on the processes of society-wide engagement in responding to actual or potential disasters. Studies find that inclusive

[3] UNESCAP (2011, 209) defines governance as "the sum of the many ways individuals and institutions, public and private, plan and manage the common affairs of the city. It is a continuing process through which conflicting or diverse interests may be accommodated and cooperative action can be taken".

decision-making processes in planning, open debate, accountability and collaboration are best able to mobilize knowledge and resources for longer-term as well as immediate responses (Healey 2007; Bulkeley 2010; UNESCAP/UNISDR 2012). Conversely, where governance institutions are weak, efforts for successful recovery are likely to be seriously impaired.

The emerging role of cities in disaster governance is already being acknowledged (Bulkeley 2010). As national populations increasingly urbanize, cities become the locus of both private and public life and can also become a political arena in which contestations over social and environmental justice can best be resolved through participatory public decision-making (Healey 2007; Healey and Upton 2010; Friedmann 2011). However, as important as these trends are, approaches toward environmental disaster preparedness and recovery remain poorly connected with the developmental roles of governments in Asia. The United Nations (2013, 3) finds, for example:

> Although increasingly risk management and reduction is mentioned in governmental development policies, plans and strategies…it is not treated as a truly multisectoral concern, and the institutional and legislative arrangements for disaster risk reduction are weakly connected to development sectors.

In addition, the capacities of city governments to create and use a governance system that can flexibly adapt to "uncertain and unpredicted conditions" are novel, with most continuing to rely on master planning and regulatory routines (Mirfenderesk and Corkill 2009, 152). The great uncertainty attending disasters is too complex to be effectively addressed by conventional approaches. Fixed rules typical of bureaucratic forms of government are likely to fail in disaster situations because they are based on assumptions of high levels of predictability rather than being able to respond to ruptures in the status quo (Baker and Refsgaard 2007).

Given these observations, the task at hand is to begin to build new types of disaster governance capacities into local political institutions. This would require a fundamental shift in urban governance at the city level along with substantial transfers of power and financial capacities from central to local governments. As stated by the United Nations (UNESCAP 2011, 209), "participatory local governance is one of the tenets of sustainable development and to be effective calls for a political 'space' which only decentralization can provide".

Although decentralization in the form of devolution of political power to local levels is a perennial theme in Asia, it has only begun to make progress over the past two decades in most countries (Bahl 2005; World Bank 2005; Laquian 2005). Indonesia has among the most devolved systems of governance in Asia, although it has only been in place for slightly more than a decade and is still very mixed in local government performance (Firman 2010). China is also using decentralization to improve environmental governance, with new environmental institutions and practices of local environmental policymaking involving private companies and citizen organizations leading to improvements in accountability (Mol 2009). At the other extreme are the many countries that continue to have highly concentrated government systems operating from capital cities, with localities taking "post office" administrative roles rather than possessing political decision-making authority (Douglass 2013).

In addition to more concerted political energy devoted to effectively devolving state power, innovations in local governance are also needed. In acknowledging "the immense progress made in many countries in East Asia to decentralize authority", Jha and Stanton-Geddes (2013, 35) find that "inadequate capacity to implement disaster risk management efforts at the local level remains high" and that "most preventive measures are embedded in the design and construction of infrastructure or other sectoral spending" rather than in community engagement.

To be effective for disaster planning, decentralization would need to include incentives to move away from many current patterns of response. Among the more important are anti-corruption mechanisms to inhibit an aid-dependency syndrome encountered in many disaster relief episodes.[4] The dynamics of disasters require all actors to be participants in producing knowledge, mobilizing resources and collaboratively taking action. For longer-term recovery and resilience, these innovations cannot be deployed only at the moment of urgency but instead need to be continuing elements of governance. It also needs to be flexible in quickly adapting to changing circumstances that cannot be well predicted in advance (Mirfenderesk and Corkill 2009).

Engagement with affected populations and flexibility in effectively adjusting to suddenly changing circumstances need to occur at the geographical scale of the problems at hand, as such disaster governance can be expected to operate at multiple scales. As was discovered in the aftermath of Hurricane Katrina in New Orleans in 2006, cross-scale linkages, each matched to the size and needs of the disaster, proved to be among the most important dimensions of much needed innovations in resilience in the face of the disaster (Baker and Refsgaard 2007). Participatory planning among organized actors that promoted trust to engage in coordinated responses was crucial for long-term recovery, as was institutional flexibility in situations of uncertainty, high levels of confusion and unpredictability in moving understandings of problems into remedial actions. One important but often neglected scale is that of neighbourhoods and smaller urban spaces.

2.7.1 Neighbourhoods

Given the context of most Asian cities previously described as being unequal in access to what Friedmann (1992) calls the basis of social power, which includes housing and living incomes, and what Nussbaum (2002) refers to as capabilities,

[4] Cohen and Werker (2008) find that some governments not only underinvest in disaster prevention if they know that they will be bailed out; they also create a racket effect of not preparing for disasters as a way of rent-seeking from humanitarian aid coming with a disaster. Rampant corruption lies at the root of these practices. Devolution of power to local forms of participatory governance is found to be an important, though not sufficient, means to limit these practices.

disaster governance needs to include an intentional focus on the urban poor, marginalized people and the otherwise most vulnerable in the spaces in which they live and earn incomes. Including social and environmental justice is paramount, and for this reason, neighbourhoods and their social institutions need to be included in any approach toward disaster governance. This would mean, for example, that instead of automatically turning toward the relocation of poorer households from flooded areas, governments would work with affected communities to first try to find ways in which people and their livelihoods, traditions and social relations can stay intact by remaining in these sites.

Because disasters impact poorer members of society more than others and because recovery is generally the most difficult for these people as well, such efforts would play "a central role in shaping a city's resilience" (Jabareen 2013, 224). Governance in this context goes beyond disaster recovery as an infrastructure and emergency service project to include broader policies toward empowering people to regain livelihoods and secure social spaces for living (UNISDR/UNESCAP 2012). If such efforts are not successful, environmental disasters can lead to downward spirals, lessening resiliency into the future.

Where they are successful, research finds that environmental degradation that contributes to natural hazards is often reduced. The UNDP (2013) reports that new initiatives for "community-driven development" (CDD) programmes in Indonesia, Laos and the Philippines are successfully enabling communities and local institutions to take the lead. In addition to CDD initiatives, the government of Indonesia passed the Disaster Management Law in 2007 that gives citizens rights to protection from and during disasters. The law is implemented through fines and jail sentences to law offenders. It also identifies rights to information, education and training for disaster risk reduction and calls for the establishment of new national as well as provincial disaster management agencies that allow for the active participation of nongovernment organizations.

Thailand's Baan Mankong programme is an example of the ways in which lower-income neighbourhoods can become active as decision-makers in integrating disaster risk reduction and preparedness into their social and economic lives (Boonyabancha 2005; Archer 2010; UNDP 2013). In a Bangkok community exposed to repeated flooding, Baan Mankong residents worked with the government to improve infrastructure and houses, which meant raising them onto stilts, while also investing in flood preparedness, including establishing a fund to which neighbourhood members contribute $1 per month that is made available to families in need when flooding occurs. As the 2011 flood came to Bangkok, the community organized itself to prepare sandbags, opened a disaster centre at the uphill temple with a kitchen to provide food and stockpiled basic medical equipment. Volunteers from the community staffed the disaster centre. Because they were from the community, they readily understood the situations of families in need.

The Baan Mankong approach is an important case of what disaster governance can accomplish at the smaller urban scales of the neighbourhood. Its approach stands in contrast to the more general pattern of slum clearance and relocation by governments and advisors who use sector approaches and disaster management understandings to prioritize canal clearance over trying to sustain the social fabric of neighbourhoods and livelihoods.

2.7.2 The City Region Scale

While many of the most serious impacts of disasters are experienced at neighbourhood or community scales, the incubation of compound disasters typically occurs at the scale of the city region. In Asia's MURs the dynamics at play that lead to flood disasters include (EEPSEA 2009; Alcamo 2009; Bates et al. 2008; Marcotullio 2007; Firman 2010):

- Land subsidence from overdrawing groundwater
- Saltwater intrusions into underground water supplies
- Deforestation in upland areas
- Massive increases in non-porous ground cover
- High population densities and loss of open spaces
- Growth of low-quality settlements along waterways
- Uncontrolled dumping of waste into waterways
- Infrastructure failures

The location of most MURs in coastal areas adds the additional challenges of sea rises and heightened vulnerability to extreme weather events. Together, these phenomena comprise a multistranded assault on urban ecologies that requires a regional scale approach to address.

The search for a form of governance that can integrate planning at the city region scale is a long-standing one (Freire 2006; Laquian 2005). Solutions are wide ranging, including the establishment of consolidated metropolitan planning authorities, federations of local governments and bilateral functional agreements to share specific services and infrastructure between contiguous countries (Vogel et al. 2010). Several metropolitan regions, such as Beijing, Hanoi, Osaka, Seoul, Shanghai, Tianjin and Tokyo, have been placed under unified governance by expanding their physical boundaries to absorb other cities and rural districts into them. In the cases of Beijing, Shanghai and Tianjin, for example, the central government created single unitary governments headed by mayors appointed by the national government.

The Tokyo Metropolitan Region is also a consolidated form of MUR governance. It has been able to implement a comprehensive development plan encompassing its core 23 wards within Tokyo City and parts of the prefectures of Saitama, Kanagawa and Chiba and the district of Tama. Significant results of this consolidation have been reported in terms of traffic congestion alleviation through region-wide transportation planning (Laquian 2005). Air pollution has also been reduced.

Although many proposals have been put forth for environmental sustainability, such as those for compact cities, linear cities, smart cities, eco-cities and, simply, green cities, these do not directly speak to issues such as how to guide urban growth in a way that minimizes the potential for flooding. In terms of disaster governance, a return to McHarg's (1995) concepts of guiding urban form in a way that best supports city region ecologies, particularly to avoid flooding, would be more relevant than many eco-city proposals are. Thus, conversion of land to urban uses would avoid steep uplands, waterways needed for natural drainage, aquifers where rain runoff can collect and be absorbed to replenish groundwater supplies and coastal zones that are critical interfaces between land and sea ecologies.

The relationship between urban form and disaster management is gaining recognition. The new master plan for Hanoi, for example, reserves 70 % of the land area to be open and green till at least 2030. This is made possible, in part, by the government decision in 2008 to incorporate a surrounding province and several districts of other provinces into a single consolidated municipality. Overnight, Hanoi became one of the largest cities in the world in terms of area, which now encompasses an area of 3,328 km^2, or three times its former geographical size and more than twice its previous population. As a master plan by international urban design teams that entails sweeping changes to the city region and has no public participation, the new plan remains controversial in that while the landscape might be flood resistant, how lower-income households fit into it is not revealed in detail.

Efforts underway in the Pearl River Delta of China between HK, Macao and Guangdong to create an "Urban Cluster Coordinated Development Program" (CCDP) as a regional scale of governance take the form of a federation of local governments, in this case ones with unique administrative status and border authorities. Among the many objectives of this rescaling of territorial government is to provide open space and agreed upon environmental conditions to control overall development cooperatively. Spatial guidance for individual cities is included in the programme to ensure that the development of cities is in accordance with regional spatial strategies. A principal intention of the plan is to divert development away from ecologically sensitive areas (Vogel et al. 2010).

From the 1970s in Indonesia, attempts were also made to create a super regional scale of planning with proposals for a broader spatial scale of planning for Jakarta. Coined as Jabotabek, it sought to combine Jakarta with parts of West Java as an appropriate environmental scale to coordinate planning in the emerging mega-urban region extending far beyond Jakarta's boundaries (Douglass 1991; Jones and Douglass 2008). However, neither the national nor provincial governments have officially adopted any of these proposals. Over the ensuing years, the growth of this city region has been among the most spectacular in Asia, which jumped from under five million in 1970 to nearly 30 million inhabitants in 2010, representing annual increases of over half a million people per year (Kurniawati 2009).

Now officially called Jabodetabek-Punjur,[5] the greater Jakarta region has generated environmental sustainability problems that also continue to increase in

---

[5]As the mega-urban region of Jakarta has expanded, so has the name for it, beginning in the 1970s with Jabotabek, then Jabodetabek and now Jabodetabek-Punjur to signal its expansion toward Bandung.

scale and impacts (Arai 2001; Firman 2004; Peresthu 2005; Tunas 2008).[6] The floods of 1996, 2002, 2007 and 2013 were the greatest and most destructive ever recorded in the city's long history.[7] Political reform and the devolution of government to provincial and regency levels have potentially generated a new era for a federated approach toward governing the Jabodetabek MUR. As annual flooding becomes more widespread and mutually destructive beyond the Jakarta DKI,[8] this might also increase incentives to begin to effectively rescale governance to make this mega-urban region more resilient in the face of increasing disaster risks.

Despite differences in institutional arrangements, all efforts to rescale urban governance have in common the attempt to achieve a high level of coordinated planning capacities at the city region scale that can overcome tendencies among local governments in the region to disregard opportunities and impacts that cross borders. However, the idea of moving upward toward city region scales of governance for disaster preparedness and resilience is still novel, and the role of disaster governance in reducing vulnerability and exposure to hazards is just beginning to be recognized.

2.7.3 Transborder Riparian Regions

Transborder riparian basins are among the most at risk of compound disasters. Asia's riparian systems are immense. They include at least 40 major transborder rivers and lakes (Fig. 2.3), totalling more than 16 million square kilometres of land area in their basins. As Asia urbanizes, cities reach ever more deeply into these regions to harness water supplies, generate hydropower, appropriate natural resources and reorganize local economies to serve urban demand. Global climate change adds to their transformations as the Tibetan-Himalayan glaciers rapidly recede and are likely to disappear and unusual weather events create extraordinary episodes of flooding and landslides that threaten life, livelihoods, social and cultural practices and biodiversity throughout each region (UNEP 2006, 2007).

---

[6] The gap between low-cost housing provision and demand continues to increase and is now reaching a deficit of 800,000 units (Widoyoko 2007).

[7] In the 2007 episode as much as 75 % of the city was flooded, displacing a recorded 430,000 people, mostly poor, from their homes (BBC 2007; Steinberg 2007). Health impacts—diarrhoea, skin and respiratory problems and dengue fever—breakdown of basic urban services and loss of livelihoods lingered long after the floodwaters subsided (Yuniar 2009). Thousands of homes were totally destroyed, and business losses were estimated to total $1 billion (Rukmana 2011). The 2013 torrential rains flooded more than 100,000 homes, left 47 people dead and shut down the entire city of 10 million people for several days (Jakarta Globe 2013). The estimated economic cost of the flood is more than $3 billion.

[8] Jakarta DKI (Daerah Khusus Ibukota) is the name given to Jakarta as a special city region with status equal to that of a province.

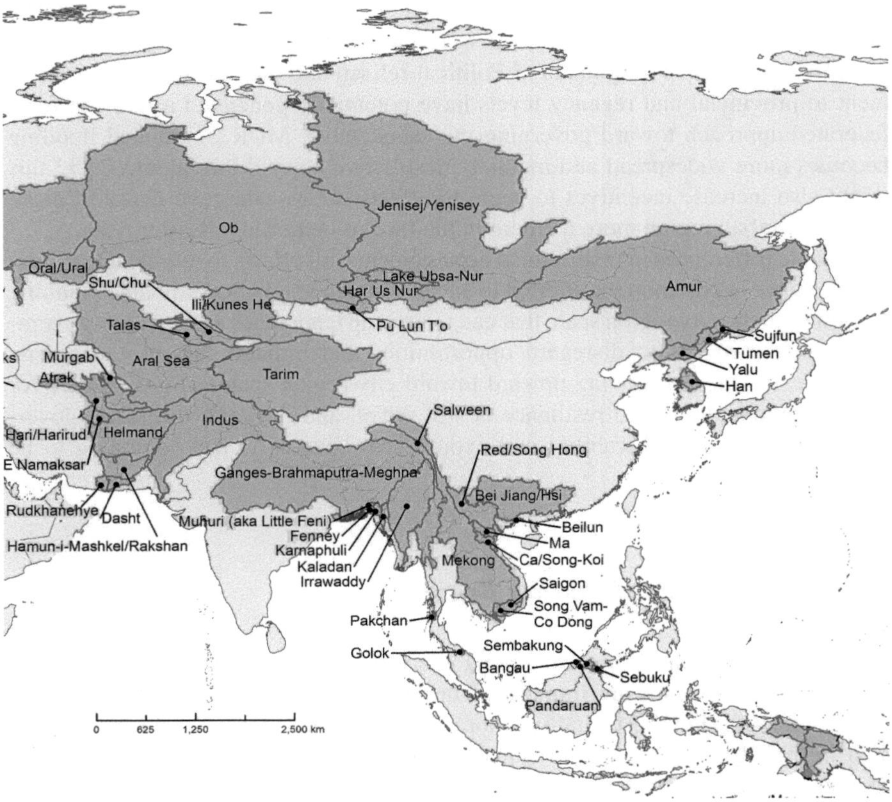

Fig. 2.3 Transborder riparian regions of Asia (*Source*: UNEP (2007))

Among the more controversial urban intrusions into riparian regions is the construction of mega-dams. Research by the independent World Commission on Dams (2000) found that the damaging ecological impacts of the larger dam projects are substantial and, in some instances, irreversible (WCD 2000; ICEM 2007). When riparian regions cross national borders, nationalism and a lack of incentives for upstream countries to be concerned about downstream disasters have made riparian governance extremely difficult (Ashayagachat 2008; Chellaney 2007; Dinar et al. 2007; Gunn and McCartan 2008).[9] These observations lead to the need for cooperative transborder mechanisms to govern riparian regions that are able to cover four dimensions of governance. Among the key components of a much needed riparian governance outlined by Douglass (2011) are:

- *Information processing and sharing*. Governments of upstream countries can be unwilling to share information with downstream countries about impending

[9] Most large dams are significantly underperforming with reduced holding capacities from silting and thus power-generating ability (Bauer and Rudolph 2001; WCD 2000).

water diversion, dam construction and waste disposal. Multiple forms of data are needed from different sources to ensure open dialogue and greater trust among stakeholders (Mirumachi and Nakayama 2007; Wyatt and Baird 2007).

- *Transborder agreements* to establish legitimacy for long-term transborder cooperation and conflict resolution. These can take many forms: treaties, compacts, memoranda of understanding, protocols and others, including personal relations of trust among national leaders.
- *Civil society participation* to ensure that local social and economic interests are included in decisions that reshape regional ecologies and human lives that depend on them.
- *Supranational governance authority* to establish a significantly autonomous and neutral source of information, agreement brokering and venue for participatory transborder regional governance.

The Mekong River Basin provides one approach to transborder riparian governance that more recently includes concerns over environmental disasters. Running from the Tibetan Plateau through six countries, the Mekong is the 12th longest river in the world (Fig. 2.4). Approximately 60 million people live in the Lower Mekong Basin where the river supplies water for drinking, irrigation, hydropower, transportation and commerce for and connected with the region's fast expanding cities. It serves millions more in China and Myanmar (UNDP 2006).

Transformations of the basin have intensified in recent decades (ADB 2008; WWF 2008). The Mekong River Basin is at high risk of flood disasters. Almost 70 % of the forest cover is gone. The 2011 floods set new records (International Rivers 2013). In the same year a study by Princeton University concluded that if the 27 new hydropower dams planned for the region were built, they "could have a catastrophic impact on the river's fishery and millions of people who depend on it" (Sullivan 2012, 1).

The Mekong River Commission (MRC) was re-established in 1995 as a means to create a form of transborder governance capable of guiding development toward more sustainable ecology and livelihoods of people in the region (Jacobs 2002). Funded by UNDP, the MRC includes provisions for cooperative natural resource planning, environmental and social cost management, databases and information systems and organizational management and cooperation. China and Myanmar are dialogue partners of the MRC to further its reach for transborder cooperation. The 2001 Work Programme represented an important shift toward creating region-wide approaches to planning rather than continue to focus on individual projects. It also included the idea of the MRC as a "learning organization" that was to engage civil society organizations in finding "bottom-up" solutions to river basin planning issues, particularly with regard to livelihoods.

The MRC is also attempting to turn toward a transborder regional development approach through "Integrated Water Resource Management" (IWRM). Defined by the Global Water Partnership as "coordination of development and management of water, land and other resources for maximizing of economic results and social welfare with no compromise on environment" (GWP 2003), the central principals of

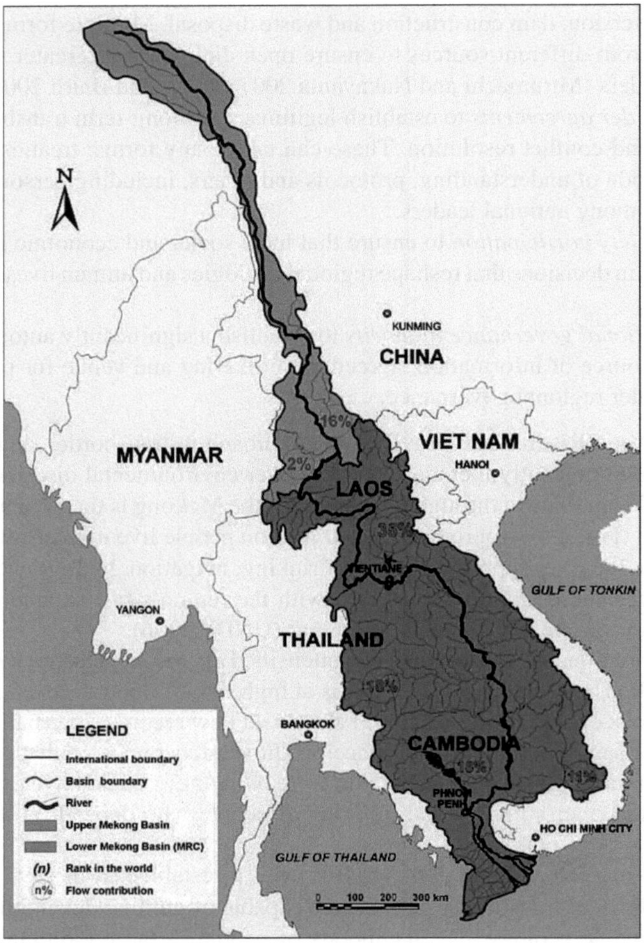

Fig. 2.4 The Mekong river basin (*Source*: The Mekong River Commission (MRC 2009))

the IWRM are participation and integration of the resources, institutions and stakeholders for sustainable water resources. Transborder collaboration across the riparian region is included in the approach (Biswas 2008).

As a donor-driven organization that has been criticized for putting urban-centred interests over those of the environmental and riparian populations, the MRC faces continuing challenges in getting governments to buy into the IWRM formulation as a new approach toward riparian governance with environmental disaster elements (Varis et al. 2008a, b; Hirsch and Jensen 2006). In response, the MRC has stepped up research and dissemination of technical reports on major trends and issues in the region and has begun requiring large-scale projects to do environmental impact statements before, during and after construction (MRC 2009). Given the often contentious political contexts of transborder riparian planning and the unlikelihood of supranational regulatory institutions to appear in the near future, the most promising areas for collaboration remain in the realm of transparency in information analysis and sharing, open forum for discussions and, particularly among powerful international funders, continued openings for "bottom-up" planning from within the regions at very local levels.

2.8 Conclusions: The Urban Matrix of Disaster Governance

> Natural disasters occur in a political space. (Cohen and Werker 2008, 795)

> Public involvement is critical in all aspects of disaster risk planning from central to local governments and to community levels…It is important to decentralize policies and customize them according to local needs and priorities. (ADB 2013)

Asia's rapid urban transition and its effects on environmental disasters call for new approaches to disaster governance, not just for cities but also for the regions incorporated into their spheres of influence. The sheer size of still expanding urban agglomerations that now include millions of people in a single contiguous space presents novel conditions for approaching the many questions about how to confront the portent of increasing disasters. The gravitation of urban populations to coastal and river basin areas with high disaster risk magnifies the need to reflexively respond to scales and impacts of disasters for which societies have little previous experience. Urbanization in Asia also adds new types of vulnerabilities in need of concerted attention, which in addition to children, elders and disabled people would include the millions of people living precariously in high disaster-prone slums.

In the emerging urban matrix of social, political and economic relations, disasters can quickly compound into other disasters that cascade through spatial systems, with events in one location impacting conditions in another in almost instantaneously unexpected ways. As cities organize distant natural resource regions and other rural sites to supply energy and resources to them, disasters can incubate unnoticed or unreported in jurisdictions with weak authority to oversee projects such as nuclear power plants or mega-dams.

Disasters emanating from such areas can boomerang back to the metropolitan regions that put them in place. They can also leap frog from city region to city region on a global scale, with the impact of an earthquake or tsunami shutting down production in one city region or country creating a production crisis in another.

While physical recovery of a disaster site might be possible in a relatively short period of time, finding resilience to both its social consequences and its compound effects can be expected to take much longer and in some cases might not occur at all. Preparing for possible disasters thus extends beyond a single event, and for this reason governance should be given attention to underscore the need for a social learning process that includes many perspectives and voices in political processes of decision-making. As summarized by Jabareen (2013, 227):

> City and community resilience is a phenomenon that is complex, non-deterministic, dynamic in structure, and uncertain in nature. It is a phenomenon that is affected by a multiplicity of economic, social, spatial, and physical factors. Its planning involves a wide range of stakeholders including civil society, local and national governments, the private sector, and various professional communities, and it therefore affects a variety of urban communities and city residents.

Though not automatic, participatory governance can provide an arena for pursuing social and environmental justice in responding to environmental disasters. This entails governance institutions and mechanisms organized at the critical scale of critical "problem sheds" (Allen 1998), three of which are the neighbourhood, city region and transborder riparian region. The case of transborder riparian regions presents formidable challenges in improving transborder governance capacities. The social and economic life of most of continental Asia depends on the ability to prevent and recover from what seems certain to be increasing environmental disasters in these regions. Asia's urban transition figures highly in the matrix of riparian region governance as cities reach ever more deeply into these regions to build dams for hydroelectricity and water for urban, industrial as well as commercial agricultural uses.

Decentralization, democratization and participatory planning are fundamental in creating arrangements that are able to work horizontally and vertically over space to link the smaller scales of daily life spaces with city- and regional-level planning processes. Success at one scale can magnify problems at another, however, and mechanisms are needed to overcome the tendency of boundaries to separate or generate competitive relations rather than integrate through collaborative agreements among them. In terms of disaster governance, an important consideration is whether decentralization of government powers within nation states would allow and encourage city regions most affected by transborder disasters to proactively engage in disaster governance across borders with other affected city regions. This is a question that has yet to have significant experiential evidence to assess.

The five types of effects of Asia's urban transition on environmental disasters at three geographical scales discussed above are illustrative of the multiplicity of ways in which our understanding of environmental disasters needs to be recalibrated in research and policy. Instead of understanding disasters as separate events, we need to understand them as ruptures that reflexively compound through chains of interdependencies that begin well before a disaster and ripple long after in unexpected ways. To cover these temporally and spatially wider complexities, we need perspectives from many disciplines and knowledge from multiple sources, most especially from people who live in localities experiencing a natural disaster. By helping to triangulate competing reconstructions of events, a scanning of many sources of knowledge is important not only for building consistent evidence-based records and coherent explanations of disasters and their impacts. Through such a process, research can also better contribute to the production of knowledge as a social learning process that can bring the knowledge and skills of academics and disaster specialists into conversation with the richly contextual knowledge of people in disaster localities. In this way, too, it can make needed contributions to inclusive processes of disaster governance.

References

Alcamo, J. (2009). Climate change and the changing frequency of floods and droughts: Scenario analysis of risk and adaptation in Europe. *Earth and Environmental Science*, 6, 292016. http://www.iop.org/EJ/toc/1755-1315/6/29. Accessed 7 June 2014.

Allen, T. (1998). Watersheds and problemsheds: Explaining the absence of armed conflict over water in the Middle East. *Middle East Review of International Affairs*, 2(1), 49–51.

Arai, K. (2001). Only yesterday in Jakarta: Property boom and consumptive trends in the late new order metropolitan city. *Southeast Asian Studies*, 38(4), 481–511.

Archer, D. (2010, September 19–23). *Empowering the urban poor through community - based slum upgrading: The case of Bangkok, Thailand*. In 46th ISOCARP Congress, Nairobi.

Asgary, A., Anjum, M. I., & Azimi, N. (2012). Disaster recovery and business continuity after the 2010 flood in Pakistan: Case of small businesses. *International Journal of Disaster Risk Reduction*, http://dx.doi.org/10.1016/j.ijdrr.2012.08.001. Accessed 7 June 2014.

Ashayagachat, A. (2008, November). Dams upriver hurting people living downstream. *Indus Asia Online Journal*. http://iaoj.wordpress.com/2008/11/14/dams-upriver-hurting-people-living-downstream. Accessed 15 Jan 2009.

Asia Development Bank (ADB). (2008). *Regional cooperation operations business plan Greater Mekong Subregion 2009–2011*. Manila: Asia Development Bank.

Asia Development Bank (ADB). (2013, April). *Disaster risk management in Asia and the Pacific*. ADB, Issues Paper. Manila: ADB.

Assmuth, T., Hildén, M., & Benighaus, C. (2009). Integrated risk assessment and risk governance as socio-political phenomena: A synthetic view of the challenges. *Science of the Total Environment*, 408, 3943–3953.

Bahl, R. (2005). Promise and reality of fiscal decentralization. In S. Ichimura and R. Bahl (Eds.), *Decentralization policies in Asian development* (pp. 1–26). Singapore: World Scientific Publishing.

Baker, D., & Refsgaard, K. (2007). Institutional development and scale matching in disaster response management. *Ecological Economics*, 63, 331–343.

Bankoff, G., & Hilhorst, D. (2009). The politics of risk in the Philippines: Comparing state and NGO perceptions of disaster management. *Disasters*, 33(4), 686–704.

Bates, B., Kundzewicz, Z., Palutikof, J., & Wu, S. (2008). *Climate change and water*. Technical Paper presented at the Intergovernmental Panel on Climate Change. London: IPCC Working Group II Technical Support Unit.

Baur, J., & Rudolph, J. (2001). *Water facts and findings on large dams*. D+C Development and Cooperation, Deutsche Stiftung für internationale Entwicklung (DSE). http://www.inwent.org/E+Z/zeitschr/de201-4.htm. Accessed 7 June 2014.

BBC. (2007, February 7). *Jakarta's deadly floods receding*. http://news.bbc.co.uk/2/hi/asia-pacific/6333945.stm. Accessed 7 June 2014.

Biswas, A.K. (2008). Integrated water resources management: Is it working? *International Journal of Water Resources Development*, 24(1), 5–22.

Boonyabancha, S. (2005). Baan Mankong: Going to scale with "slum" and squatter upgrading in Thailand. *Environment & Urbanization*, 17(1), 21–46.

Bulkeley, H. (2010). Cities and the governing of climate change. *Annual Review of Environment and Resources*, 35, 229–253.

Chellaney, B. (2007, June 26). China aims for bigger share of South Asia's water lifeline. *Japan Times*, http://www.japantimes.co.jp/opinion/2007/06/26/commentary/china-aims-for-bigger-share-of-south-asias-water-lifeline/ Accessed 7 June 2014.

Cohen, C., & Werker, E. (2008). The political economy of "natural" disasters. *The Journal of Conflict Resolution*, 52(6), 795–819.

Demographia World Urban Areas. (2013, March). *Largest urban agglomerations in the world*, 9th annual edition. www.demographia.com/db-worldua.pdf. Accessed 7 June 2014.

Dinar, A., Dinar, S., McCaffrey, S., & McKinney, D. (2007). *Bridges over water: Understanding transboundary water conflict, negotiation and cooperation*. New Jersey: World Scientific.

Displacement Monitoring Centre (DMC). (2013, May 13). *Displaced by disasters: 32.4 million people uprooted in both rich and poor countries*. http://blog.internal-displacement.org/category/disasters/. Accessed 1 Aug 2013.

Douglas, I. (2009). Climate change, flooding and food security in South Asia. *Food Security*, 1, 127–136.

Douglass, M. (1991). Planning for environmental sustainability in the extended jakarta metropolitan region," Ch. 12. In N. Ginsburg, B. Koppel and T.G. McGee (Eds.), *The extended metropolis: Settlement transition in Asia*. Honolulu: UH Press.

Douglass, M. (2010). Globalization, mega-projects and the environment: Urban form and water in Jakarta. *Environment and Urbanization*, 1(1), 45–65.

Douglass, M. (2011). Cross-border water governance in Asia. In S. Cheema (Ed.), *Cross-border governance in Asia and the Pacific* (pp. 122–168). Tokyo: United Nations University Press.

Douglass, M. (2013). Decentralizing governance in a transborder urban age: East Asia and the Busan–Fukuoka "Common Living Sphere". *Pacific Affairs*, 86(4), 731–758.

EEPSEA. (2009). *Climate change and Southeast Asia – Compilation of vulnerability database*. http://www.eepsea.cc-sea.org/pages/resource/sociecon.html. Accessed 17 Jan 2014.

Firman, T. (2004). New town development in Jakarta metropolitan region: A perspective of spatial segregation. *Habitat International*, 28, 349–368.

Firman, T. (2010). Multi local-government under Indonesia's decentralization reform: The case of Kartamantul (The Greater Yogyakarta). *Habitat International*, 34, 400–405.

Freire, M. (2006, April 6–9). *Urban planning: Challenges in developing countries*. Paper presented at the International Congress on Human Development, Madrid.

Friedmann, J. (1992). *Empowerment: The politics of alternative development*. New York: Basil Blackwell.

Friedmann, J. (2011). *Insurgencies: Essays in planning theory, London and New York*. London: Routledge.

Global Water Partnership (GWP). (2003). *Integrated water resources management toolbox, Version 2*. Stockholm: Global Water Partnership Secretariat.

GreenAsh. (2013). *Natural disaster risk levels of the world's largest cities*. http://greenash.net.au/thoughts/2013/03/natural-disaster-risk-levels-of-the-worlds-largest-cities/. Accessed 8 May 2013.

Grimm, N., Faeth, S., Golubiewski, N., Redman, C., Wu, J., Bai, X., & Briggs, J. (2008). Global change and the ecology of cities. *Science*, 319, 756–760.

Gunn, G., & McCartan, B. (2008, August 31). Chinese dams and the Great Mekong floods of 2008. *Japan Focus*. http://www.japanfocus.org/-B-McCartan/2865. Accessed 14 Apr 2014.

Healey, P. (2007). *Urban complexity and spatial strategies: Towards a relational planning for our times*. New York: Routledge.

Healey, P., & Upton, R. (Eds.) (2010). *Crossing borders: International exchange and planning practices*. Oxon: Routledge.

Hirsch, P., & Jensen, K. (2006). *National interests and transboundary water governance in the Mekong*. Australian Mekong Resource Centre. http://www.mekong.es.usyd.edu.au/projects/mekong_water_gover-nance.htm. Accessed 3 Oct 2008.

International Centre for Environmental Management (ICEM). (2007). *Pilot strategic environmental assessment in the hydropower sub-sector-Vietnam*. Final report—Risks to biodiversity from hydropower in the 6th Power Development Plan. Indooroopilly: ICEM.
International Rivers. (2013, March 28). *The lower Mekong dams factsheet text: A transboundary water crisis*. http://www.internationalrivers.org/resources/the-lower-mekong-dams-factsheet-text-7908. Accessed 6 June 2014.
Jabareen, Y. (2013). Planning the resilient city: Concepts and strategies for coping with climate change and environmental risk. *Cities*, 31, 220–229.
Jacobs, J. (2002). The Mekong River commission: Transboundary water resources planning and regional security. *The Geography Journal*, 168(4), 354–364.
Jakarta Globe. (2013, January 24). *Jakarta flooding highlights prevention gaps*. http://www.thejakartaglobe.com/archive/jakarta-flooding-highlights-prevention-gaps/
Jha, A., & Stanton-Geddes, Z. (2013). *Strong, safe, and resilient: Strategic policy guide for disaster risk management in East Asia and the Pacific*. Washington, DC: The World Bank.
Jones, G., & Douglass, M. (Eds.). (2008). *The rise of mega-urban regions in Pacific Asia – Urban dynamics in a global era*. Singapore: Singapore University Press.
Kawata, Y. (2011). Downfall of Tokyo due to devastating compound disaster. *Journal of Disaster Research*, 6(2), 176–184.
Kurniawati, D. (2009, July 24). The floods: A swelling city is at the root of the problem. *Jakarta Globe*. http://www.thejakartaglobe.com/waterworries/the-floods-a-swelling-city-is-at-the-root-of-the-problem/319991. Accessed 14 Apr 2014.
Laquian, A. A. (2005). Metropolitan governance reform in Asia. *Public Administration and Development*, 25, 307–315.
Marcotullio, P. J. (2007). Urban water-related environmental transitions in Southeast Asia. *Sustainability Science*, 2(1), 27–54.
McGee, T.G. (1991). The emergence of *desakota* regions in Asia: Expanding a hypothesis. In N. Ginsburg, B. Koppel, & T.G. McGee (Eds.), *The extended metropolis: Settlement transition in Asia* (pp. 3–35). Honolulu: University of Hawaii Press.
McHarg, I.L. (1995). *Design with nature*. London: Wiley.
Mekong River Commission (MRC). (2009). *The Mekong River commission for sustainable development*. http://www.mrcmekong.org/. Accessed 15 Jan 2009.
Ministry of Economy, Trade and Industry (METI). (2011). *Japanese industry-lasting change in manufacturing industry*. Tokyo: METI.
Mirfenderesk, H., & Corkill, D. (2009). Sustainable management of risks associated with climate change. *International Journal of Climate Change Strategies and Management*, 1(2), 146–159.
Mirumachi, N., & Mikiyasu Nakayama, M. (2007). Improving methodologies for transboundary impact assessment in transboundary watercourses: Navigation channel improvement project of the Lancang-Mekong River from China-Myanmar boundary marker 243 to Ban Houei Sai of Laos. *International Journal of Water Resources Development*, 23(3), 411–425.
Mol, A. (2009). Urban environmental governance innovations in China. *Current Opinion in Environmental Sustainability*, 1, 96–100.
Mulvihill, P., & Ali, S.H. (2007). Disaster incubation, cumulative impacts and the urban/ex-urban/rural dynamic. *Environmental Impact Assessment Review*, 27, 343–358.
Neumayer· E., & Plümper, T. (2007). The gendered nature of natural disasters: The impact of catastrophic events on the gender gap in life expectancy, 1981–2002. *Annals of the Association of American Geographers*, 97(3), 551–566.
Nicholls, R.J., Hanson, S., Herweijer, C., Patmore, N., Hallegatte, S., Corfee-Morlot, J., Chateau, J., & Muir-Wood, R. (2007). *Ranking port cities with high exposure and vulnerability to climate extremes exposure estimates*. Paris: OECD.
Nussbaum, M. (2002). Capabilities and social justice. *International Studies Review*, 4(2), 123–135.

Peresthu, A. (2005). *Jakarta's "exurbia" kampongs*. http://www.etsav.upc.edu/urbpersp/num01/inf01-1.htm. Accessed 14 Apr 2014.

Rees, W. (1996). Revisiting carrying capacity: Area-based indicators of sustainability. *Population and Environment*, 17(3), 195–215.

Rukmana, D. (2011). *Jakarta annual flooding in 2011*. http://indonesiaurbanstudies.blogspot.com/2011/02/jakarta-annual-flooding-in-february.html. Accessed 14 Apr 2014.

Satterthwaite, D., Huq, S., Pelling, M., Reid, H., & Romero, P. (2007). *Adapting to climate change in urban areas: The possibilities and constraints in low and middle-income nations* (International Institute for Environment Working Paper). London: International Institute for Environment and Development.

Schneider, A., Friedl, M., & Potere, D. (2009). A new map of global urban extent from MODIS satellite data. *Environmental Resources*, 4, 1–11.

Sen, A.K. (1990). Food, economics, and entitlements. In J. Staatz & C. Eicher (Eds.), *Agricultural development in the third world* (pp. 189–205). Baltimore: Johns Hopkins University Press.

Steinberg, F. (2007). Jakarta: Environmental problems and sustainability. *Habitat International*, 31, 354–365.

Sullivan, J. (2012). *Current stories plans for dams on Mekong River could spell disaster for area fisheries*. http://www.princeton.edu/main/news/archive/S33/58/78C70/index.xml?section=topstories&via=Princeton. Accessed 14 Apr 2014.

Tunas, D. (2008). *The spatial economy in the urban informal settlement*. Leuven: Faculteit Sociale Wetenschappen, Katholieke Universiteit.

Turner, B.A. (1976). The organizational and interorganizational development of disasters. *Administrative Science Quarterly*, 21, 378–397.

United Nations (UN). (2011). *Global assessment report on disaster risk reduction 2011*. New York: United Nations.

United Nations (UN). (2013). *Global assessment report on disaster risk reduction 2013*. New York: United Nations.

United Nations Development Programme (UNDP). (2006). *The challenges of governance*. New York: United Nations.

United Nations Development Programme (UNDP). (2012). *The millennium development goals report 2012*. http://www.un.org/en/development/desa/publications/mdg-report-2012.html. Accessed 14 Apr 2014.

United Nations Development Programme (UNDP). (2013), *Disaster risk reduction, governance & mainstreaming*. New York: United Nations.

UNESCAP. (2012). *The state of Asian cities 2010/11*. Bangkok: UNESCAP.

United Nations Environment Programme (UNEP). (2006). *GEO yearbook 2006 – An overview of our changing environment*. Nairobi: UNEP.

United Nations Environment Programme (UNEP). (2007). *Global outlook for ice and snow*. Nairobi: UNEP Division of Early Warning and Assessment.

UNESCAP. (2011). *The state of Asian cities 2010/11*. Bangkok: United Nations Economic and Social Commission for Asia and the Pacific.

UNESCO (United Nations Educational, Scientific and Cultural Organization). (2006), *Water: A Shared Responsbility – the United Nations World Water Development Report 2*. www.unesco.org/water/wwap. Accessed 22 August 2015.

UNISDR/UNESCAP. (2012). *Asia-Pacific disaster report 2012 – Reducing vulnerability and exposure to disasters*. Bangkok: UNESCAP.

Varis, O., Keskinen, M., & Kummu, M. (2008a). Mekong at the crossroads. *Ambio*, 37(3), 146–149.

Varis, O., Rahaman, M.M., Virpi, S. (2008b). The rocky road from integrated plans to implementation: Lessons learned from the Mekong and Senegal river basins. *International Journal of Water Resources Development*, 24(1), 103–121.

Vogel, R.K., Savitch, H.V., Xu, J., Yeh, A.G.O., Wu, W., Sancton, A., Kantor, P., Newman, P., Tsukamoto, T., Cheung, P.T.Y., Shen, J., Wu, F., & Zhang, F. (2010). Governing global city regions in China and the West. *Progress in Planning*, 73(2010), 1–75.

Widoyoko, D. (2007). *Good governance and provision of affordable housing in DKI Jakarta, Indonesia*. Loughborough: Loughborough University Water, Engineering and Development Centre.

World Bank. (2005). *East Asia decentralizes*. Washington, DC: IBRD.

World Commission on Dams (WCD). (2000). *Dams and development: New framework for decision-making*. London: Earthscan.

World Wildlife Fund (WWF). (2008). *Mekong–Protecting the river of life from source to sea*. http://www.worldwildlife.org/what/wherewework/mekong/. Accessed 7 June 2014.

Wyatt, A.B., & Baird, I.G. (2007). Transboundary impact assessment in the Sesan river basin: The case of the Yali Falls Dam. *International Journal of Water Resources Development*, 23(3), 427–442.

Yuniar, Y. (2009, March 28). Indonesia dam bursts, killing scores. *Asia News*. http://online.wsj.com/article/SB123812301079354491.html. Accessed 14 Apr 2014.

Chapter 3
Building Cities in a Subduction Zone: Some Indonesian Dangers

Anthony Reid

Abstract This chapter seeks to assess the future risks to Indonesian mega-cities by examining the record of tectonically generated disasters in the past. It argues that there are two clear trends, to some extent balancing each other. On the one hand, the quantifiable impact of natural disasters is growing as populations increase and concentrates ever more in endangered coastal mega-cities. On the other, global advances in communications, and scientific understanding of the dangers of eruptions, tsunamis and earthquakes, provide unprecedented opportunities for preparedness. A third factor appears to be cyclical. Indonesian cities were (badly) planned and built, and attracted half the Indonesian population to endangered coastal locations, during a century which was unusually mild in terms of both volcanic eruptions and tsunamis. Since 2004 it has become clear that mega-tsunamis must recur every few centuries to release the build-up of pressure in Indonesian subduction zones. The periodicity of mega-eruptions is harder to predict, but even a repetition of the 19th century pattern would bring not only unprecedented death and destruction in the 21st, but incalculable disruption to agriculture and air transport.

Keywords Volcanoes • Tsunamis • Indonesia • Urbanisation • Tectonic risks • Subduction

3.1 Introduction

The twentieth century wrought a profound demographic as well as political transformation of Indonesia, as in Southeast Asia more generally. Although one of the world's least urban regions in 1900, Indonesia became almost 50 % urban during a century when the overall population more than doubled. The vast movement of people flowed overwhelmingly towards a small number of coastal cities whose location had been selected to suit the commercial and strategic needs of a trading company not very secure out of sight of its ships. Jakarta, Surabaya, Semarang,

A. Reid (✉)
Australian National University, Canberra, Australia
e-mail: anthony.reid@anu.edu.au

M.A. Miller, M. Douglass (eds.), *Disaster Governance in Urbanising Asia*,
DOI 10.1007/978-981-287-649-2_3

Padang, Makassar, Manado, Ambon and Kupang were transformed in the twentieth century from colonial ports to vast urban agglomerations in unusually endangered coastal locations. The century in which the transformation took place was one in which tectonic dangers were little considered, because their effects were relatively mild in that period and not well understood. A longer-term analysis shows that the twentieth century's relative mildness was a misleading guide, and the certainty of major disasters should be built into planning for the future.

3.2 Are We Living in Particularly Dangerous Times?

The recent round of natural disasters in the Asian maritime region has appeared exceptionally severe in the memory of those living today. The appalling earthquakes and tsunamis in Aceh and the Indian Ocean on 26 December 2004 and in Northeast Honshu on March 2011 were destructive on a scale unknown in human memory. Since they occurred in a rich country, the Tohoku earthquake and tsunami were ranked the most expensive natural disaster in human history by insurers (at US$210–300 billion) (Swiss Re 2013). The tsunami from the Aceh earthquake, slightly bigger at 9.2 magnitude, had ten times as many casualties (over 230,000) and is considered Indonesia's worst ever in terms of the number of deaths. Cyclone Nargis in the Irrawaddy Delta in 2008 (138,000 deaths) appeared to have no Burmese or Southeast Asian precedent and can only be matched globally by other cyclonic floods further west in the Bay of Bengal (1970, killed perhaps 500,000; 1991, killed 139,000).

Typhoons appear to be on an unprecedented scale in the Philippines, with no record of winds as high as Typhoon Haiyan's 315 km per hour or of a toll of death and destruction on the scale wrought on Tacloban and its surrounds on November 2013. The Philippines had also not experienced for centuries a volcanic eruption on the scale of Mount Pinatubo in 1991, the largest the world has seen in the past century in terms of explosive matter cast around Luzon and circling the planet. The floods that affected Bangkok and all the deltaic areas of Indo-China in 2011 were rated by Swiss Re as the most destructive of property of any freshwater flood in human history with US$47 billion in losses—even though it occurred in the poor global south where property is valued far less by the Zurich bean-counters than in North America and Europe. Despite that bias in their data, Swiss Re rates the top eight most endangered cities in the world all to be in our eastern Asian region headed by (in order) Tokyo, Manila, the Pearl River Delta, Osaka and Jakarta.[1]

[1] Swiss Re calculates in terms of the number of people exposed to risks rather than the degree of risk—which is of course far harder to quantify.

3.3 Are Natural Disasters Intensifying?

Of course human memory is short, while capitalism and demography ensure that the numbers keep increasing. The task of the historian is to situate the limited past of memory within the much longer past that is knowable. In order to help understand how these seemingly unprecedented disasters fit in the longer term, I want to introduce three different patterns that are interacting here—getting worse through scale, getting better through preparedness and cyclical movements in the tectonic subduction.

3.3.1 Getting Worse Through Scale

As Michael Douglass in this volume points out, Asian cities are growing at an astonishing pace, and the sheer demographic scale ensures that the numbers affected by particular disasters grow steadily greater. The flooding of coastal cities in cyclonic and other storm surges is the factor most directly aggravated by rapid urbanisation. The sea level is rising through global warming and the rapid melting of the ice-caps, while siltation of rivers and the asphalting of surfaces increase the tendency of rivers to flood. Most critically, Southeast Asian coastal cities are sinking at an alarming rate. Jakarta, the most perilous, is reckoned to be sinking at 5–10 cm a year and some districts as much as 26 cm a year as a result of overbuilding and the extraction of the water table below (Berg 2012; Wardany 2010). Its northern areas are already below the sea level. The same applies to the northern port areas of Semarang and Surabaya. Bangkok has recently been estimated to be sinking at 2–5 cm a year and Saigon at 2 cm a year. These Asian megacities are being critically endangered many times faster than Venice, though with only a fraction of the funding and technology to do something about it (UNOCHA 2010; WWF 1986).

When this factor of ever-worse flooding is linked to the ever-increasing agglomerations of people in these coastal cities, there is no doubt that the numbers affected every year by devastating floods can only increase into the tens of millions. The man-made causes of this crisis, notably the depletion of the aquifers, worsen this problem. It has been argued, however, that the tectonic interaction of plates may also be contributing to Jakarta's and perhaps even Bangkok's subsidence, which would put this problem in my third, cyclical, box.

With hindsight it is tragic that the twentieth-century planners and politicians did not move their administrative capitals to safer places before the massive expansion in their populations began. To do so now would be painful and politically almost unthinkable. Yet government has been the massive driver of urbanisation, and future administrative infrastructure is the easiest for planners to control. Only the military dictatorship in Burma was able to do what now appears to have been the right thing for the wrong reasons and fortunately before the massive urbanisation that capitalism will ensure in that newly opened country. The last dictator, General Than Shwe, chose to move his capital to centrally placed Naypyidaw in 2005, not because of planning reasons but due to paranoia about foreign conspiracies in Rangoon, cosmological advice from his soothsayer and emulation of past capital-building emperors. Yet Napyidaw immediately became one of the fastest growing cities in the world and was already Burma's third largest city by 2012. Unfortunately, it is less easy for democracies to make such radical decisions, but nevertheless the governance of growth requires new thinking about the siting of administrative centres.

3.3.2 Getting Better Through Preparedness

The human death toll of the tsunami in Japan was only a tenth that of the tsunami in Sumatra, and the death toll of the earthquake in California (Northridge 1994, 6.7 M) was less than a thousandth that of a comparable earthquake in Haiti (2010, 7.0 M). Under modern conditions, there is a great deal which rich countries can do to prepare themselves against disaster by introducing effective building codes and more ingenious engineering, by building sea and river dykes against flooding and by evacuating populations from regions facing imminent typhoon, volcanic eruption or even a tsunami. The best hope of avoiding terrible human loss is that the cities most exposed to disaster, almost all of them in Asia, will grow sufficiently in wealth, education and sophistication of governance to be able to emulate today's rich countries in their preparedness.

Even poor countries benefit greatly from the increase in global scientific and monitoring capacity. An excellent case in point was the eruption of Mount Pinatubo in western Luzon in 1991. This was measured as a 6 on the Volcanic Explosivity Index (VEI), the biggest eruption that Asia or the world had experienced since Krakatau in 1883. It had a huge impact globally in slowing planetary warming by darkening the skies with ash, and in Luzon itself it destroyed a quarter of a million homes and livelihoods. It behaved very cooperatively for the volcanologists, emitting a series of escalating blasts which convinced people that they should believe in science and get out. The first 5,000 people were evacuated in March 1991, then another 25,000 in a 20 km radius around the volcano on June 9, the evacuation of the US Clark Air Base on June 10 and finally another 30,000 people on June 12 before the eruption proper began 3 days later (Tantingo 2011; Rantucci 1995). Hence, only 800 people were killed directly by the eruption, compared with at least 35,000 by Krakatau (1883) or even the death toll (5,100 and 1,500, respectively) of the much smaller eruptions of Kelud, Java (1915), and Agung, Bali (1963). The longer-term effects of all the destruction of crops and buildings, displacement of people and disruption of climate are much more difficult to calculate.

One might therefore see our present predicament as a race between scale and governance: between the threats to Asia's cities becoming rapidly worse and the capacity to prepare for them growing steadily better. Governance is crucial and far more needs to be done internationally and domestically to prepare for future disasters.

3.3.3 Cyclical

But there is a third pattern of particular relevance to the tectonic subduction zones which govern much of Indonesia, New Guinea and the Philippines. This is the necessary periodic recurrence of a build-up of pressure as one tectonic plate grinds slowly below another and its release in massive earthquakes. The Indo-Australian plate moves north at about 6 cm a year and subducts under Sumatra, Java, Bali and the Lesser Sundas, while the Philippine plate to the east of the Philippines subducts under the eastern Philippines, southern Japan and the Marianas. The subducting plate generates molten magma squeezed out at depths of around 100 km and rising to accumulate in magma chambers below the earth's crust, where pressure gradually builds. Much is released underwater or leaks out gradually in some areas such as Hawaii. Explosive volcanic eruptions, the time and place of which are extremely hard to predict, are however a dangerous feature of the Indonesian section of the ring of fire, the area which places the largest populations in danger of volcanic disaster (de Boer and Sanders 2002).

There is more predictability about the places, if not the times, of the most dramatic release of pressure on the subducted plates in mega-earthquakes. The earthquakes (with 9 or above magnitude) and accompanying tsunamis in our own time bring the knowledge that similar events must have happened in the past, even if no records of them survive, and will happen again in the future. Unfortunately, the record of past events in Southeast Asia and the Pacific before 1900 is extremely inadequate, and hence, we risk being as unprepared by science for future mega-disasters as we were for the Sumatra tsunami of 2004. Though the record in Japan is far better, the geologists had been confidently predicting a tsunami-generating large earthquake in the Nankai Trough off the coast of Shikoku, rather than in the Tohoku area thought to have been less seismic because it was an older and deeper subduction—similar to that in Java.

The remainder of this chapter will focus on this cyclical pattern in Indonesia and the urgent need for research to establish from past events where the most exposed areas are and the approximate periodicity of mega-events.

3.4 The Long Twentieth Century: A Seemingly Quiet Tectonic Period

For Indonesians, the succession of disasters since 2004, summarised in Table 3.1, seems extraordinary.

Before this disastrous series, western Indonesia's last earthquake above 8 on the Richter scale was before the seismograph and Richter scale were invented around 1900. The earthquake and tsunami that devastated Padang and coastal Nias in 1861 have been reconstructed as a likely 8.4 magnitude (Newcomb and McCann 1987). It appears, therefore, that 143 years passed without a major tsunami-generating

Table 3.1 Indonesian geological disasters since 2004

Date	Magnitude	Place	Direct fatalities
26 December 2004	9.2 M+tsunami	Aceh-Nias	167,800
28 March 2005	8.6 M+tsunami	Nias	1,313
26 May 2006	6.3 M	Yogyakarta	5,749
17 July 2006	7.6 M+tsunami	Pangandaran, West Java	730
12 September 2007	8.4 M	Mentawai, Padang	25+
30 September 2009	7.6 M	Padang	1,117
25 October 2010	7.7 M+tsunami	Mentawai	400
26 October 2010	VEI 4 eruption[a]	Merapi, Java	324
11 April 2012	8.6 earthquake	Aceh	5
September 2013	VEI 2 eruptions	Sinabung, North Sumatra	31
13 February 2014	VEI 2 eruption	Kelud, East Java	4

[a]VEI = Volcanic Explosivity Index, offering a scale from 1 to the biggest recorded in modern times, Tambora at 7, discussed below

quake in the subduction zones of Sumatra or Java. Until the 2004 onwards disasters, neither scholars nor planners had given Indonesia's vulnerability the attention it deserved. After the Krakatau eruption of 1883, there had been more than a century of relative geological calm, without any sufficiently huge disaster to draw attention. If there were other major events, they seem to have escaped scientific observation or its communication to the broader world, prior to 1900 when much of Indonesia remained little known and during 1941–1970 when political upheavals disrupted such work. The Indonesian death toll from earthquakes, tsunamis and volcanic eruptions in the whole twentieth century was recorded (no doubt minimally) as 32,000. By hindsight, this appears to have been a misleading time to have begun to understand geological normality. It looks increasingly like a kind of lull between spasms. Each of the eruptions in the nineteenth century—Krakatau (1883) and Tambora (1815)—and the tsunami of 2004 killed more people and created more homelessness and destruction than all the recorded events of the twentieth century put together (ADRC 2000; Reid forthcoming).[2]

By comparison with this moderate geological death toll, the staggering damage of the dramatic political events of mid-century explains why the scholars, planners and politicians were preoccupied with other matters. At least 300,000 died in Indonesia during the Pacific War, 60,000 or more during the Indonesian revolution of 1945–1950, 40,000 in the rebellions of the 1950s, roughly half a million in the massacres of the Left after 1965 and around 65,000 after the Indonesian invasion of East Timor in 1975 (Clodfelder 1992, pp. 1127–1140).

[2]The best cumulative data set online is that of NOAA, National Geophysical Data Centre.

3.5 Projecting Further into Indonesia's Tectonic Past

To confirm the seeming quiet in the long twentieth century was a lull, not a norm; one must work harder on the earlier record, when reporting was woefully inadequate both from indigenous and foreign sources. Taking volcanoes first, the twentieth century's worst in the Southeast Asian region was Pinatubo (1991), VEI 6, with huge damage to property but relatively few casualties. In Indonesia, the most damaging volcanoes since Krakatau (1883) have all been rated as only 4 or less. Judging by death tolls, the worst were Kelud (East Java) in 1919, with 5,100 killed after its crater lake burst, though probably tens of thousands of indirect excess deaths were caused by crop failures (Reid 2013); Gunung Agung (Bali) in 1963, with 1,584 deaths; and Merapi (Central Java) in 1930, with 1369 deaths.[3] Galunggung (West Java) in 1982 was the first to alert the airline industry to the dangers of volcanic ash, which knocked out engines on two 747s out of Singapore but killed only a reported 72 people on the ground.

By contrast the nineteenth century had the two biggest eruptions the world has known in modern times, in Tambora (Sumbawa) in 1815 (VEI 7) and Krakatau, in the Strait between Java and Sumatra, in 1883 (VEI 6). The estimates of direct deaths on the ground were around 90,000 and 36,000, respectively, but the indirect effects around the world were far greater, with global temperature dropping by up to 10° in Europe in 1816, crop failures around the world and a resulting cholera epidemic. In Indonesia, Tambora produced total darkness for 3 days within a 300 km radius and ash up to 1 m deep in Sumbawa, 30 cm in Lombok and 10 cm in Bali (de Boer and Sanders 2002). Recent studies of the demographic effects of Krakatau have shown over 100,000 "missing" people throughout West Java by 1885, in comparison with Java's overall growth rate of 1.6 % (Brata et al. 2013; Reid 2013). In addition to these two, the explosion of Mount Galunggung (West Java) in 1822, rated VEI 5 by the amount of debris it cast into the atmosphere, was also bigger than anything in Indonesia's twentieth century. Some 4,000 people were thought to have been killed directly.

The Southeast Asian data prior to the nineteenth century is very limited. But the recent progress in charting global climate changes has revealed new evidence of periodic mega-eruptions in Indonesia that must have been responsible for ash deposits in the two polar ice-caps, as well as global cooling, crop failures and famines in the historical record of the Northern Hemisphere. The most definite of these now appears to be the eruption of Lombok's volcano, Samalas, on the edge of what is left as Mount Rinjani, in 1257. This was a bigger explosion than Tambora and appears to have caused even more havoc around the world (Emile-Geay et al. 2008; Hilts 2012; Lavigne et al. 2013). No doubt, therefore, it must have virtually depopulated Bali and Lombok at that time. The sources of other dramatic incidents of global cooling and of ash deposits in the polar caps, in AD 535 and 1452 and in the 1590s, have yet to be clearly identified, but Indonesia is the chief suspect. Another

[3] http://en.wikipedia.org/wiki/List_of_volcanic_eruptions_by_death_toll

great eruption in Central Java is thought to have ended the Mataram civilisation that built Borobudur and Prambanan around AD 920.

The picture for earthquakes and tsunamis is similar. As mentioned above, the Sumatra subduction has produced four major quakes since 2004 of greater magnitude than anything Sumatra had experienced in the twentieth century. The worst quakes Indonesia faced in that first century when measurements were possible were all in less populated areas of East Indonesia—an 8.5 in the Banda Sea (1938), an 8.2 in Papua (1996) and an 8.0 in North Maluku. Fortunately, the death toll from the resultant tsunamis was in the hundreds at most. Bigger death tolls came from tsunamis produced by lesser earthquakes but in more populous areas. For instance, 1,200 died in Lembata in 1979, 1,170 in Flores in 1992, 600 in the Makassar Strait in 1969 and 223 in East Java in 1994.[4]

In the nineteenth century, we are reliant on subsequent estimates by historical geologists judging from a scatter of evidence often from distant reporting sites, and as usual Indonesia has been far less systematically worked on in this way than other key subduction zones. It was the work of Newcomb and McCann in 1987 that first drew attention to the misleadingly quiet record of the twentieth century. Before their study, "Sumatra was characterised as relatively aseismic as inferred from the lack of great earthquakes in the instrumental record of this [twentieth] century" (Newcomb and McCann 1987, p. 421). They were able to estimate figures for the historic earthquake and tsunami events that devastated Dutch posts on the west coast of Sumatra in 1797, 1833 and 1861 as well beyond anything the twentieth century offered and thus were the first to point to the potential for the disaster we witnessed in 2004. By comparison, they found little evidence for major earthquakes and tsunamis in Java even before the twentieth century and thus did nothing to dispel the unfortunate idea that Java might be aseismic (Newcomb and McCann 1987). The theory accompanying this judgement was that Java, like the Tohoku area of northern Japan, was an older subduction and therefore may have created a smoother interaction than occurred in younger collisions. The Tohoku earthquake and tsunami blew one great hole in this theory, and the two destructive Java tsunamis of 1994 and 2006 blew another, but the work on the ground is still dangerously inadequate with regard to Java.

In East Indonesia, comparable reconstruction also badly needs to be done and can rely on a more consistent stream of reports from the Dutch spice-collecting ports of Banda, Ambon and Ternate than elsewhere. It is clear, however, that the tsunamis that hit Ambon in 1674 and Banda in 1899, each killing over 2,000 people on a small island, and the tsunami wave heights that were reported higher than 25 m in 1674, 1820 and 1871 were beyond anything the better-measured twentieth century suffered.[5]

In the wake of the 2004 and 2005 mega-earthquakes and destructive tsunamis in Sumatra, geological research on that coast has intensified. Geologists from the Earth Observatory Singapore and Caltech have been able to date the extent of coral

[4] NOAA, National Geophysical Data Centre, Tsunami events search, 2013.

[5] NOAA, National Geophysical Data Centre, Tsunami events search, 2013.

uplift during past subduction thrust mega-events in Sumatra, including those known to historians in 1797 and 1833. They also found much bigger ones in the fourteenth century that were previously unknown to the slender historical/archaeological record (Meltzner et al. 2010; Sieh 2012; Sieh et al. 2008). The remarkable series of uplifts of 1–4 m in the coral on a number of occasions between 1350 and 1450 coincides with at least one tsunami of 2004 scale showing in the sand deposit layers in both northwest Sumatra and southwest Thailand (Bondevik 2008; Monecke et al. 2008). The historians will need to reassess the divide between Buddhist Sumatra and Islamic Sumatra, now known to have been accentuated by at least one massive earthquake of around 9 magnitude and an attendant tsunami on a scale of that of 2004.

While geology is thus making some progress in demonstrating the necessary regularity in tectonically induced disasters in the Indonesian subduction zones, history can also add precious additional information when the opaque sources are combed for clues. I believe I have identified two seventeenth-century tsunamis in Java in 1618 and Aceh in 1660. On 5 January 1660, according to a Dutch company representative in Banda Aceh, "in less than three hours beginning at nine in the evening the water carried over the whole land, even streaming over the walls of the forts along the beach as well as over the lofts or dwellings of most of the inhabitants, of whose houses more than a hundred were carried away by its force, and more than a thousand souls were drowned as well as innumerable cattle".[6] The Pidië coast (around modern Sigli) was also badly affected, with 14 vessels wrecked and many men drowned. The Dutch report characteristically paid more attention to the loss of VOC textile stocks, as well as the damage to its rivals in the market, than to the effect on the government of the queen, who was lucky to survive though losing many of her servants and slaves. The 1660 date coincides with evidence from Siberut (northern Mentawai) of a substantial coral uplift in that year.[7] This dramatic event undoubtedly set Aceh's city and port back for many years. Indeed, its population and its importance in Asian trade would never be quite the same, though this trauma has never before been recognised as among the reasons for the setback. The VOC post was withdrawn, and we hear little about the city over the next decade.

Java is still more crucial because no tsunamis on the exposed south coast have been definitely identified before 1994. Nevertheless, the supernatural power of the southern ocean was even more important as a theme of palace literature and ritual than that of the Merapi volcano, these being the twin poles, north and south, around which the cosmology of the Central Javanese states has evolved. The chronicles of the Mataram kingdom of the seventeenth and eighteenth centuries and the continuing

---

[6] Jan van Groenewegen to Batavia, 28 January 1660, VOC 1233 (KA 1123), f. 502r (my translation). I gratefully acknowledge the labours of Takeshi Ito in transcribing and editing this material, now newly published by Brill (Ito 2015).

[7] Correspondence with Kerry Sieh and Belle Philibosian, November–December 2012. They pointed out, however, that other evidence does not so far suggest that an event in Siberut would have such major effects as far away as Banda Aceh. Clearly, this is important new evidence in that direction, however.

ritual of its modern successors in Yogyakarta and Surakarta make clear that both volcano and southern ocean are supernaturally dangerous unless propitiated and harnessed by rightful kings, but it is the Queen of the South Seas (Ratu Kidul) who is more important. She is believed to have given the conquering founders of the Mataram dynasty, the shadowy Senopati (d. 1601) and mighty Sultan Agung (r. 1613–1646), their power to rule by her mystic union with them, though ordinary mortals who provoked her would be carried to a watery grave.

The most chronologically reliable of the early Javanese chronicles, the *Babad ing Sangkala*, does provide the most explicit evidence for a tsunami with words translated by Merle Ricklefs as: "In Mataram, they moved [the court] to Karta, indeed, when *disappearing, all was turned into sea*" (Ricklefs 1978, pp. 32–33). The italicised words repeated three times in slightly different form for emphasis are a chronogram (a date in words) for the Javanese year 1540, which began in February 1618 CE, at the beginning of Sultan Agung's campaigns to conquer Java. In other words, there is historic evidence to locate a powerful earthquake and tsunami on the south coast of Central Java, which adds to the urgency of scientific research on the ground (Reid 2012).

3.6 Pre-colonial Cities Avoid Tsunami Coasts

Prior to the twentieth century, Indonesians (like Japanese) were wise enough to build one-storey dwellings of wood, bamboo and palm leaf, which caused little danger during an earthquake. As William Marsden remarked after complaining how much damage Sumatran quakes did to European brick buildings in Sumatra, "the usual manner of building renders them but little formidable to the natives" (Marsden 1811, p. 30). The only buildings of brick and stone on an Indian model were Hindu and Buddhist temples of the period 600–1200 CE, which raises questions about this strategy. Such sacred buildings were erected on high ground well away from the tsunami-endangered west coast of Sumatra and south coast of Java. Any temples erected at coastal harbours would have been swallowed by tsunamis. Yet, even so, earthquakes invariably destroyed them sooner or later. When Raffles and his collaborators in 1813 visited the monuments of Prambanan and Borobudur in Java, subsequently rebuilt many times, they noted their "desolation", with "large hillocks of fallen masses of stone" (Raffles 1817, Vol. II, p. 11). Indian traders and craftsmen with little knowledge of Southeast Asian conditions may have been partly responsible for this miscalculation, yet the contrast is striking with the earliest mosques, also initially inspired by foreigners but invariably built of wood, thatch and flexible materials without minarets. The point may be that mosques were built for the faithful to assemble, whereas the Hindu and Buddhist temples were frequented by gods rather than men. Their destruction allowed the next generation to gain further merit by rebuilding them, as has been noted of Burma.

The most tsunami-affected coasts do appear to have been avoided by their Indonesian inhabitants, who built their major centres in the interior. The west coast of Sumatra was virtually deserted until pepper growing on virgin soil began there to meet the European demand in the sixteenth century. Even as late as 1830, the coastal plain of today's province of central Sumatra held fewer than 100,000 people, only one tenth that of its uplands. Today, the coast has over two million people, slightly more than the interior (Reid 1998, p. 61–89). The people of Nias, probably the most vulnerable to tsunamis of all complex Indonesian societies, spurned their coasts completely before sea-based Dutch infrastructure arrived in the second half of the nineteenth century. They fished only in freshwater rivers and built their villages on hilltops where possible (Gruber 2007). Their origin myths, like those of similarly threatened Mentawai, are concerned to explain earthquakes, tsunamis and the subsidence of the coasts beneath the seas (Hämmerle 1999; Loeb 1935; Schefold 1988).

The subduction zone along the south coast of Java was also virtually deserted before the Dutch established a post in Cilacap in 1840. Given the evidence above, it seems likely that fear of the southern ocean, embodied in the Ratu Kidul myth, was well grounded in experience of disasters like that of 1618 and probably an earlier event during the early Mataram civilisation that built Borobudur and Prambanan in the first millennium. In Maluku, also, the remembered pre-colonial origin villages of each society are in the interior, with the European presence gradually drawing or forcing removal to the coasts. The great monuments of Indonesia's Hindu-Buddhist past are all well away from the coasts.

3.7 Colonial Priorities, Not Tectonic or Climatic Threats, Determined the Siting of Southeast Asia's Primate Cities

The Portuguese, Spanish, Dutch and English arrived successively in Southeast Asia in heavily armed ships, and all their forts and trading posts were built compactly within sight of those ships. Inevitably, they experienced earthquakes that wrought havoc with the brick-and-mortar buildings they had erected as protection against attack and fire. In 1645, the St. Andrew's Day earthquake, subsequently estimated at 7.9 magnitude, famously "reduced the proud and magnificent city of Manila to a pile of ashes and rubble" (Phelan 1959, p. 101).[8] Less well remembered is the earthquake that damaged virtually every building of the Dutch headquarters of Batavia (Jakarta) in 1699, "so heavy and strong that nothing comparable had ever been known to have occurred here". Some 50 stone and brick buildings collapsed, killing 28 of the Dutch inhabitants. The most serious disruption, still being blamed for the difficulties of the canal system 12 years later, was from the huge number of uprooted trees that came down from the uplands and blocked the rivers.[9] The minor Sumatran outposts of Padang and Bengkulu were the worst affected, Padang being virtually

[8] A fuller modern analysis of the quake can be found in Bankoff (2007, pp. 411–427).

[9] The quotation is from the official Dutch report to the VOC in Coolhaas, *Generale Missiven* VI, pp. 49–50, pp. 830 & 831–832. This disaster was noted also by Chinese traders between Batavia and Nagasaki, who reported to the Nagasaki authorities that "a heavy earthquake hit the castle of

destroyed in 1797 and heavily damaged in 1833 and 1861. The Europeans built a little better after these and other disasters and in the nineteenth century moved their dwellings to somewhat safer and healthier suburbs on higher ground. Yet the strategic and commercial imperatives of seaborne shipping were too important for the Europeans until the twentieth century for them to consider whole new administrative centres.

Only when the colonial peace and the construction of railways allowed the colonisers to imagine their empires as coherent countries in the early twentieth century could this distortion be reconsidered. That was the time when Britain opted to move the Indian capital from Calcutta to Delhi (1911), the British East Africa capital to Nairobi (1907) and the South African administrative capital to Pretoria (1910). No such shifts occurred in Southeast Asia, unfortunately, where traditional interior capitals like Mandalay, Chiang Mai and Yogyakarta declined into elegant marginality. The closest to colonial-built interior cities of this railway era were the Chinese tin-mining centres of Malaya—Kuala Lumpur, Ipoh and Taiping—and Bandung in upland West Java. The Dutch made some tentative gestures to decentralise, moving the Governor General's secretariat and the departments of agriculture, arts and education to Buitenzorg (Bogor) and the railway departments to Bandung around 1900 The capital Batavia (now Jakarta) was itself planned to be relocated to Bandung in the 1930s when it was still a manageable city of 200,000, but only the military headquarters had moved at the time of the Japanese invasion. In the turbulent post-war revolutionary period, the coastal ports again became protected colonial enclaves, with refugees rapidly expanding their populations. However, their population take-off truly began when the state became the great hope of employment after independence and the bureaucracy ballooned.

It was in the relative geological lull that I have described in the long twentieth century that these colonial enclaves on the coast were transformed into huge and endangered cities. Colonial Southeast Asia had been demographically somewhat distorted in its lack of urbanisation, with the urban proportion declining through most of the nineteenth century as ever-increasing population was stuffed into shared peasant poverty. It made up for this "urban deficit" with steady growth after 1900 and breakneck urban crowding after 1950, much of it in makeshift buildings and shanties ill equipped to withstand disasters. The long century of political upheaval rather than natural disasters between 1885 and 2000 saw the Indonesian population grow from about 25 million to 205 million and its urban (predominately coastal) population from a little over a million to 90 million. Wealth, hospitals, communications, schools, universities and entertainment were now concentrated in the colonially sited coastal cities, which were also safer places in times of turmoil. Life was simply better there.

As we noted, even apart from their murderous traffic, Jakarta and Manila are now among the most endangered big cities in the world. But there are also now large cities exposed to tsunamis in places that pre-colonial populations had the sense to

---

Kelapa [Batavia], causing some casualties. An earthquake of this magnitude seldom takes place anywhere. The country was greatly disturbed" (Ishii 1998, p. 237).

avoid. Since 1870, the importance of colonial ports as hubs for each province has pushed the population to cluster dangerously on coastal plains. In the interval between Padang's earthquakes of 1833 and 2009, the population of that West Sumatran littoral had soared from 80,000 to over two million, a high proportion of them living in substandard but "modern" brick and concrete dwellings subject to earthquake damage, whereas in past events only a handful of Europeans and Chinese were affected by the collapse of such buildings. Hence, the 2009 September earthquake, although registered as magnitude 7.6, centred off the coast south of Padang, caused a death toll of over a thousand with 135,000 homes destroyed and 1.25 million people affected. The much bigger shocks of 1833 and 1861, by contrast, had caused only a handful of deaths.

In Java, similarly, the exposed southern coast was almost unpopulated until the nineteenth century, but the Dutch foundation of a major base at Cilacap in 1840 and the subsequent location of an oil refinery there have seen the population of the southern littoral grow to about two million today. Both Padang and Cilacap seem likely to face a major tsunami in the relatively near future, as Kerry Sieh has been warning (Sieh 2012). Other coastal cities facing a tsunami threat are Banda Aceh and Bengkulu in Sumatra, Maumere and Ende in Flores, Kupang in Timor and Ambon and Ternate in Maluku.

3.8 Conclusion

Modern Indonesia's urban infrastructure was situated and built during a century of turbulent politics but relative geological calm, which cannot be expected to continue in the twenty-first century and beyond. Tectonic and climate research on a global scale, rather than the efforts of Southeast Asian historians, increasingly make clear that the heavily populated islands of the Sunda Arc have generated many of the world's most devastating tectonic events, in the form of both volcanic eruptions and tsunami-generating earthquakes. There is an imperative to extend this understanding of the past, with its implications for the future, through cooperation between science, history and archaeology. The international community needs to join with the immediately affected Indonesian community to extend this research and generate better planning for infrastructural developments in the future—including the relocation of capitals. Adequate disaster preparedness must include purposeful education and the development of national and international governance preparation for evacuation and response. I hope I live long enough to see our military establishments having their joint exercises as simulated responses to the certainty of natural disasters on this tectonic frontier, rather than highly tendentious and improbable military threats.

References

Abeyasekere, S. (1987). Jakarta: A history. Singapore: OUP.

Asian Disaster Reduction Center (ADRC). (2000). Data book on Asian natural disasters in the twentieth century. Kobe: Asian Disaster Reduction Center.

Bankoff, G. (2007). Fire and quake in the construction of Old Manila. The Medieval History Journal, 10 (1&2), 411–427.

Berg, N. (2012, April 26). Jakarta is sinking itself into the ocean. The Atlantic Citylab. http://www.citylab.com/weather/2012/04/jakartas-sinking-itself-ocean/1857/. Accessed 13 Nov 2014.

Bondevik, S. (2008). Earth science: The sands of tsunami time. Nature, 455.

Brata, A., Rietveld, P., de Groot, H., & Zant, W. (2013). The Krakatau eruption in 1883: Its implications for the spatial distribution of population in Java. Economic History of Developing Regions, 28(2), 27–55.

Clodfelder, M. (1992). Warfare and armed conflicts: A statistical reference to casualty and other figures, 1618–1991 (Vol. II, pp. 1127–1140). Jefferson: McFarland.

de Boer, J.Z., & Sanders, D.T. (2002). Volcanoes in human history: The far-reaching effects of major eruptions. Princeton: Princeton University Press.

Emile-Geay, J., Seager, R., Cane, M.A., Cook, E.R., & Haug, G.H. (2008). Volcanoes and ENSO over the past millennium. Journal of Climate, 21, 3134–3148.

Gruber, P. (2007). Adaptation and earthquake resistance of traditional Nias architecture. http://www.nirn.org/pdf/01_070614_gruber.pdf. Accessed 21 July 2012.

Hämmerle, J. (1999). Asal usul masyarakat Nias: Suatu interpretasi. Gunungsitoli: Penerbit Yayasan Pusaka Nias.

Hilts, C. (2012, August 6). London's volcanic winter. Current Archeology. http://www.archaeology.co.uk/articles/features/londons-volcanic-winter.htm. Accessed 12 November 2012.

Ishii, Y. (Ed.) (1998). The junk trade from Southeast Asia: Translations from the Tosen Fusetsugaki, 1674–1723. Singapore: ISEAS.

Ito, T. (Ed.). (2015). Aceh Sultanate: State, society, religion and trade. The Dutch sources, 1636–1661 (2 Vols.). Leiden: E.J. Brill.

Lavigne, F., Degeai, J.-P., Komorowski, J.-C., Guillet, S., Robert, V., Lahitte, P., et al. (2013, September 4). Source of the great A.D. 1257 mystery eruption unveiled, Samalas volcano, Rinjani Volcanic Complex, Indonesia. Proceedings of the National Academy of Sciences. http://www.pnas.org/content/early/2013/09/26/1307520110. Accessed 20 Oct 2013.

Loeb, E. (1935). Sumatra: Its history and people. Vienna: Institut für Völkerkunde der Universität Wien.

Marsden, W. (1811, Repr. 1966). The history of Sumatra (3rd edn.). Kuala Lumpur: Oxford University Press.

Meltzner, A.J., Sieh, K., Chiang, H.-W., Shen, C.-C., Suwargadi, B.W., Natawidjaja, D.H., et al. (2010). Coral evidence for earthquake recurrence and an A.D. 1390–1455 cluster at the south end of the 2004 Aceh-Andaman rupture. Journal of Geophysical Research, 115, B1040. doi:10.1029/2010JB007499.

Monecke, K., Finger, W., Klarer, D., Kongko, W., McAdoo, B., Moore, A., et al. (2008). A 1,000-year sediment record of tsunami recurrence in northern Sumatra. Nature, 455, 1232–1234.

Newcomb, K.R., & McCann, W.R. (1987). Seismic history and seismotectonics of the Sunda Arc. Journal of Geophysical Research, 92(B1), 421–439.

Phelan, J.L. (1959). The hispanization of the Philippines: Spanish aims and Filipino responses 1565–1700. Madison: University of Wisconsin Press.

Raffles, T.S. (1817, Repr. 1965). The history of Java, 2 vols. Kuala Lumpur: Oxford University Press.

Rantucci, G. (1995). Geological disasters in the Philippines: The July 1990 earthquake and the June 1991 eruption of Mount Pinatubo. Description, effects and lessons learned. Rome: World Conference on Natural Disaster Reduction & Italian Ministry of Foreign Affairs.

Reid, A. (1998). Inside-out: The colonial displacement of Sumatra's population. In P. Boomgaard, F. Columbijn, & D. Henley (Eds.), Paper landscapes: Essays in the environmental history of Indonesia (pp. 61–90). Leiden: KITLV Press.

Reid, A. (2012). Historical evidence for past tsunamis in the Java subduction zone (Asia Research Institute Working Paper no. 178). Singapore: Asia Research Institute. www.ari.nus.edu.sg/docs/wps/wps12_178.pdf.

Reid, A. (2013). Population history in a dangerous environment: How important may natural disasters have been? Masyarakat Indonesia, 39(2), 505–526.

Reid, A. (Forthcoming). History and seismology in the ring of fire: Punctuating the Indonesian past. In D. Henley & H. Schulte-Nordholt (Eds.), The longue durée in Southeast Asian history. Leiden: Brill/KITLV.

Ricklefs, M.C. (1978). Modern Javanese historical tradition: A study of an original Kartasura chronicle and related materials. London: SOAS.

Schefold, R. (1988). Lia: Das große Ritual auf den Mentawai-Inseln (Indonesien). Frankfurt: Dietrich Reimer Verlag.

Sieh, K. (2012). The Sunda megathrust: Past, present and future. In P. Daly, R.M. Feener, & Reid A. (Eds.), From the ground up: Perspectives on post-tsunami and post-conflict Aceh (pp. 1–24). Institute of Southeast Asian Studies.

Sieh, K., Natawidjaja, D., Meltzner, A.J., Shen, C.C., Cheng, H., Li, K.S., et al. (2008). Earthquake supercycles inferred from the sea-level changes recorded in the corals of West Sumatra. Science, 322, 1674–1677.

Swiss Re. (2013). Mind the risk. A global ranking of cities under threat from natural disasters. http://media.swissre.com/documents/Swiss_Re_Mind_the_risk.pdf. Accessed 16 Sept 2013.

Tantingo, R. (2011). Pinatubo: The volcano in our backyard. Quezon City: Center for Kapampangan Studies, Holy Angel University.

UN Office for the Coordination of Humanitarian Affairs (UNOCHA). (2010, January 13). Thailand: Act now to stop Bangkok sinking, urge scientists. IRIN News. http://www.irinnews.org/report/87715/thailand-act-now-to-stop-bangkok-sinking-urge-scientists. Accessed Sept 2013.

Wardany, I. (2010, April 23). Jakarta Sinking as fast as 10 cm per Year. The Jakarta Post, http://www.thejakartapost.com/news/2010/04/23/jakarta-sinking-fast-10cm-year.html. Accessed 24 Apr 2014.

World Wildlife Fund (WWF). (1986). Mega-stress for mega cities: A climate vulnerability ranking of major coastal cities in Asia. Gland: WWF.

Chapter 4
Muddy Resistance: Community Empowerment in Mudflow Disaster Governance in Porong, Sidoarjo, Indonesia

Rita Padawangi

Abstract Mainstream disaster governance strategies and mechanisms place an emphasis on material and quantifiable losses in order to determine the amount of material or financial compensation. Calling for a humanistic approach to respond to disasters, this chapter examines how space becomes tools of community empowerment in a continuing disaster. Data used in the analysis are obtained through ethnographic interviews and participant observation of social movements on Lapindo mudflow as well as news articles related to the disaster in the span of several years until the year 2014. The mudflow as a unique prolonged disaster in the context of disjointed, seemingly negotiated government bureaucracies that tend to favour corporations with powerful actors sheds light on the importance of building community resilience by focusing on livelihoods rather than by emphasising on compensation and other temporary interventions. Embracing environmental disaster spaces as part of everyday life that encompasses culture and economy is an alternative disaster governance approach that puts people as the priority, as active members of society rather than victims. While the conditions on the ground are very much nuanced, the focus on human flourishing is in line with empowerment for the long term by constructing social identity, interactions and relationships as resilient communities.

Keywords Mud volcano • Mudflow • Lapindo • Porong • Sidoarjo • Disaster governance • Empowerment

The fieldwork for this paper is funded by the by the Singapore Ministry of Education AcRF Tier 2 grant for the project 'Aspirations, Urban Governance, and the Remaking of Asian Cities' (MOE2012-T2-1-153).

The author would like to thank Rudy Hermawan, Daris Ilma and Anton Novenanto for their support in the research, for their time to review the paper before publication and for providing meaningful comments and suggestions for improvement.

R. Padawangi (✉)
Asia Research Institute, National University of Singapore, Singapore, Singapore
e-mail: ritapd@nus.edu.sg

M.A. Miller, M. Douglass (eds.), *Disaster Governance in Urbanising Asia*,
DOI 10.1007/978-981-287-649-2_4

4.1 Introduction

Sanggar Alfaz was a house owned by Cak Irsyad's family. Right beside the house was a levee, more than 3 m high, that separated the settlement from a pool of hot mud. In front of the house, a bamboo hut was equipped by a television set for visitors to relax while watching shows. It also had a terrace where musical instruments were kept, among which there were traditional drums, self-made instruments from big mineral water bottles and a gong. Inside the house, the living room had been converted into a small library with a computer and rows of children's books. Together with several people who volunteered, Cak Irsyad turned the house into an open house for children to play in 2009, after the mudflow disaster hit the subdistrict of Porong, Sidoarjo, East Java. The mud eruption made many children lose their outdoor play spaces, received less attention from parents and made many lose their friends, because families had to evacuate and communities were dispersed to different resettlement areas (Fig. 4.1).

Most residents in the area found out about the mud volcano eruption in Porong on the fateful day, 29 May 2006, when they were finally informed that volumes of hot mud had started to flow out from the ground. Started with a flow of methane-filled mud eruption as high as trees, it steadily swallowed the surrounding villages and piled up as high as 20 m, with the speed escalated from 5,000 to 120,000 m^3 per day for the first 11 weeks (Batubara and Utomo 2012; Mazzini et al. 2007). The

Fig. 4.1 An art installation by Yogyakarta-based Taring Padi group in the Porong mud volcano (Author 2014)

eruption occurred within the drilling site of Lapindo Brantas, an Indonesian company, which had bought pieces of land in Porong for their gas exploration.

How does space become tools of community empowerment in an escalating disaster? By examining grassroot activism in an environmental disaster that steadily grows as a predator of life spaces, this chapter critically questions the reduction of 'natural' disasters into material and quantitative interpretations that has widely dominated disaster mitigation and scholarship discourses. Although this chapter uses the case of the mud volcano eruption in Porong, Indonesia, the focus of the analysis is the community and their lives rather than the disaster in itself. The analysis is based on ethnographic interviews and participant observation of community activities and events in 2007, 2013 and 2014, as well as news articles from local and international agencies. As environmental disasters are socially and politically constructed, an emphasis on the disaster as an event risks neglecting the human side of disaster-affected spaces, which is rightfully the primary reason why disasters become important subjects.

4.2 Locating Life Spaces in Disaster Governance Literatures

Focusing on the communities and the livelihoods in spaces affected by disasters may sound simple and basic, but it is common to find policies on disaster risk reduction and disaster response that do not thoroughly take livelihoods into consideration or, if so, only superficially. Environmental disasters are usually represented in terms of the number of affected people and financial loss to describe the magnitude of the social and financial rupture (Café 2012; Donaldson et al. 2013; Jigyasu 2004; Walker et al. 2011). Disaster risk assessment and analysis measure the impacts of natural and technological hazards, which are aimed to inform public policies and the insurance industry (Tierney 1999; Penning-Rowsell and Pardoe 2012).

The attention to life space is scarce in disaster governance policies and literatures. The understanding of life space, defined as territorial base of household political economies that are engaged in the production of life and livelihood, is only possible through a comprehension of the everyday life of members in particular communities (Friedmann 1992; Wisner 1993). The focus of abstractions of disasters into quantifiable measurements has been at the centre of policy analysis that has direct impact on the people, both in immediate times of disaster and in the subsequent process of rebuilding, if necessary. Moreover, quantitative representations of disaster impact or risk assessment are often uncritically viewed as accurate, even by social scientists, despite the processes to derive these estimates that might deserve further scrutiny (Hilgartner 1992; Stallings 1995; Tierney 1999). As a result, disaster management policies often focus on compensation and rebuilding of dwellings. Community mobilisation is often treated as the territory of humanitarian assistance or outsourced to organisations that are usually independent of the government.

The domination of quantifiable measurements in the policy domain is inextricable from the dominant paradigm in the organisation of society. Friedmann (1992) pointed out that increasingly there had been a fascination towards quantifiable measurements because the numbers make complicated urban issues comprehendible. In other words, it simplifies the complexities of the problem in order to make them understandable across a broad spectrum of people (Friedmann and Hudson 1974). On the one hand, the built environment is produced under the capitalist development regime, in which spaces are treated as abstract nominal values, which becomes a form of capital (Harvey 1992, 2002). The process of making the spatial value abstract undervalues or even devalues social relationships as secondary spatial attribute (Lefebvre 1991).

On the other hand, the social construction of precision and reliability of natural sciences and engineering vis-à-vis social sciences perpetuates the socially constructed superiority of engineering options and economic outcomes in dealing with disasters (Johnson and Penning-Rowsell 2010). The dominance of expert-reliant solutions is sustained through education system, in which experts are trained and knowledge are presumably developed (Gramsci 2010). Despite social movement efforts, if any, that might attempt to reclaim political, intellectual and moral leadership, the physical consequences of disasters embody the belief that social and material existence of human beings is affected by a greater force that are out of human control (Café 2012; Laclau and Mouffe 2001; Rosol 2013; Torfing 1999). Post-disaster disempowerment may occur as a result of ruptures to everyday social and political life and even systems (Café 2012; Fischer 2003; Gilbert 1995; 1998). Another possibility is to have hopeful social change following a disaster that destabilises existing societal systems.

Although there have been contributions from psychology in the risk assessment discourse (Tierney 1999), these are inadequate to gain understanding of the human and communities as social and cultural beings. Gotham and Greenberg (2014) argued that there has been a lack of thorough socio-spatial analysis of the production of risk, in spite of critical contributions from risk society theorists in highlighting political and economic structures as well as power relations in the study of disasters. They made a case in differentiating disaster and crisis to better understand these structures and relations. 'In societies in which risk and inequality are minimal, disasters remain contained and do not result in full-blown social, economic, and environmental crises' (Gotham and Greenberg 2014, p. 6).

A notable exploration into power relations and daily life in constructing disaster vulnerability was previously offered by Wisner (1993), who argued for the importance of understanding processes of marginalisation in order to link disaster occurrence and social theory. The recognition of marginalised groups—based on class, gender, age, ethnicity and disability—is a step towards applied disaster governance policies that are sensitive towards life spaces. Yet, a humanistic view of life spaces would require attention towards human actions and social relationships that are empowering members of society to be agents of social change. Interpreting these life spaces before, during and after disasters by means of thorough insight of the everyday lives of the residents is a meaningful contribution to the disaster governance literatures, particularly when these life spaces are situated within the broader social, political and economic structures as well as power relations. Environmental disaster governance literatures have looked at some of these efforts as coping and survival mechanisms, but gaps remain in understanding how disaster governance can sustain, revive or transform life spaces in empowering rather than prescriptive ways.

4.3 The Disaster Governance of Lapindo Mudflow

While many environmental disasters, such as volcanic eruptions, earthquakes or floods, come in relatively short spans of time, the mud volcano in Porong, Sidoarjo, emerged rather slowly but surely. The case in Porong is also much longer than usual mud volcano eruptions that are expected to last only a few days (Mazzini et al. 2007). The hot mud has been flowing out for 9 years from a pit that was drilled for natural gas exploration. Analysis on the causes of the mudflow appeared in the following years, debating whether the eruption was caused by the Lapindo Brantas Indonesia drilling—which would make it a 'man-made' disaster—or triggered by the Yogyakarta earthquake 2 days before the major mud eruption in Porong (Davies et al. 2007, 2011; Mazzini et al. 2012; Rudolph et al. 2011). Several researchers recently suggested to not categorise the disaster as a mud volcano but as an eruption of the naturally overpressured deep hydrothermal system beneath East Java as a whole (Fitrianto 2012; Mazzini et al. 2012). Publications by Indonesian scientists have firmly called the mudflow a natural mud volcano (Istadi et al. 2009; Sawolo et al. 2009). This chapter does not focus on how textual arguments have been circulated to legitimise or delegitimise the hypothetical causes of disaster, but the important point from the scientific inquiries is that they remain inconclusive, although the official discourse had leaned towards viewing the disaster as natural rather than man-made (Istadi et al. 2009). While the cause of disaster has been subjected to debates throughout the 9 years of the disaster, social consequences of contradictive hypothetical views have been instantly materialised in the lives of the residents and were systematised through government policies.

The government-funded Sidoarjo Mudflow Agency (*Badan Penanggulangan Lumpur Sidoarjo/* BPLS) that was formed as a disaster management agency for the Lapindo mudflow was largely ineffective (Drake 2013). The first year of the mudflow was marred with instant anger and frustration of the residents as their houses were drowned in the mud. More than 18,000 households had to take refuge in evacuation camps (BBC Indonesia 2010a). Initially, the residents of Renokenongo village had refused to evacuate, even when the mud had already invaded their homes up to knee level. However, their hope of the mudflow to be overcome before their homes were fully submerged was eroded when there was another gas explosion (Farida 2014). A slow witnessing of property and community loss, with almost zero likelihood to return to the pre-disaster setting and uncertainty of livelihoods in the future, becomes a process of forced disempowerment. Moreover, social institutions are forced to close down with no certainty of revival. One of the most disheartening losses is the closure of schools, which resulted in a dispersal of the children, who were forced to be separated from their friends, especially when their homes were also gone (Drake

2013). Economic survival was a struggle, as many residents were forced to look for alternatives—most likely less reliable sources of income (Farida 2014).

The representation of disaster damages as monetary values is a product of capitalist production of space, in which urban spaces, the built environment and activities associated with them are considered as properties that can be monetised. In fact, using monetary values to illustrate resources needed for rebuilding and to compensate damages to livelihoods are commonly found in disaster management practices. Literatures on how insurance companies define and measure damages are abounding, particularly in relation to flood disasters. While urban theories such as the right to the city and social- and spatial-justice literatures are strongly critical against capitalist logic as the sole reasoning of valuing urban spaces, they rarely touch on compensation-based disaster management practices.

The compensation system's inability to take into account the intangible values has resulted into at least three responses. Firstly, the system translated those intangible values into no value. Secondly, the intangible values are valued lower than material properties that are easily monetised. Thirdly, the attempts to insist valuation on intangible values are perceived as negotiation efforts to increase financial compensation to be more than what was previously valued. In other words, the emphasis on monetary representation of damages reduces human interests solely into moneymaking and perpetuates the image of the victim as helpless and powerless. In practice, it is true that there is a tendency of disempowerment due to an overwhelming helplessness that leads to fatalism following a tragedy. In their experiences of six major disasters in ten years in the Philippines, Carcellar et al. (2011) identified the vulnerabilities that cause this helplessness to include 'limited financial access, insecure land and house tenure, high risk locations and a lack of organization' (p. 371). However, the landscapes of victimhood, as will be shown in this chapter, are more complex than a general disempowerment of the people.

Aspects of community life that are often abandoned in disaster mitigation policies are often generalised into the 'intangible' aspects. However, these aspects are not intangible; rather, they are invaluable or unquantifiable. Neighbourly relations, for example, are not intangible, because in reality, they are practised in everyday life. Arts and culture also have their material manifestations although they may not be quantifiable or are neglected in the disaster impact valuation process. Rather than sustaining disaster governance approach that favours compensation system, one that is focused on building resilience, sustaining livelihoods and empowering communities are likely to be suitable in the long term. Empowerment, in this case, is interpreted as a focus on strengthening human agency as members of society, which allude to presenting maximum opportunities for human flourishing (Friedmann 1992).

4.4 The Social Impact of the 'Natural' Mud Volcano

After the eruption of the mud volcano on 29 May 2006, the landscape of Porong has been gradually but surely changing. Mud discharge started with the volume of 50,000 m^3 per day, but the following year reached 180,000 m^3 per day (Mazzini et al. 2009). Although calculation from satellite images indicated an average of 90,000 m^3 per day in 2008, the mud discharge volume in reality has unevenly spread throughout the years (Istadi et al. 2009; Rudolph et al. 2011). By December 2008, the inundated area had reached 1,500 ha. Several villages that were not drowned were also affected by gases that were released from the mud.

Six presidential decrees had been issued regarding the mudflow disaster. The first was Presidential Decree 14/2007, followed by 48/2008, 40/2009, 68/2011, 37/2012 and 33/2013. The contents of these decrees are defining the affected areas and therefore making them eligible for compensation from the state. The five decrees after the first one in 2007 were revisions on the boundaries and areas eligible for being purchased to compensate for their property loss. The categories of impact are differentiated into at least three categories:

Firstly, the direct and material impact, which included mudflow, water flow, poisonous gas flow, flammable gas flow, landslide that affected houses and public facilities, water supply contamination and blockage of irrigation canals.

Secondly, the less direct impact, with similar characteristics of the first category, but without the inundation of homes. With homes being intact, people are perceived to have considerably less degree of damages.

Thirdly, the indirect impact, which included wider areas that were economically affected by the irrigation canal blockage and hot mud flow.

In 2008, more than 6.4 km^2 have been officially 'included in the map' and were therefore eligible for state compensation. Forty-two percent of those were agricultural land and 45.6 % were categorised as gardens associated with landed houses.

Despite the seemingly rural profile of the 'map', Porong had been a strategic district for trade in East Java. Its location between two major cities, Malang and Surabaya, had made Porong the main route for trades, even after the mudflow disaster started. The market in Porong had been an important trade centre in Sidoarjo because of its location that was not too far from agricultural production as well as fisheries (Batubara and Utomo 2012). The market was converted into a temporary refugee centre during the early days of the mud volcano eruption. Tanggulangin, a nearby centre for local handbag craft production, suffered tremendous loss as tourist visits decreased after the disaster started.

Presidential Decree 14/2007, chapter 15 (1), declared that the 'social and societal problems' will be handled 'through the purchase of land and buildings affected by the Sidoarjo mud through stages of payment, according to the map of affected area on 22 March 2007, with the proper land titles that state the size of the land and location'. The payment started with 20 % as down payment and the rest would be given as instalments to the affected people within two years' time. Every year in the two subsequent years after the decree was issued, there had been amendments that focused on the definition of the affected area and the responsible parties to provide compensation. Presidential Decree 40/2009 had expanded the areas eligible for compensation by adding several districts outside the 'affected area map' that expe-

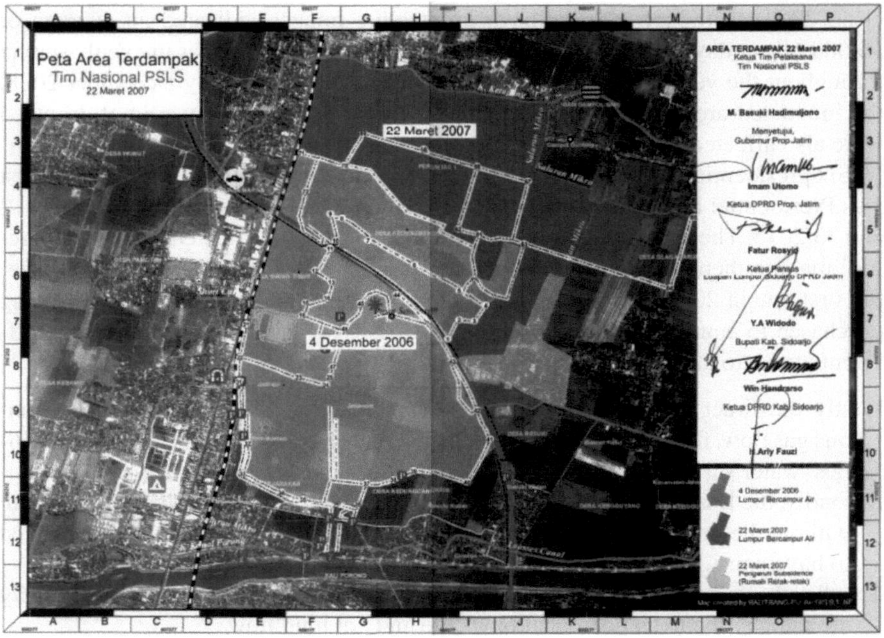

Fig. 4.2 Map of the affected area (22 March 2007) as the basis of compensation framework in the Presidential Decree 14/2007 (Presidential Decree 14/2007)

rienced impacts of 'dangerous gases' and 'mud eruption' which made the places 'inhabitable' (Figs. 4.2 and 4.3).[1]

The definitions of affected areas as declared through the presidential decree and its amendments were inseparable from the continuous expansion of the mud flow. One immediate strategy to contain the mud flow was to build a levee to surround the source of the eruption, for which the landowners that would be submerged within the levees had to be compensated. Despite many protests from the villagers, who felt that the compensation scheme was still unclear, the levee was constructed with the protection of East Java Provincial Police that blockaded popular demonstrations (Liputan6 2006). A diversion canal to channel the mud to Porong River was also constructed as part of the disaster management plan, following recommendation of the BPPT. Five years after the mudflow started, the levees were expanded again to allow three villages (Pejarakan, Besuki and Kedungcangkring) to be submerged.

The focus of the presidential decree to measure damages primarily on the material aspect is consistent with the monetisation of disaster impact discourse, but the mudflow case has exposed at least three weaknesses of such valuation. Firstly, many

[1]The term "inhabitable" was first used in the 2009 Presidential Decree. Presidential Decree 40/2009 used the term "inhabitable area" to replace the term "affected area" in Presidential Decree 14/2007.

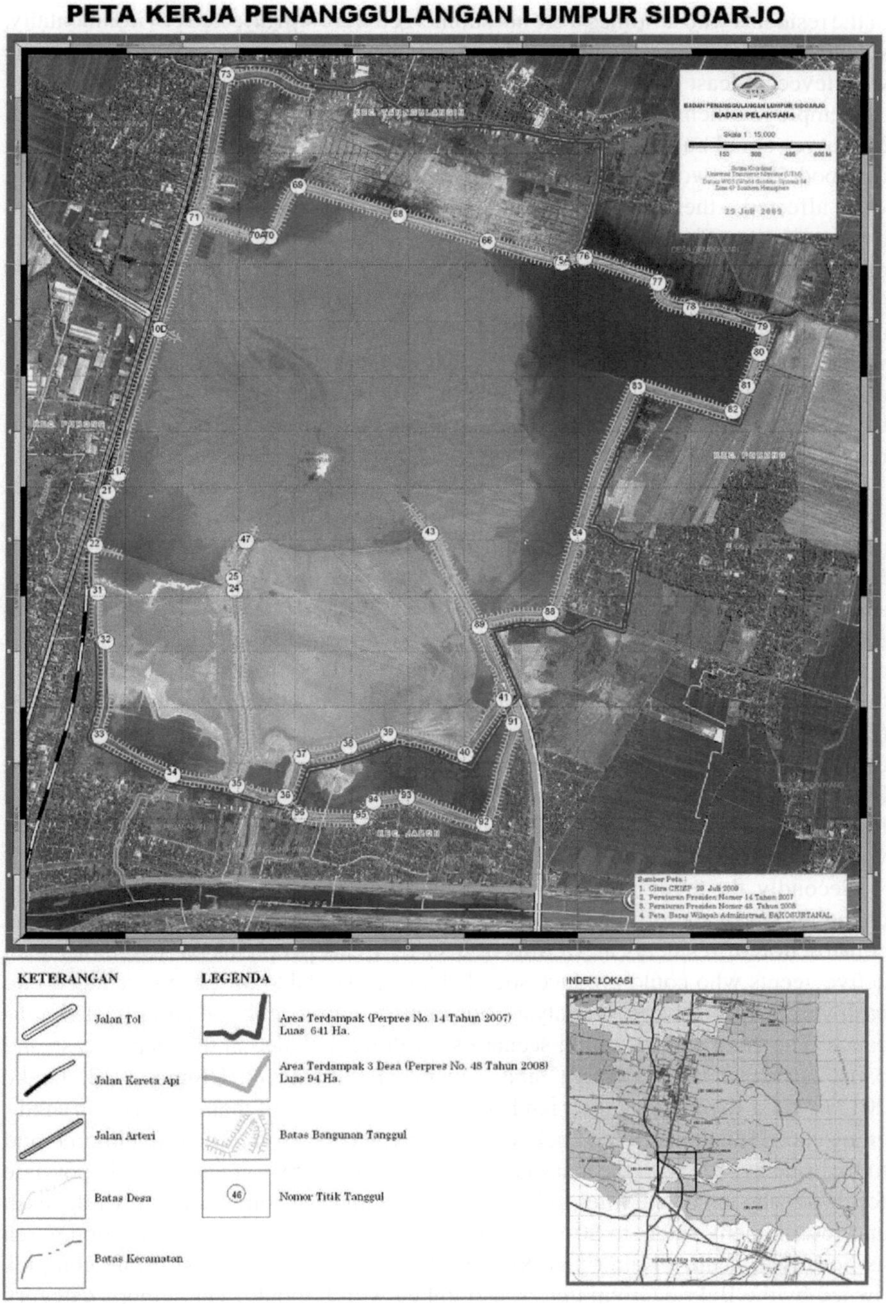

Fig. 4.3 Map of the affected area as of 29 July 2009 (Badan PenanggulanganLumpur Sidoarjo/ BPLS)

of the residents whose homes were not submerged were still affected environmentally, socially and economically. In the case of Ketapang village across the street from the west levee, at least many residents became unemployed because three big factories that employed them had been drowned in mud and were completely dysfunctional. Moreover, small businesses that have grown around the factories, such as those selling food for the workers, were also wiped out. Factories that were not submerged were affected—their productions were down by at least 40 % because the access roads were cut off. A portion of the Malang-Surabaya highway was also drowned in mud and had not been reconnected even until 8 years after. Malang's economic growth fell from 7.2 % in 2005 to 5.3 % in 2006, while Sidoarjo's fell from 6.7 % in 2005 to 4.6 % in 2006 (McMichael 2009). Small and medium enterprises were worse hit than the large manufacturers that had more distribution networks, which means that the economic impact of the mud volcano was much wider than the identified area in the government's map and was unequal among different scales of businesses (McMichael 2009).

Besides employment, schools were submerged and children lost their friends because they had to move to different schools or areas. There were also water supply problems due to groundwater contamination (Purwaningsih and Notosiswoyo 2013). However, the demands of the affected residents whose homes were not physically submerged were not met. Nine villages that sent delegations to the National Parliament and to the government in Jakarta in January 2008 were snubbed in the first amendment (48/2008) that chose to compensate villages that would be affected by the new diversion canal rather than those that were socially and economically affected (Batubara and Utomo 2012). The social, environmental and economic losses by the disaster are not limited to those that are directly drowned in mud, but also included livelihoods of those who are dependent on the functions that were drowned. The losses were also felt by those that would be affected by future expansion of the mud volcano, of which the longevity can be for many years to come (Rudolph et al. 2011).

Secondly, the focus on financial compensation takes away the attention from the changing landscape of the city of Porong. In other words, the social and political construction of residents as victims of disaster treats people as passive rather than active agents who could produce social change. The value of space is reduced into cash values without a good analysis, nor inquiry, into its cultural and social significance. The three compensation scenarios for those included in the map—'cash and carry,' 'cash and resettlement' and 'others' (Batubara and Utomo 2012; Drake 2013)—refer to replacing their land and houses, but none of them focus on strengthening social relationships and networks that are necessary to rebuild livelihoods and to live side by side with the disaster zone. 'Cash and carry' scenario is only possible for those who have land titles to be exchanged with cash compensation, while 'cash and resettlement' refers to 20 % cash compensation and 80 % compensation given as houses in nearby Kahuripan Nirwana Village, Sidoarjo, a real estate complex owned by the Bakrie Group that is also linked with Lapindo. The 'others' category refers to those without land titles, which is common for the residents there (Tempo.co 2008). None of these scenarios have been fully completed; until the end of 2008, approximately, only 300 out of 4,000 sign-ups for 'cash and carry' had been fully compensated, while only 400 out of 2,500 'cash and resettlement' had received their compensations. As of February 2010, only 20 % of the compensation had been completed, and in December 2010 only 7,280 cases out of the total 13,237 that had completed their paperwork had been granted compensation (BBC Indonesia 2010a, c).[2] In addition to the problematic execution of this compensation model, it is aspatial, due to its ignorance towards the meaningful values of space and social relations, demonstrated by the reduction of space into cash values and land barter to a real estate complex. Rather than using the maps to make the scheme spatial, the compensation schemes made the maps aspatial.

It should also be noted that the compensation was given as property transaction. Although the presidential decrees mandated the affected households to be compensated, they also forced the victims to sell their land to get the compensation. In other words, the presidential decrees frame land acquisition as compensation that benefit the residents. Consequently, the affected areas were converted into exchangeable property to be acquired by either Minarak Lapindo Jaya Company or the government, which comes with real social, cultural and spatial consequences along with displacement. However, the lack of attention in the decrees towards the transformation of life spaces of communities due to the disaster-induced landscape change of Porong has perpetuated a devaluation of community life in favour of monetary property values.

Thirdly, the emphasis on the material aspect was inseparable from the interventions that focused on reducing the subsequent impacts on them. The strategies to build the levee and the channel, to let the mud flow into the Porong River, were to reduce the possibility of further property damages in the future. On the one hand, these interventions had been useful in terms of curbing the spread of the mud. On the other hand, these solutions were infrastructure projects that reified the reliance on particular developers, companies and institutional brokerages that started the disaster to begin with. The selection of construction companies was under criticisms and led groups of residents to protest against the projects. Protests also questioned the prioritisation of spending funds on levees rather than addressing residents' compensation that was seriously lagging behind (Liputan6 2006). The risks of levee breakdown during the rainy season continue to define the surrounding areas with uncertainty and continuous steady threat of mud invasion (BBC Indonesia 2010b). Furthermore, the channelling of mud into the Porong River expands the disaster space beyond its main eruption point, which would have implications on the spatial attributes of the river and the new channel. It had been reported that fisheries in the Porong River (as a result) were seriously affected by the mud, causing shrimp and fish farmers to suffer economic loss. In 2007–2008, 60 % of shrimp ponds in Sidoarjo were hit by massive death of shrimps and Sidoarjo fishery production declined by more than 25 %, presumably caused by substances contained in the

[2] Minarak Lapindo Jaya, the subsidiary of Lapindo that was established specially to manage compensations, claimed that they had no money to purchase all the victims' land in lump sum and offered payment as instalments instead. Communities rejected the request on the basis that they needed to buy new homes and could not wait any longer.

mud that affected the river's aquaculture (Fitrianto 2012). The implication of the mud in changing the spatial characteristics and meanings of the Porong River was largely overlooked. The development of the channel also indicates the pseudo-merging of the mud volcano space and the river space, which potentially expands the social and cultural disaster-affected dynamics as well.

4.5 Muddy Strategies

There are at least three kinds of community responses that are spatially contextual, in their efforts to live with the mudflow disaster, all of which are empowering in different ways. Firstly, the reality of the levee and the mud has evolved to be the spatial reality of the residents' everyday lives. Although they may come across as commercialisation of victimhood, to many, these efforts are coping mechanisms. The government has started to promote the mudflow disaster area as an official geological tourism site since 2010, which includes BPLS' effort to highlight the new island emerging at the end of Porong River because of mud accumulation. But the 'mud tourism' economy in the levee area has been socially constructed by various actors and groups. Visitors now can encounter a handful of vendors selling DVDs of mudflow documentaries and offering motorbike rides to take visitors closer to the centre of the mud volcano. The western levee wall along the railway features an open site that is used as a parking spot for motorcycles, cars and buses. If there is no particular event on the levee, motorcycles can go all the way up to the walls, along with the motorbike ride runners. Upon parking one's vehicle, a man dressed casually would approach the driver to obtain a parking fee. Negotiating this fee might be tougher for tourists compared with locals who are familiar with the place or who knows people with leadership positions in the communities (Fig. 4.4).

Most of these vendors were former residents of the submerged villages. For example, Kusdyanto, now 48 years old, was a former pedicab driver who had lost his home and the pedicab—which were drowned in mud (Himawan 2013). Another vendor, a woman in her forties, brought her three children to the levee when she went around to sell DVDs and to offer motorbike rides. Her house was submerged and her husband, who used to work in one of the factories, lost his job, and the family had never received their compensation beyond the first 20 % that they received in the beginning. These stories are consistently found among the vendors, who also have their routine social gathering sessions that are used for sharing problems and experiences as well as to exchange information on the latest developments (Farida 2014). Along with the formation of social relationships around the mud volcano and economic survival, the vendors have socially constructed the mud volcano into an entrepreneurial space. While cities often outlaw street vendors to roam freely in public spaces, the levee and the mud volcano have become an open space for interpretation and appropriation. From the state's point of view, these appropriations may come across as anarchic. However, in practice these actions and actors have their ways of forming relationships, managing territories and forming consensus on

Fig. 4.4 'Masuk Area Wisata Lumpur Panas' (Entrance to the Hot Mud Tourism Area) signage outside the parking lot (Author 2014)

the use of space, including the state's official intention to make the mud volcano a tourist attraction.

The second mode of community response is the social mobilisation against Lapindo and Aburizal Bakrie's cronies, the owners of Lapindo shares. In the first couple of years of the mudflow disaster, this group utilised radical mobilisation strategies that mainly featured demonstrations against the Bakrie family and the presidential decrees that protected the cronies. Two of the most vocal groups are *Pagar Rekontrak*[3] and *Korban Lapindo Menggugat* (KLM, Lapindo Victims United). Those in this effort considered Lapindo and its owners as the ones responsible for botched drilling safety procedures that caused the mud volcano to erupt (Batubara and Utomo 2012; Farida 2014). The group also refused to be included in the map of compensation-eligible places. Their standpoint viewed the government's policy to have Lapindo buy the land from the communities as unfairly beneficial for Lapindo, as the community members would be forced to sell their unusable land to the company that would eventually be the main landowner. If there is indeed a wealth natural gas beneath the mud volcano to be explored in the future (Mazzini et al. 2012), then the mudflow would be a tool to impose 'forced eviction' of the communities for Lapindo's benefit, as the residents could not sell the land to anyone

[3] *Pagar Rekontrak* is an acronym for *Paguyuban Rakyat Renokenongo Menolak Kontrak* (Association of Renokenongo People Against Contract), which later was renamed *Pagar Rekorlap* (*Paguyuban Warga Renokenongo Korban Lapindo*/Association of Renokenongo Victims of Lapindo) (Batubara and Utomo 2012).

else. With the slogan 'Menolak Masuk Peta, Menolak Menjual Desa, Menolak Pengeboran' (Refuse inclusion in the map, Refuse to sell our village, Refuse drilling), they refused to sell their land for the purpose of expansion of the mud levee and the gas exploration drilling by Lapindo (Liputan6 2012).

These social mobilisations expanded the Lapindo mudflow spaces beyond their locality, particularly to the national capital city, in two ways. Firstly, groups with social mobilisation strategies made outreach efforts to nongovernmental organisations (NGOs), advocacy groups and networks outside Porong and Sidoarjo to spread their resistance message. Secondly (which is related to the first), central spaces were appropriated in the national capital as an extension of their protests in Porong. Various groups of residents have marched in Jakarta, the national capital, to protest at the Presidential Palace and at Aburizal Bakrie's office. Protesters tossed mud balls, made of the Lapindo mud they brought from home, to the Ministry for Social Welfare, Bakrie's ministerial office. Groups have also marched to the main demonstration site of Jakarta, *Bundaran Hotel Indonesia*, with self-made banners with slogans that demanded persecution of Lapindo and anyone associated with the disaster.

The same group also chained themselves in front of the Presidential Palace. When approached during the demonstration, one of the members of the protest responded with a frustrated tone, 'I lost everything. My home, my land. I don't have anything anymore.' Groups from other villages in Porong have also come to Jakarta and camped at the Proclamation Monument for a few days. These visits to Jakarta did not yield in any meetings with Bakrie, ministers or the president, but these groups were able to connect with and to form coalitions with Jakarta-based nongovernmental organisations (NGOs), particularly Kontras and Komnas HAM (National Human Rights Commission) that work on human rights violations in Indonesia (BBC Indonesia 2010a). Social mobilisation experience in Jakarta may empower but may also overwhelm the victims. As the city of Jakarta and its residents continue their everyday activities and problems with or without the presence of Lapindo mudflow victims' street protests, those who were involved in the march might internalise this continuously busy life of the city as being ignorant of their cause (Drake 2013). But, for those who were involved in the meetings with NGOs in Jakarta, social mobilisations beyond the small town and semirural area of Porong expanded the spaces of disaster to construct public spaces of discussion and advocacy that bring together active victims of disaster with members of the public that do not have firsthand experience of the mudflow. These social mobilisations have also appropriated politically strategic public spaces in Jakarta as megaphones to express their concerns. The presence of Lapindo mud in Jakarta, whose residents specially brought to display across the Presidential Palace and to throw at Bakrie's office, becomes an artefact that anoint both Bakrie, the President of Indonesia and the city of Jakarta as actors and entities that bear responsibilities over the suffering of the people in Porong (Fig. 4.5).

The third kind of response features local coping efforts that lead to external outreach on sharing empowerment activities and initiatives beyond the locality. These efforts are inseparable from the first two, as they often overlap and collaborate. This

Fig. 4.5 Protest and body chaining demonstration across the Presidential Palace, 22 November 2007 (Author 2007)

group also refused to be included in the map and also shared similar concerns with the radical movements. However, the empowerment activities in this category evolved from rather mundane initiatives that built up to life-changing experiences.

An example of this would be Cak Irsyad and *Sanggar Alfaz*. Cak Irsyad's community was torn because of the mudflow. The levee was built right beside his house. He was one of the people who resisted being 'included in the map' (*menolak masuk peta*). Obviously disappointed by how the government handled the situation, Cak Irsyad opened his house as a gathering place for children, which later developed together with the community to become an art centre where children could learn music and theatre. The *sanggar* (open house) started to connect with other *sanggars* in other cities, from which they started to collaborate for public performances, starting with the third anniversary of the mudflow on 29 May 2009 (Novenanto et al. 2014). Performance during mudflow anniversaries and the sanggar's anniversary became their annual activities, through which collaborations in the process, practice and choreography were conducted by volunteers from the communities. During the seventh anniversary of the mudflow in 2013, children from *Sanggar Alfaz* performed traditional-style dance choreography based on the mudflow story, and in 2014 they prepared choreography of the *Timun Mas* legend to reappropriate the local folklore as an empowering story of good, small people against a big, ugly and evil giant.[4] In Cak Irsyad's own words in June 2013, 'This disaster had somehow led me to connect with other people, outside my community. I get to know other people who have this *sanggar* initiative… This whole experience changed me.' His narrative presents an experience of growing social space as a result of disaster.

Starting with the name *Sanggar Anak Alfaz* (*Sanggar Alfaz* for children), the name was changed in 2011 to *Sanggar Alfaz* to invite all age groups. *Sanggar Alfaz* started community gardening to produce vegetables to be shared. Supported by financial assistance and training from colleagues in the network, they started a small printing business with the hope of providing employment opportunities for local youths. *Sanggar Alfaz* also became a gathering place for women in the East Besuki community every Saturday night. Called *Jimpitan Sehat*, the weekly gatherings were spaces to share information about health, welfare programmes and benefits. The feeling of helplessness and disempowerment due to slow disaster uncertainties have paved the way for Cak Irsyad's initiatives to expand the reach and diversity of his social network. Sanggar Alfaz became known and respected well outside Porong and participated in children's activities in the network of various sanggars.[5]

Despite being grateful of inspirational experiences post-mudflow, Cak Irsyad was clearly upset and saddened about the situation, a feeling that was shared by many members of the community. Not only were the communities separated but children lost their friends due to their parents' move elsewhere. People were also divided in the way that they thought they should respond to the mudflow. The resistance to be included in the map was arguably stronger in the early years of the disaster, but as years went by, families started to yearn to move on with their lives. The capacity to cope with disaster was overwhelmed by the feeling of insecurity in the prolonged disaster situation without any signs of ending. Those who were tired of the situation and wanted to give up by receiving compensation started to feel aggravated by those who were still insistent on their refusal cause. This tension was further fuelled by allegations that those who refused to be included in the map were only holding on to demand more compensation money, a 'boomerang effect' that was widely circulated among officials and groups of local academics (Drake 2013).

Findings from the aforementioned three strategies to live with the mud are threefold. Firstly, the strategies are overlapping and are related to each other. News and updates spread relatively quickly among communities in Porong and its surrounding villages, especially with visible activities around the embankments (Farida 2014). With the scale of disaster impact, media spotlight and the continuously changing landscape of Porong, people are considerably aware of the social movements and initiatives after the mud volcano eruption. In the past several years, these groups have collaborated and took turns to organise annual commemorations of the mudflow disaster on a stage on the levee. The event typically features music and

[4] The folklore, about a girl who defeated an evil giant by spreading shrimp paste that turned into a mud lake that drowned the giant, had been previously used by geologist Awang Harun Satyana to justify the mudflow as a natural disaster to legitimise the innocence of Lapindo Brantas (Nurcahyo 2014).

[5] Information based on discussion with members of Sanggar Alfaz, Porong and Sanggar Merah Merdeka, Surabaya.

dance performances, as well as poem reading and stand-up comedy. But there has always been a main feature of the event that became the highlight of the commemoration. Led by *Sanggar Alfaz*, the commemoration event in 2013 brought members of the communities to build an effigy (locally known as *ogoh-ogoh*) of Aburizal Bakrie. The highlight of the event was to throw the effigy into the mud as an expression of *ruwatan* or getting rid of bad spirits.

This effigy stood in the mud even during the 2014 commemoration event, which was led by KLM. The commemoration in 2014 also featured Survivor, an installation of 110 mud sculptures by sculptor Dadang Christanto that represented the victims of the mudflow who are sorrowful, separated and drowned (Tempo.co 2014a). Loss of their everyday activities and social relationships was represented by the broken furniture and tools, such as stoves, televisions, buckets and some dolls that were put in the sculptures' arms. Although the submerged land presumably had to be sold to Lapindo, even today, community groups are able to construct an event stage, plant sculptures and hold gatherings there. These collaborative annual events become climaxes of space occupation to represent the victims, which are inseparable from its everyday appropriation by the mounting disaster tourism activities and the growth of related underground economy (Figs. 4.6 and 4.7).[6]

Fig. 4.6 The Aburizal Bakrie effigy in 2014, with the hardened mud beneath it, is still a spectacle for mud volcano visitors (Author 2014)

[6] Local activists have mentioned the existence of pickpockets in the mud tourism area, who mainly target visitors. These pickpockets become more rampant during big events such as the annual commemoration.

Fig. 4.7 Survivor by Dadang Christanto, art installation to commemorate the 8th anniversary of the mudflow in 2014 (Author 2014)

Secondly, community efforts to overcome challenges in the mudflow-affected spaces have had their traces since before the disaster. In Besuki village, where Sanggar Alfaz is located, the Surabaya-Gempol toll road has divided the village into East and West in 1982, following at least a decade of industrial expansion over their agricultural land. Weekly communal prayer activities paused for a while because of the difficulty in crossing the toll road, especially at night. This was worsened by the construction of concrete walls on both sides of the road in 1995, in response to the people's effort to cut off barbed wire barricades to cross the road. Despite the construction of pedestrian bridge, the people preferred to tear down the concrete walls for easier crossing (Novenanto et al. 2014). These efforts have shown the resilience of the community to overcome challenges they were facing in spite of physical barriers, although there was also a limit as in practice the community was inevitably split into two. To some, the barbed wire cutting and the tearing of the concrete wall could be interpreted as lawlessness, but these actions compellingly demonstrate the power of human agency in finding alternative solutions to challenges considerably long before the mudflow disaster.

Thirdly, these three kinds of resistance strategies may also change as the disaster lingered on for many years. After the issuance of Presidential Decree 37/2012, relocation issues heated in East Besuki and *Jimpitan Sehat* participants decreased, particularly after relocation was discussed during one gathering (Novenanto et al. 2014). Entering the eighth year of the mudflow, Cak Irsyad's community was subjected to compensation and moved to another location that was originally an agricultural land. In the process, Cak Irsyad was denied ownership of land in the new place, with the reason that there was not enough land for everyone, and had to buy a house elsewhere, separated from his neighbours. While the actual reason for leaving him behind may never be officially or legally revealed, many friends believed that he was singled out because of his outspokenness.[7]

Korban Lapindo Menggugat (KLM) also underwent change from their first stance of refusal to be included in the map. KLM has strong links to the Urban Poor Consortium (UPC) and Urban Poor Linkage (UPLink), an NGO and a network that are actively defending eviction-prone urban poor communities in several cities in Indonesia, on the basis of the urban poor's right to the city. Despite being one of the promoters of the 'refusal to be included in the map' (*menolak masuk peta*) movement, KLM is now advocating for 'fair compensation' of the affected communities. KLM's main focus—to demand 'speedy and fair compensation'[8]—was adopted as one of the political promises of Indonesia's president-elect, during the signing of political promise as the main highlight of the eighth anniversary of the mudflow in 2014. This change was brought by the long wait of many mudflow refugees who are often left without proper housing, having to spend financial resources to rent homes while waiting for compensations that never came. In addition, the perceived characteristics of leaders' attitude and leadership style also affected the shift towards compensation-focused campaign. The presidential candidate, who later won the election, was widely seen as a clean politician and had established good working relationship as a governor of Jakarta with eviction-threatened communities in the capital city that were in the same network as KLM.

4.6 Conclusion: The Resilience of Life Spaces

As this chapter is written, the mudflow still continues. Less than four months after the 8th anniversary of the mudflow, the 110 Survivor sculptures are now partially submerged. Despite the pressure from the Constitutional Court on the state to pressure Lapindo to fulfil their obligations, Minarak Lapindo Jaya has stated that they did not have the money to fully compensate the victims at the designated deadline on 30 June 2014. On 18 May 2014, several victims threw mud on the Bakrie effigy, and on 27 August 2014 groups of affected residents blocked access for BPLS regular mud removal from the volcano in a protest against the severely delayed compensation by Lapindo. Mud has crawled up to the chests of Survivor sculptures, which continue to be silent witnesses of the continuing disaster in recent months (Tempo.co 2014b, c, d). In November 2014, mud victims expressed their disappointment and vowed to block BPLS' access to the levee again when they were informed that the national budget 2015 has not included enough funds to cover the remaining 781

[7] Information based on discussions with members of Sanggar Alfaz, Porong and Sanggar Merah Merdeka, Surabaya.

[8] As stated in the speech at the 8th anniversary of the mud volcano.

billion rupiah compensation that has not been paid, despite victim group representatives having had meetings with the Ministry of Public Works and relevant institutions on the government's commitment to cover the outstanding financial compensation (Korbanlumpur.info 2014).

Finally, in December 2014, the government announced the provision of advance money (*dana talangan*) from the national budget adjustment as a loan to Lapindo to pay off the outstanding compensation. The money was scheduled to be disbursed to the victims by March 2015, although it was later postponed to May 2015 (Tempo.co 2014h, 2015). Lapindo is required to pay off the debt to the government in 4 years; otherwise, the government will assume ownership of all Lapindo's assets as the loan's collateral. Lapindo's assets are largely the land that has been bought from the victims through the compensation scheme, with the belief that the land value would increase when the mud 'surely' stops in the future (Tempo.co 2014f). The government's move is reflective of the President's 'speedy and fair compensation' pledge in his political campaign promise. In addition, the new administration's Minister of Public Works, Basuki Hadimuljono, was a member of the old administration's team that monitored the mudflow in its early days in 2006. *Dana talangan* has been both applauded and criticised: It's applauded as the state's presence in the lives of the victims after many years of absence (Khudori 2015). Yet Novenanto (2015a, b, c) argued that the state has always been present in the form of political alliances and policies, and the only consistent absence of the state, even with the *dana talangan*, is on the punishment to Lapindo for neglect in permit processing and disaster management. *Dana talangan* is also criticised for early loan approval before the final valuation of Lapindo's assets is done. The basis of the criticism is the worry that the value of the assets may be lower than that of the loan and thus makes the loan a bailout of the businessman through national budget (Khudori 2015). There are also worries that the victims to be compensated by this *dana talangan* is subjected to brokers (*calo*) in Lapindo as has happened in the process before (Tempo.co 2014e, g). Although many remain hopeful that the scheme will settle the tension and enable victims through financial means to rebuild livelihoods, the focus on land value rather than life space remains unchallenged and is inseparable from the prevailing hegemony of material compensation as a cure for disaster victims.

This chapter has shown that a humanistic approach through focusing on life spaces to examine community resistance strategies in a prolonged disaster has demonstrated a distinctively empowering angle in examining the impact of disasters on livelihoods without neglecting the power of human agency. It is obvious that disaster governance mechanisms that are fixed on material replacement are insufficient in comprehending the extent of the disaster, but the focus on the quantification of disaster impact remains the most appealing as it is the one that has been repeatedly done. In the case of the mudflow, the two groups that refused to be included in the map have adopted compensation strategy in the eighth year. The change of strategy has been the one that was more pragmatic rather than idealistic, given the prolonged disaster experience that may not end in the near future and the prospect of an incoming progressive president.

Through the years, the levee and the mud volcano had been socially constructed into an unrestricted space, in which Porong residents could articulate the space as they wish. Entrepreneurial vendors socially constructed the mud volcano as a tourism area and offered motorcycle rides to bring visitors closer to the centre of the mud volcano. Furthermore, the production of DVDs sold by vendors on the levee was supported by the local youth group (*Karang Taruna*), although criticisms on the extortion of contributions through indeterminate sale price of the DVDs remain. The disaster has also left a space that is open for interpretation, which is almost anarchic. The areas drowned by mud had transformed into a nonregulated space that can be appropriated by many different groups for different purposes.

While the vendors interpreted the levee and the mud as a business opportunity as part of their economic survival, activists could throw an effigy into the mud that would not be removed even after one year when the mud beneath it was hardened and could be walked upon. Despite its nature of mocking Aburizal Bakrie, a powerful businessman and politician who was at one point an aspiring presidential candidate, the effigy could last as a monument that to a certain degree reflected the social consensus about him being the man behind the disaster. The mud volcano also became the site for artists to make and to display artwork, even those crafted by artists from outside Porong such as the Survivor and the drowned hand installations during the eighth anniversary of the disaster.

Current disaster governance mechanisms and scholarship have both focused on immediate solutions to the problems faced by disaster-affected communities. The persistence of material and quantitative measurements in valuing (or devaluing) life spaces as private properties is proven insufficient in understanding the needs of communities for post-disaster resurgence, let alone in understanding the dynamics and possibilities of change in long-term disasters such as the Lapindo mudflow. The mudflow case in Porong, Sidoarjo, is a reflection of inequalities and marginalisations in society, demonstrated by the impunity of Lapindo up until now and the neglect towards the victims. The victims may have different experiences of the disaster and encounters with the mud. To some of them, as examined in this chapter, the disaster had been a transformative incident that has been incorporated into their life spaces and social identity. With the long-term disaster still looming, the analysis in this chapter has shown that focusing on disaster as the everyday life would yield a more useful examination of the potential ways in which local initiatives allow communities to empower themselves to embrace the transformative identity as active members of a disaster-affected society.

References

Batubara, B., & Utomo, P. W. (2012). *Kronik Lumpur Lapindo: Skandal Bencana Industri Pengeboran Migas di Sidoarjo*. Yogyakarta: INSIST Press.

BBC Indonesia. (2010a). *Komnas HAM Usut Lumpur*. http://www.bbc.co.uk/indonesia/berita_indonesia/2010/02/100212_lumpursidoarjo.shtml. Accessed 13 Aug 2014.

BBC Indonesia (2010b). *Warga Mengungsi Karena Lumpur Lapindo*. http://www.bbc.co.uk/indonesia/berita_indonesia/2010/12/101224_lapindoflood.shtml. Accessed 13 Aug 2014.

BBC Indonesia. (2010c). *Pembayaran Ganti Rugi Lumpur Lapindo 'Mandek'*. http://www.bbc.co.uk/indonesia/berita_indonesia/2010/12/101230_lapindo.shtml. Accessed 13 Aug 2014.

Carcellar, N., Rayos Co, J. C., & o Hipolito, Z. (2011). Addressing disaster risk reduction through community-rooted interventions in the Philippines: Experience of the homeless People's Federation of the Philippines. *Environment and Urbanization*, *23*(2), 365–381.

Café, D. P. (2012). The social construction of disaster: Ondoy (Ketsana) in the context of Sagrada Familia and Inquirer.net. *Asian Social Science*, *8*(10), 45–57.

Davies, R. J., Swarbrick, R. E., Evans, R. J., & Huuse, M. (2007). Birth of a mud volcano: East Java, 29 May 2006. *GSA Today, 17*(2), 4–9.

Davies, R. J., Mathias, S. A., Swarbrick, R. E., & Tingay, M. J. (2011). Probabilistic longevity estimate for the LUSI mud volcano, East Java. *Journal of Geological Society London*, 168(2), 517–523.

Donaldson, A., Lane, S., Ward, N., & Whatmore, S. (2013). Overflowing with issues: Following the political trajectories of flooding. *Environment and Planning C: Government and Policy*, 31(4), 603–618.

Drake, P. (2013). Under the mud volcano. *Indonesia and the Malay World*, 41(121), 299–321.

Farida, A. (2014). Reconstructing social identity for sustainable future of Lumpur Lapindo victims. *Procedia Environmental Sciences*, 20, 468–476.

Fischer, H. W. (2003). The sociology of disaster: definitions, research questions, & measurements. Continuation of the discussion in a post-September 11 environment. *International Journal of Mass Emergencies and Disasters*, 21(1), 91–108.

Fitrianto, A. R. (2012). Shrimp farmers' innovation in coping with the disaster (A case study in Sidoarjo mud volcano disaster toward shrimp farmers' responses). *Procedia Economics and Finance*, 4, 168–176.

Friedmann, J. (1992). *Empowerment: The politics of alternative development*. Cambridge, MA/ Oxford: Blackwell.

Friedmann, J., & Hudson, B. (1974). Knowledge and action: A guide to planning theory. *Journal of the American Institute of Planners*, 40(1), 2–16.

Gilbert, C. (1995). Studying disaster: A review of the main conceptual tools. *International Journal of Mass Emergencies and Disasters*, 13(3), 231–240.

Gilbert, C. (1998). Studying disaster: Changes in the main conceptual tools. *What is a Disaster*, ed. E.L. Quarantelli. London and New York: Routledge, 11–18.

Gotham, K. F., & Greenberg, M. (2014). *Crisis cities: Disaster and redevelopment in New York and New Orleans*. Oxford: Oxford University Press.

Gramsci, A. (2010). *Prison notebooks*. New York: Columbia University Press.

Harvey, D. (1992). Social justice, postmodernism and the city. *International Journal of Urban and Regional Research*, 16(4), 588–601.

Harvey, D. (2002). The urban process under capitalism. *The City: Land Use, Structure, and Change in the Western City*, 2, 5.

Hilgartner, S. (1992). The social construction of risk objects. In J. F. Short & L. Clarke (Eds.), *Organizations uncertainties and risk* (pp. 40–53). Boulder: Westview Press.

Himawan, A. (2013). *Kisah Lebaran dari Pinggir Danau Lumpur Lapindo: Kisah Penjaja VCD Film Lumpur Lapindo*. Kontan Online. http://nasional.kontan.co.id/news/kisah-lebaran-dari-pinggir-danau-lumpur-lapindo. Accessed 13 Aug 2014.

Istadi, B. P., Pramono, G. H., Sumintadireja, P., & Alam, S. (2009). Modeling study of growth and potential geohazard for LUSI mud volcano: East Java, Indonesia. *Marine and Petroleum Geology*, 26(9), 1724–1739.

Johnson, C. L., & Penning-Rowsell, E. C. (2010). What really determines policy? An evaluation of outcome measures for prioritising flood and coastal risk management investment in England. *Journal of Flood Risk Management*, 3(1), 25–32.

Jigyasu, R. (2004). Sustainable post disaster reconstruction through integrated risk management–the case of rural communities in South Asia. *Journal of Research in Architecture and Planning*, 3, 32–43.

Khudori. (2015). *Talangan APBN Buat Lapindo*. http://www.tempo.co/read/kolom/2015/02/17/1953/Talangan-APBN-buat-Lapindo. Accessed 30 Apr 2015.

Korbanlumpur.info. (2014). *Korban Lapindo Ancam Blokade Tanggul*. http://korbanlumpur.info/2014/11/korban-lapindo-ancam-blokade-tanggul/. Accessed 18 Nov 2014.

Laclau, E., & Mouffe, C. (2001). *Hegemony and socialist strategy: Towards a radical democratic politics*. London: Verso.

Lefebvre, H. (1991). *The production of space*. Oxford: Blackwell.

Liputan6. (2006). *Warga Jatirejo Memprotes Pembangunan Tanggul Permanen*. http://news.liputan6.com/read/131067/warga-jatirejo-memprotes-pembangunan-tanggul-permanen. Accessed 13 Aug 2014.

Liputan6. (2012). *Korban Lapindo Menolak Masuk Peta Terdampak*. http://news.liputan6.com/read/391794/korban-lapindo-menolak-masuk-peta-terdampak. Accessed 13 Aug 2014.

Mazzini, A., Svensen, H., Akhmanov, G.G., Aloisi, G., Planke, S., Malthe-Sørenssen, A. et al. (2007). Triggering and dynamic evolution of the LUSI mud volcano, Indonesia. *Earth and Planetary Science Letters*, 261(3), 375–388.

Mazzini, A., Nermoen, A., Krotkiewski, M., Podladchikov, Y., Planke, S., & Svensen, H. (2009). Strike-slip faulting as a trigger mechanism for overpressure release through piercement structures. Implications for the Lusi mud volcano, Indonesia. *Marine and Petroleum Geology*, 26(9), 1751–1765.

Mazzini, A, Etiope, G., & Svensen, H. (2012). A new hydrothermal scenario for the 2006 Lusi eruption, Indonesia. Insights from gas geochemistry. *Earth and Planetary Science Letters*, 317–318, 305–318.

McMichael, H. (2009). The Lapindo mudflow disaster: Environmental, infrastructure and economic impact. *Bulletin of Indonesian Economic Studies*, 45(1), 73–83.

Novenanto, A. (2015a). Negara Absen dalam Kasus Lapindo, Apa Iya? (Pra Semburan). http://korbanlumpur.info/2015/03/negara-absen-dalam-kasus-lapindo-apa-iya-1/. Accessed 30 Apr 2015.

Novenanto, A. (2015b). Negara Absen dalam Kasus Lapindo, Apa Iya? (Timnas). http://korbanlumpur.info/2015/03/negara-absen-dalam-kasus-lapindo-apa-iya-2/. Accessed 30 Apr 2015.

Novenanto, A. (2015c). Negara Absen dalam Kasus Lapindo, Apa Iya? (BPLS). http://korbanlumpur.info/2015/03/negara-absen-dalam-kasus-lapindo-apa-iya-3/. Accessed 30 Apr 2015.

Novenanto, A., Amiruddin, L., & Ilma, D. (2014). Pemanfaatan Sanggar Alfaz sebagai strategi pemuda Besuki Timur mengatasi bencana industri Lumpur Lapindo. *Youth Studies Center (YouSure) Universitas Gadjah Mada*.

Nurcahyo, H. (2014). Rekayasa Dongeng dalam Bencana Lumpur Lapindo. Asosiasi Tradisi Lisan Jawa Timur.

Penning-Rowsell, E. C., & Pardoe, J. (2012). Who benefits and who loses from flood risk reduction? *Environment and Planning C: Government and Policy*, 30(3), 448–466.

Presidential Decree 14/2007. *Government of Indonesia*. http://prokum.esdm.go.id/perpres/2007/perpres_14_2007.pdf. Accessed 26 Sept 2014.

Presidential Decree 48/2008. *Government of Indonesia*. http://www.kemendagri.go.id/produk-hukum/2008/07/17/peraturan-presiden-nomor-48-tahun-2008. Accessed 26 Sept 2014.

Presidential Decree 40/2009. *Government of Indonesia*. http://sipuu.setkab.go.id/PUUdoc/16767/PERPRES0402009.pdf. Accessed 26 Sept 2014.

Presidential Decree 68/2011. *Government of Indonesia*.

Presidential Decree 37/2012. *Government of Indonesia*.

Presidential Decree 33/2013. *Government of Indonesia*.

Purwaningsih, E. & Notosiswoyo, S. (2013). Hydrochemical study of groundwater in Sidoarjo mud volcano area, East Java Indonesia. *Procedia Earth and Planetary Science*, 6, 234–241.

Rosol, M. (2013). Vancouver's "EcoDensity" planning initiative: A struggle over hegemony? *Urban Studies*, 1–18. doi: 0042098013478233.

Rudolph, M.L., Karlstrom, L., & Manga, M. (2011). A prediction of the longevity of the Lusi mud eruption, Indonesia. *Earth and Planetary Science Letters*, 308(1), 124–130.

Sawolo, N., Sutriono, E., Istadi, B. P., & Darmoyo, A. B. (2009). The LUSI mud volcano triggering controversy: Was it caused by drilling? *Marine and Petroleum Geology*, 26(9), 1766–1784.
Stallings, R. A. (1995). *Promoting risk: Constructing the earthquake problem*. New York: Aldine de Gruyter.
Tempo.co. (2008). *Tiga model Ganti Rugi Ala Lapindo Tak Beres*. http://www.tempo.co/read/news/2008/12/02/058149149/Tiga-Model-Ganti-Rugi-ala-Lapindo-Tak-Beres. Accessed 13 Aug 2014.
Tempo.co. (2014a). *Makna 110 patung di peringatan Lumpur Lapindo*. http://www.tempo.co/read/news/2014/05/28/058580852/Makna-110-Patung-di-Peringatan-Lumpur-Lapindo. Accessed 26 Sept 2014.
Tempo.co. (2014b). *Ganti rugi tak jelas, korban Lapindo mengamuk*. http://www.tempo.co/read/news/2014/05/18/058578539/Ganti-Rugi-Tak-Jelas-Korban-Lapindo-Mengamuk. Accessed 26 September 2014.
Tempo.co. (2014c). *Korban Lumpur Lapindo usir pekerja BPLS*. http://www.tempo.co/read/news/2014/08/27/058602598/Korban-Lumpur-Lapindo-Usir-Pekerja-BPLS. Accessed 26 Sept 2014.
Tempo.co. (2014d). *Lumpur Lapindo meninggi, 110 patung tenggelam*. http://www.tempo.co/read/news/2014/09/03/058604152/Lumpur-Lapindo-Meninggi-110-Patung-Tenggelam. Accessed 26 Sept 2014.
Tempo.co. (2014e). *BPLS Akan Bagikan Ganti Rugi Korban Lapindo*. http://www.tempo.co/read/news/2014/12/19/173629731/BPLS-Akan-Bagikan-Ganti-Rugi-Korban-Lapindo. Accessed 30 Apr 2015.
Tempo.co. (2014f). Kasus Lapindo, Duit Negara Rp 10 T, Ical Rp 3,8 T. http://www.tempo.co/read/news/2014/12/19/078629771/Kasus-Lapindo-Duit-Negara-Rp-10-T-Ical-Rp-38-T. Accessed 30 Apr 2015.
Tempo.co. (2014g). *Jokowi Jangan Bayar Ganti Rugi Via Lapindo, Kenapa?*http://www.tempo.co/read/news/2014/12/23/058630486/Jokowi-Jangan-Bayar-Ganti-Rugi-Via-Lapindo-Kenapa. Accessed 30 Apr 2015.
Tempo.co. (2014h). *DPR Setuju, Dana Talangan Lapindo Dibayar Maret*. http://www.tempo.co/read/news/2014/12/29/087631602/DPR-Setuju-Dana-Talangan-Lapindo-Dibayar-Maret. Accessed 30 Apr 2015.
Tempo.co. (2015). *Menteri PU Targetkan Dana Talangan Lapindo Cair Mei*. http://www.tempo.co/read/news/2015/03/29/090653789/Menteri-PU-Targetkan-Dana-Talangan-Lapindo-Cair-Mei. Accessed 30 Apr 2015.
Torfing, J. (1999). *New theories of discourse: Laclau, Mouffe and Zizek*. Oxford: Blackwell.
Tierney, K. J. (1999). Toward a critical sociology of risk. *Sociological Forum*, 14(2), 215–242.
Walker, G., Whittle, R., Medd, W., & Walker, M. (2011). Assembling the flood: Producing spaces of bad water in the city of Hull. *Environment and Planning A*, 43(10), 2304–2320.
Wisner, B. (1993). Disaster vulnerability: Scale, power and daily life. *GeoJournal*, 30(2), 127–140.

Chapter 5
Disaster Governance in War-Torn Societies: Tsunami Recovery in Urbanising Aceh and Sri Lanka

Malin Åkebo

Abstract The chapter focuses on disaster governance in the context of the earthquake and tsunami that severely hit the war-torn areas of Aceh and Sri Lanka in December 2004. It explores how the tsunami response and recovery actions were influenced by the ongoing armed conflicts and how the process of tsunami recovery, in its own turn, shaped the politics of the violent conflicts. The chapter takes a closer look at the urban dimension of tsunami recovery and at urban–rural intersections. While the tsunami on the one hand opened up a window of opportunity for reaching a peaceful solution to the violent conflicts, at the same time, as illustrated in the chapter, the natural disaster also contributed to consolidate dividing lines between geographical localities, urban and rural societies and identity groups. The chapter points to the importance of recognising competing governance systems and how prior tensions, cleavages and political power struggles might be reproduced and accentuated in the light of environmental disasters. In conclusion, the chapter underlines the essential importance of taking context into account and of recognising the political processes at play so as to understand the response and recovery from environmental disasters in different conflict settings and societies.

Keywords Disaster governance • Tsunami • Conflict • Conflict resolution • Aceh • Sri Lanka

5.1 Introduction

The 2004 Indian Ocean tsunami and undersea earthquake occurred on an incredible scale, directly impacting a dozen of countries connected by the Indian Ocean, killing nearly 300,000 people and displacing more than 1.8 million (Sarvananthan and Sanjeewanie 2008). The experiences from the 2004 tsunami serve as a painful

---

M. Åkebo, PhD (✉)
Department of Political Science, Umeå University, Umeå, Sweden
e-mail: malin.akebo@umu.se

M.A. Miller, M. Douglass (eds.), *Disaster Governance in Urbanising Asia*,
DOI 10.1007/978-981-287-649-2_5

reminder of the massive destruction and suffering that environmental disasters can produce in contemporary societies. It also underlines the essential importance of increasing our insights into how societies respond to and recover from such disasters and what challenges they may encounter along the way.

This chapter focuses on disaster governance in the context of the earthquake and tsunami that severely hit the war-torn areas of Aceh and Sri Lanka in December 2004. The densely populated coastal lines of Aceh and Sri Lanka were the areas worst hit by the tsunami. While a dozen countries were affected, a vast majority of all casualties came from these two regions. In addition, Aceh and Sri Lanka also share common characteristics and experiences because they are both protracted identity-based violent conflicts. The tsunami struck Aceh during the violent conflict—which had raged for almost three decades—between the government of Indonesia and the Free Aceh Movement (Gerakan Aceh Merdeka, GAM). Likewise in Sri Lanka, since the early 1980s, a long-drawn war had persisted between the government and the separatist Liberation Tigers of Tamil Eelam (LTTE). In this chapter, I explore how the tsunami response and recovery actions in Aceh and Sri Lanka were influenced by the ongoing armed conflicts and how the process of tsunami recovery, in its own turn, shaped the politics of the violent conflicts. The main argument is that the tsunami response and recovery to large extent reflected competing systems of governance in the two war-torn societies.[1]

As has been noticed both in media reporting and in scholarly works, in the wake of the tsunami the conflicts in Aceh and Sri Lanka embarked on different trajectories. Aceh, on the one hand, experienced a peaceful ending of the protracted violent conflict with the signing of the Helsinki Peace Agreement in August 2005. In Sri Lanka, on the other hand, we saw a deterioration of conflict, deepening polarisation and eventually a resumption of war between the government and the LTTE. However, although the outcome of these two conflicts differs, a closer look at the processes of tsunami response and recovery nonetheless suggest that the picture is more ambiguous. While the tsunami on the one hand opened up a window of opportunity for reaching a peaceful solution to the violent conflicts, at the same time, the natural disaster also contributed to spur new conflicts, to intensify strives for recognition and to provoke additional attempts to increase control over people, territories and resources. This includes the distribution of the large influx of humanitarian relief and development aid. In different ways, the tsunami thus contributed to consolidate dividing lines between geographical localities, urban and rural societies and identity groups.

In this chapter, the tsunami recovery will be analysed through the lenses of the protracted violent conflicts and the changing dynamics and interactions between multiple stakeholders in each conflict setting. I treat the tsunami as a structural factor that in various ways influenced the capacity, interest and perspectives of state and non-state actors as well as relationships within and between groups and societies. Furthermore, I stress the importance of agency for understanding how and why the natural disaster impacted the way it did in each particular setting. In analysing

[1] The notion of competing systems of governance has been inspired by Goodhand and Klem (2005).

these processes, I explore how key actors—including governments, rebel groups, international donors and civil society organisations—responded to and reinterpreted their interests, objectives and perspectives in the light of the natural catastrophe. The chapter is based on the assumption that the impact of environmental disasters must be understood by taking into account the social context in which they occur. In other words, as Choi (2009) puts it, 'natural processes are never divorced from their social milieu' (p. 194).[2] This line of reasoning provides important insights into our understanding of the impact of natural disasters such as the tsunami in war-torn societies. Essentially, it recognises that the social conditions that existed prior to an environmental disaster are important for understanding the characteristics and severity of the disaster's effects. This is because the impact of a natural phenomenon is assumed to be mediated through pre-existing structures of society. Accordingly, exploring linkages between nature and society is important for understanding the consequences of a natural disaster in specific social contexts (Bastian 2009; Klem 2006).[3] While these understandings about the interaction between societal structures and natural phenomenon can be generally applied, it might be particularly important to bear in mind when analysing natural disasters in deeply divided societies experiencing protracted violent conflicts. As Choi (2009) stresses, natural phenomena can impact people disproportionally depending on prior conditions of social relations. In conflict-prone societies, it can thus be assumed that pre-existing societal structures and conflicting lines will lead to people experiencing environmental disasters differently, since 'social conditions of inequality and marginalization ... play into the experiences and impact of a disaster' (Choi 2009, p. 195). Also Goodhand and Klem (2005) argue that when a natural disaster occur in war-torn societies, these societies have a higher level of pre-existing vulnerability compared to more peaceful regions and that these vulnerabilities are often focused on the geographical areas that have been most affected by the violent conflict. With regard to the present study, these insights suggests that the pre-existing social conditions in Aceh and Sri Lanka must be taken into account in order to understand the impact of the tsunami and the dynamics of its recovery. Thus, it points to the importance of recognising competing governance systems and how prior tensions, cleavages and political power struggles might be reproduced and accentuated in the light of environmental disasters. In the cases of both countries, this entails taking a closer look at changes and continuities in centre–periphery relations, regional dynamics and urban–rural intersections. Empirically, the chapter focuses on response and recovery in the immediate and midterm aftermath of the tsunami catastrophe in Aceh and Sri Lanka. The analysis primarily builds on written documentations, including scholarly articles and books, and reports from research institutes, international organisations, INGOs and NGOs. In addition, the study has

[2] This perspective can be contrasted to those that view natural disasters as single events and that in terms of recovery focuses solely on rebuilding what was destroyed by the disaster.

[3] To emphasise the difference, a distinction is sometimes made between 'hazards' and 'disasters'. The former refers to the natural phenomena and the latter to the interaction between the natural phenomena and society (Bastian 2009, p. 222).

also been informed by interviews on the Aceh and Sri Lankan peace processes conducted during research trips to Indonesia, Sri Lanka and Singapore between 2010 and 2012.

The outline of the chapter will be as follows: First, a greater emphasis will be placed on the two case studies of Aceh and Sri Lanka, respectively. Thereafter, similarities and differences between the two tsunami-affected conflict contexts will be explored in order to draw comparative insights from them. In conclusion, the chapter underlines the essential importance of taking context into account and of recognising the political processes at play so as to understand the response and recovery from environmental disasters in different conflict settings and societies.

5.2 Tsunami, Conflict and Peace in Aceh

The epicentre of the 9.0-magnitude earthquake that triggered tsunami waves across the Indian Ocean in December 2004 was located just 150 km outside Aceh on the Indonesian island of Sumatra. Not surprisingly, the remote province on the westernmost corner of the Indonesian archipelago was the area that was most brutally hit by the natural disaster. An approximate 167,000 people were killed in Aceh alone and more than 500,000 were left homeless (Nazara and Resosudarmo 2007, p. 1). In some areas, the water flooded several kilometres inland and created enormous suffering and destruction. It has been estimated that the natural disaster destroyed some 3,000 km of road in the Aceh province, damaged 2,000 school buildings and made more than 60,000 ha of farming land unusable (Barron 2008, p. 59). Within minutes, families and communities were split apart and thousands of people lost their assets and their livelihoods as the waves washed over the coast.

When the tsunami hit Aceh, an ongoing violent conflict between the government of Indonesia and the GAM movement had raged since the mid-1970s.[4] It is difficult to assess the number of casualties from the conflict, but it has been estimated that somewhere between 12,000 and 20,000 were killed (Aspinall 2009, p. 2) and that 100,000 people were forcibly displaced (Nazara and Resosudarmo 2007, p. 4). The dynamics of the long-drawn conflict in Aceh has resulted in specific societal structures and relationships that are necessary to take into account in order to understand the disaster governance in this particular setting. In this section, first, I will briefly explore the impact of the tsunami, taking into account pre-existing structures and conditions in Aceh and the conflict context. Thereafter, I will examine the dynamics of tsunami response and recovery and explore linkages between the tsunami recovery and the politics of peace and conflict in Aceh.

[4] For detailed accounts of the conflict, see, for example, Aspinall (2009), Miller (2009) and Reid (2006).

5.2.1 Conditions in Aceh and the Impact of the Tsunami

When the Indian Ocean tsunami struck in 2004, Aceh was one of the poorest regions in Indonesia. Almost one-third of the population was living below the poverty line, which can be compared to the rest of Indonesia's provinces with an average poverty level at 17 % (World Bank 2008). The modest economic development in Aceh can be understood both as one of the contributing causes to the protracted violent conflict and as a consequence of the conflict. Aceh is rich on natural resources, including significant oil and gas reserves, flourishing fishing waters and fertile soil. Nevertheless, historically, most of the revenues have been allocated directly to Jakarta (Miller 2006; Aspinall 2009). In addition, due to the long-drawn conflict, Aceh was for decades practically isolated from the outer world. When the Indonesian government declared martial law in Aceh in 2003—after the breakdown of peace talks and of what initially appeared to be a promising ceasefire agreement—this isolation of the region was sustained and even intensified. No foreigners, journalists or civil society were allowed to enter the province (Åkebo 2013). In addition, both GAM associates and civil society activists in opposition to the Jakarta rule were forced to exile. The isolation of Aceh, combined with the sense of insecurity that the war produced, contributed to the slow economic development in the province. These conditions had negative impacts on business and trade and resulted in few foreign investments. When the tsunami hit Aceh, an estimated 78 % of private livelihoods, including trade, farming and fisheries, were destroyed in the already poor province (Schulze 2005). The material costs of the environmental disaster were estimated at $ 4.5 billion, which at the time corresponded to 97 % of Aceh's gross domestic product (GDP) (ibid, p. 2). Consequently, large parts of the population that were already living under harsh conditions faced an imminent risk of falling even deeper into poverty.

The characteristics and severity of the tsunami varied across geographical areas in Aceh. Three-quarters of Aceh's coastline was destroyed by the disaster, and the damages were predominantly concentrated on the western parts of the province. The areas that were worst hit were the urbanised coastal districts of Aceh Jaya, Aceh Barat (West Aceh) and Aceh Besar, including the larger cities of Banda Aceh, Meulaboh and Calang. As Schulze (2005) describes it, 'Calang was virtually wiped off the map and 80 % of Meulaboh was washed into the sea' (p. 2). Important to note in this respect is that the patterns of tsunami destruction and the patterns of the violent conflict differed in Aceh. While the tsunami mainly affected the coastal areas, the violent conflict was on the other hand predominantly concentrated to the rural interior (Miller and Bunnell 2013).[5] Throughout the armed conflict, the GAM movement remained strongest in the rural districts of Pidie, North Aceh and East Aceh. GAM's strongholds in North and East Aceh were only marginally affected by the tsunami with minor flooding (Schulze 2005). Overall, in the immediate wake of

[5]Approximately 70 % of the population is living in the rural inlands (Miller and Bunnell 2013, p. 84).

the tsunami, the violence in Aceh dropped compared to the levels prior to the natural catastrophe. Nevertheless, the conflict-related violence continued in the hinterlands. During the first week after the disaster, violent acts were reported, and it gradually increased during the first six months (Barron et al. 2005; Uppsala Conflict Program 2008). The post-tsunami violence was concentrated on certain geographical areas. Sixty percent of conflict-related incidents took place in the districts of North Aceh, South Aceh, East Aceh and Bireuen, while the urban centres were practically spared from GAM–government conflict-related incidents (Barron et al. 2005).

5.2.2 Dynamics of Tsunami Recovery

As the previous section implies, the tsunami left the relatively poor war-torn province in an extremely strained position and with huge needs of relief and recovery. In relation to these needs, the response and solidarity that was shown both from within Indonesia and from the international community was impressive. Thousands of volunteers offered to participate in providing assistance, and vast amounts of funding were pledged for humanitarian relief and for the rebuilding of houses and infrastructure. Since the tsunami hit hard on the urban centres along the coast, the provincial government and many district governments had been seriously affected by the disaster. Consequently, they also lacked the capacity to organise the enormous relief and reconstruction work that needed to be done (Aspinall 2005; Mc Gibbon 2006). In this context, the Indonesian government—led by Susilo Bambang Yudhoyono who had won Indonesia's first direct presidential election three months prior to the disaster—insisted to coordinate the process of assistance and to control the distribution of aid (Aspinall 2005).[6] During the incipient democratic transition in Indonesia following the breakdown of the Suharto regime in 1998, the country had experienced years of instability and uncertainty. Parallel to democratic elections and increased freedom for media and civil society, these years had also been pervaded by patrimonialism, ineffective administration, corruption and regional violent conflicts. At the time the tsunami hit Aceh, several politicians in the province including the governor were either in jail or facing trials due to corruption allegations (ibid.). The government shift that brought Yudhoyono to power in 2004 arguably marked a pivotal moment in Indonesia's transition to democracy. These changes proved to be favourable for the positive developments of the elite-level negotiations between GAM and the government in the aftermath of the tsunami. Still, despite this progress in elite-level negotiations, the process of tsunami recovery was still not spared from problems and tensions. As emphasised by Schulze (2005), during the first year after the environmental disaster, there were problems with civil–military antagonism, lack of coordination between donors and Indonesian government agencies, insufficient consultation with the local community and

[6]The Minister for people's welfare, Alwi Shihab, had been appointed to Aceh for this purpose (Aspinall 2005).

corruption among officials. As will be further outlined in this section, the process contributed to revitalise tensions between urban and rural parts of Aceh and created cleavages between the tsunami-affected and the conflict-affected population.

When the tsunami struck, the new government in Jakarta welcomed the support from foreign militaries and humanitarian workers to the previously more or less isolated Aceh region (Mc Gibbon 2006). The new openness of Aceh to the outer world has been described as one of the most striking effects of the tsunami, and it dramatically changed the situation on the ground in the province. International aid workers and journalists arrived to assist and to report to the world about the suffering of the Acehnese people. As Barron (2008) illustratively notes, within a year as many as 3,645 NGOs had registered at the United Nation's compound in Banda Aceh. Vast amounts of humanitarian aid also began flowing into the region. The US$ 5.3 billion that was pledged from abroad and the additional US$ 2.7 billion pledged from within Indonesia even exceeded the estimated damages caused by the tsunami (ibid, p. 59). When the international aid workers arrived, they were first introduced to Aceh's urban centres, in particular the provincial capital city of Banda Aceh and the city of Meulaboh in West Aceh (Miller and Bunnell 2013). Indeed, these larger coastal cities had been hit hard by the tsunami and the humanitarian needs were urgent. However, it soon became clear that many of the disaster recovery programmes in Aceh would be concentrated particularly to the urban centres (ibid.).[7] Access to remote rural areas outside of Banda Aceh and other city centres was limited. In addition, it was costly to transport the materials necessary for reconstruction work outside the urban centres. Thus, as a result, areas within a close distance to Banda Aceh that could be reached rather easily by road received more attention. As concluded in a World Bank report from 2005, 'almost three quarters of the houses completed in 2005 will either be in the Banda Aceh and Aceh Besar area or on the east coast near the Medan road', while 'very few houses will be built in the more inaccessible regions of the west coast or the islands' (p. 36–37).[8] Furthermore, as Schulze (2005) emphasises, since media attention was predominantly directed towards developments in Banda Aceh, focusing on programmes in the provincial capital city also served the objective of many NGOs and INGOs—to promote their own projects for the benefit of their donors. It should also be noted that in 2005, 90 % of the staff within the Reconstruction and Rehabilitation Agency (Badan Rehabilitasi dan Rekonstruksi; BRR)—the agency that was established by the government in April 2005 to coordinate the distribution of all the aid flowing in to the region—were located in Banda Aceh (Nazara and Resosudarmo 2007).

Furthermore, as previously mentioned, the armed conflict in Aceh continued for several months in the aftermath of the tsunami. While the Indonesian armed forces engaged in humanitarian relief, they simultaneously also searched for GAM fighters (Åkebo 2013). When GAM fighters came out from the jungle to search for their

---

[7] This tendency can also be seen in other countries experiencing disasters (Miller and Bunnell 2013).

[8] A second earthquake in March 2005 hit particularly hard on the islands of Simeulue (Aceh) and Nias (North Sumatra).

families and friends, many were either shot or arrested (Merikallo 2006). The military was keen to remain its control over the military campaign against GAM in the rural areas and did not want international relief workers to travel to those areas where the conflict still raged (Miller and Bunnell 2013). Thus, travel restrictions were put on foreign workers in order to hinder them from moving outside of the main urban centres. In addition, the military was reluctant to have outsiders inquiring into their economic businesses in Aceh. During the first months after the tsunami, there were reports of the military taking control over the end-point distribution of tsunami aid particularly along the west coast (Aspinall 2005).

The priority place given to the urban centres in the tsunami recovery accentuated and deepened cleavages between Aceh's rural and urban areas, in particular between Banda Aceh and the rural interior (Miller and Bunnell 2013).[9] During the Suharto regime, Aceh's urban centres had been under the rule of the central government, and many urban elites in the province had developed close ties with Jakarta. The Aceh independence struggle had its origins in the rural interior where the violent conflict between GAM and the Indonesian army (Tentara Nasional Indonesia, TNI) was primarily being located. While Banda Aceh became pivotal in the independence struggle in the beginning of the 2000s, in the aftermath of the tsunami, prior tension between urban and rural Aceh started to reappear. Geographic disparities in the recovery efforts became a noteworthy feature in post-tsunami Aceh. Many tsunami-affected coastal villages in the rural parts of the province only received small amounts (if any) of the massive influx of aid for tsunami recovery (Miller and Bunnell 2013). At the same time, for visitors travelling to the city of Banda Aceh a few years after the tsunami, there were almost no signs left of the natural disaster that levelled large parts of the city in late 2004. The continuation of violence on the ground in the rural areas was one of the contributing factors for the failure of reaching out with humanitarian assistance to many of the affected villages along the coast or at least significantly delaying the assistance (Miller and Bunnell 2013). In contrast, no GAM–government-related conflict incidents were reported in the capital city of Banda Aceh during the first six months after the tsunami, and GAM's position was at the time reportedly located far from urban centres such as Lhokseumawe, Langsa and Meulaboh (Barron et al. 2005).

Furthermore, even though many people also suffered deeply from the protracted violent conflict, the aid was predominantly programmed in the areas that had been affected by the tsunami. Conflict-affected areas were on the other hand less likely to get assistance. While there are some areas in Aceh that were both affected by the violent conflict and by the tsunami (and thus in the need of a combination of conflict and tsunami recovery), the most seriously conflict-affected areas could be found in the highlands that were not hit by the tsunami. Nevertheless, there was a great need of development aid and reconstruction in these areas too. For example, thousands of houses, schools and other facilities had been destroyed in the conflict (Schulze 2006). Still, they received little or no money from the massive flow of post-tsunami aid. In 2005, conflict-affected districts were 44 % more likely to be poor than the

[9] For an in-depth analysis of urban–rural intersections in Aceh, see Miller and Bunnell (2013).

average subdistrict in Aceh (Barron 2008, p. 60). The BRR that was set up by the central government to deal with the post-tsunami reconstruction only had an initial mandate to focus on tsunami-affected areas (ibid.). The work on post-conflict recovery issues was on the other hand coordinated from the provincial governor's office and Jakarta-based ministries (World Bank 2005). With a few exceptions, programmes within the multilateral donor fund in Aceh were also predominantly focusing on tsunami-affected areas, and the same was true for most of the bilateral aid agencies and NGOs too (Barron 2008). According to the World Bank analyst James Barron (2008), inequalities caused by disproportional distribution of aid resulted in significant tension and 'development disputes' at the local level (p. 60).

Thus, discrepancies between the tsunami-affected and the conflict-affected areas created particular dynamics in post-tsunami Aceh (Barron et al. 2005). This could also be seen in terms of displacements in the aftermath of the natural disaster. During the violent conflict, many people had been forced to leave the rural inlands to find a safer place in the urban centres along the coast.[10] The tsunami reversed these moving patterns. Instead, many tsunami victims who lost their homes near the coast fled to the rural interior to settle down with friends or family. For some, this meant fleeing back to the conflict-torn inland that they once had escaped from (Miller and Bunnell 2013). Since the aid that came to Aceh after the tsunami was earmarked for tsunami victims and concentrated on the urban coastlands, this also meant that those who had fled to the rural areas risked missing out such aid. In general, displaced persons who were staying with family or friends were not officially registered as displaced, and consequently they were not included in the aid delivery programmes (Schulze 2005). The majority of displaced persons fall into this category and accordingly received little or no assistance (Aspinall 2005). In addition, within minority ethnic groups that are living in the inlands, including Gayo, Alas, Batak, Mandailing, Pakpak and Java, people have historically perceived themselves as being marginalised in comparison with the coastal areas. These groups also viewed the tsunami recovery as predominantly beneficial for the coastal population (Thjin 2005).

While the developments on the ground in the aftermath of the tsunami displayed various tensions and geographical divergences, in particular between urban and rural areas, at the elite-level progress was nevertheless being made. Within weeks after the tsunami, the Indonesian government and GAM held five rounds of peace talks mediated by the former Finnish President Martti Ahtisaari.[11] In August 2005, the conflicting parties agreed to settle for a memorandum of understanding (MoU). This MoU can be described as a comprehensive peace accord that dealt with the future relationship between the Indonesian government and GAM, including political, security and socioeconomic matters (Åkebo 2013). However, when considering

[10] Many people who had been forced to leave Aceh during the conflict, including GAM affiliates or sympathisers, were also able to return. The substantial return of GAM affiliates both at the elite level and at the local level was confirmed in interviews by the author in Aceh in 2010.

[11] The peace talks were held in a mansion outside of Helsinki. Ahtisaari was involved by virtue of his position as chairman of the NGO Crisis Management Initiative (CMI).

the settlement in Aceh, it is important to note that peace attempts had been initiated prior to the tsunami. In parallel with the military operations during the martial law, low-profile initiatives were taken in order to resume peace negotiations. These attempts were further stepped up after the government shift in October 2004, and few days prior to the tsunami both parties had agreed to meet and talk (Åkebo 2013; Aspinall 2009). Thus, the agreement to search for a peaceful solution was initiated before the tsunami. Nevertheless, it can be argued that the environmental disaster created a sense of urgency that had not been present during previous peace attempts in Aceh.

With regard to the present study, the MoU importantly also included a clause that stipulated that the GAM movement would be allowed to nominate representatives fully participating in the BRR that had been set up to manage the post-tsunami reconstruction (MoU 2005, section 1.3). This has been described as important for building confidence in the process. A central part of the MoU was also the political arrangement for the governing of Aceh, which was established in April 2006 through the Law on Governing Aceh (LoGA). Nevertheless, among former GAM affiliates, there were differences between those who managed to benefit from the processes of tsunami recovery and those who did not. Many of the former GAM elites managed to gain significant political power in post-tsunami Aceh after the signing of the MoU between the government and GAM in August 2005. In the first local elections, a former GAM associate became the new governor, and affiliates also took power in many local districts. Former GAM elites also managed to win profitable reconstruction contracts (Barron 2008). Similar politics–business connections could also be found among politicians affiliated with the central government in Jakarta (Aspinall 2005).

The LoGA also seems to have had implications on rural–urban intersections in Aceh. As part of Indonesia's decentralisation policy, the LoGA devolved state power and resources to the provincial government administration in Banda Aceh (Miller and Bunnell 2013). Thus, resources from the central government level are being redistributed at the provincial government down to the district sublevels. As Miller and Bunnell (2013) argue, this has implied that the decentralisation of state power to Aceh has benefitted the urban centres, in particular Banda Aceh, at the expense of rural Aceh. As such, it has also contributed to lay the foundation for growing tensions and conflict between the urban and rural Aceh (ibid.)

5.3 Tsunami, Conflict and Peace in Sri Lanka

Sri Lanka was the second worst affected when the Indian Ocean tsunami suddenly struck on Boxing Day 2004. Two-thirds of the country's coastline was hit by the waves that caused devastating human, physical, socioeconomic and environmental destruction. It has been estimated that the tsunami killed 31,000 people in Sri Lanka. Another 4,000 people were reported missing and more than 550,000 were forcibly displaced (Mayer and Salih 2006, p. 552). In contrast to the Aceh case, the geographical areas that were affected by the tsunami in Sri Lanka to large extent overlapped with the areas that had suffered the most from the brutal civil war. When the tsunami reached the Sri Lankan coast, the war between the government and the LTTE had persisted for two decades and killed 70,000 people and left more than 800,000 forcibly displaced (Frerks and Klem 2005a, p. 2).[12] The tsunami struck particularly hard in the coastal districts of the Eastern, the Northern and the Southern provinces,[13] of which the former two were also war-torn areas.[14]

In this section, I will examine the disaster governance in the context of the violent conflict in Sri Lanka. This section will follow the same structure as Aceh's case: First, I take a brief look at the impact of the tsunami through the lenses of pre-existing social conditions in Sri Lanka and the conflict context. Second, the dynamics of tsunami response and recovery will be explored in addition to how the politics of the tsunami recovery took place between key actors in Sri Lanka.

5.3.1 Conditions in Sri Lanka and the Impact of the Tsunami

When the tsunami hit the Sri Lankan coastal line, it soon became evident that the natural disaster would have the greatest impact on people who were already marginalised in society. In general, disasters in Sri Lanka can often be linked to issues of land use and land ownership (Bastian 2009; CPA 2006). People living in low-land areas close to the water are frequently suffering from floods, droughts and landslides. Likewise when the tsunami hit, many of the victims were unauthorised residents living in flat, low-level areas close to the shore or to the railway (Frerks and Klem 2005a). While a majority of Sri Lanka's population lives in the rural areas, the country is urbanising and the coastal areas are the most densely populated zones (World Bank 2012). An estimated 25 % of the population lives within 1 km from the coast on an area that constitutes merely 5 % of the island's total area (ibid). The capital city of Colombo, which is the island's urban hub, was only marginally affected by the tsunami. However, beside the Colombo metropolitan urban area, Sri

[12] When the war ended in 2009 with the government's military defeat over the LTTE, an estimated 84,000 had been killed in the conflict between 1983 and 2009 (Uppsala Conflict Data Program 2012).

[13] The nine provinces are Sri Lanka's first level of administrative units, and each province is divided into a number of districts. In 1988, the Northern and Eastern provinces were temporarily merged into the North Eastern province as a result of the Indo–Lanka accord signed between the Sri Lankan and Indian governments the year before. The merger was incorporated into law through the 13th Amendment to the Constitution. In the Indo–Lanka accord, it was stated that a referendum would be held on permanently institutionalising the merger of the two provinces. However, this was never put in practise, and instead, the North Eastern province was ruled directly from Colombo. In 2006, the Supreme Court ruled the merger as unconstitutional and the two provinces were separated again in 2007 (ICG 2008). Hence, this separation occurred after the tsunami and accordingly after the time period that is at focus in this chapter.

[14] The Western province—where Colombo, the capital city is located—was also affected but not to the same extent as the Eastern, Northern and Southern provinces (Sarvananthan and Sanjeewanie 2008).

Lanka's urban landscape is generally characterised by small urban centres along the coast. Many of these densely populated communities were severely affected by the tsunami.[15] The people living in Sri Lanka's urban areas have steadily been pushed closer to the water, primarily due to market forces and the greater demand for land by 'capital' (Bastian 2009, p. 223.). In these densely populated coastal areas, poor people who do not posit land of their own have often occupied land illegally. These people became extra vulnerable to the tsunami disaster (Bastian 2009). Many of the coastal communities that were affected by the tsunami were already poor in comparison to other parts of the country. Within these communities, large segments of the population were depending on fishing and saw both their houses and their livelihoods being swept away with the giant waves.[16]

The long-drawn war in Sri Lanka has certainly affected the whole country, but the consequences have been particularly severe for the population in the northern and eastern territories that were claimed both by the government and by the LTTE in the latter's quest for a separate Tamil state.[17] The Tamil-dominated areas in Sri Lanka have generally been less wealthy, both as a result of the massive destruction caused by the war and as a consequence of the decades of neglect from the central government in Colombo (Orjuela 2011). While Sri Lanka has maintained a rather high level of economic growth in comparison to other war-torn states, the wealth has nonetheless been unevenly distributed between different parts of the island (ibid). The war have had a devastating impact on the urban economies in the north and east, including the larger cities of Ampara, Trincomalee and Jaffna. In these cities, much of the infrastructure was either damaged or destroyed. In addition, security measures had disrupted people's livelihoods (World Bank 2012). Because of the already weak local economy, the northern parts of the island also experienced the greatest economic losses as a consequence of the tsunami (Sarvananthan and Sanjeewanie 2008). After the tsunami, the unemployment rate in the north and east was, for example, estimated to be double the national average and the school drop-out rate was four times higher than the national average (Asian Development Bank et al. 2005, p. 5).[18] Thus, the pre-existing social conditions in the war-torn north and east made people in these areas particularly vulnerable to the environmental disaster. As we will see in the remainder of this case study, the competing systems of governance in Sri Lanka and prior cleavages in the society also seem to have influenced the patterns of tsunami recovery and distribution of the tsunami aid to different groups and localities.

[15] Many of these areas are not formally defined as urban in the national census, but they nevertheless show 'urban characteristics such as high population and building density' (World Bank 2012, p. 7).

[16] It has been estimated that one fourth of the population in Sri Lanka is living below the poverty line. Half of them are small-scale farmers (IFAD 2011).

[17] For a detailed account of the conflict, see, for example, Orjuela (2004)and Stokke and Uyangoda (2011).

[18] Ethnic and language barriers, as well as prevailing tensions between different identity groups, also restricted people's mobility and access to jobs and education across cities and regions (Shaw et al. n.d.).

5.3.2 Dynamics of Tsunami Recovery

In the immediate wake of the tsunami, for a brief period of time, the environmental disaster resulted in measures of cooperation across the island. In order to meet the urgent need for relief, people in divided communities assisted each other across ethnic lines. The government and the LTTE coordinated relief and recovery activities in the northeast (Asian Development Bank et al. 2005; Frerks and Klem 2005a; ICG 2008), and there were also anecdotes of the army and the LTTE working together on the ground.[19] Thus, the hopes were high that the natural disaster finally would lead the way to a peaceful solution to the violent conflict. However, as oftentimes had happened before, rather quickly tensions started to rise again. The initial steps of cooperation were exchanged with hostility and deepened polarisation. It soon became clear that in the context of the natural disaster, various actors in Sri Lanka would continue trying to pursue their own interests and objectives. Such behaviours were not only characteristic of the antagonistic relationship between the government and the LTTE. Also in Sri Lanka's electoral and political system, power considerations have commonly been the primary political driven force, while ideology and party programmes have come in second. In the Sri Lankan society, political power has not only been a means of influence but also wealth and employment (de Silva 1998; Orjuela 2004). Thus, the supremacy given to the personal interest of those in power has been 'an entrenched feature of Sri Lankan politics' (Mulligan and Shaw 2007, p. 73).

At various political levels, the political patronage was also present in the tsunami disaster governance, with corruption and political inference influencing the distribution of aid (Mulligan and Shaw 2007). In the wake of the tsunami, President Kumaratunga established an overarching coordination mechanism for tsunami relief. On the one hand, this response reflected the overcentralised characteristics of the Sri Lankan state. A strong administration in the capital city of Colombo with limited power for the provinces is a distinct feature of the Sri Lankan state (de Silva 1998). The unwillingness to devolve power to the provinces has resulted in low capacity and knowledge at the local levels. In addition, there have been difficulties in creating effective mechanisms for coordination of policy between different government agencies. These features of the state also hampered the tsunami reconstruction process (CPA 2006; Mulligan and Shaw 2007).[20] While the centralisation of the state is a characteristic in Sri Lanka, at the same time the state has also been described as fractured. In the wake of the tsunami, local politicians managed to take control over considerable resources to the advantage of their own electoral constituencies (Goodhand and Klem 2005). This was particularly true in the south.

[19] Interview with Paikiasothy Saravanamuttu, Executive Director CPA (Centre for Policy Alternatives), Colombo, 17 November 2010

[20] In addition, as Mulligan and Shaw (2007) argue, the context of political uncertainty in Sri Lanka due to the war and continuous power shifts with new governments 'overturning the arrangements of their predecessors' (p. 73) contributed to a poor investment climate which had negative implications on the reconstruction process.

For example, as Hyndman (2009) notes, the southern district of Hambantota—at the time being the electoral base of Sri Lanka's Prime Minister (who later became the president)—was given a priority place in urban development and housing programmes (i.e. the 'New Town' settlement). On the other hand, many Muslim ministers were seen to be less effective in managing to get control over shares for their constituencies in the east (Goodhand and Klem 2005).

The diverging patterns of aid distribution in different regional contexts in Sri Lanka were thus symptomatic of the structures and procedures of governance in Sri Lanka (Frerks and Klem 2005b). Most notable was the disproportionate distribution of aid between the northeastern areas and the southern areas. The former areas had been affected both by the tsunami and by the war, while the latter was affected by the tsunami. In addition, the Northern and Eastern provinces experienced the worst destruction from the tsunami; they account for an estimated two-thirds of all casualties and 60 % of the people being displaced (Mulligan and Shaw 2007, p. 69). For example, up to 80 % of the coastal population was affected in the northern Tamil-dominated Mullaitivu district as well as in the multiethnic Ampara district (Asian Development Bank et al. 2005). When extensive reconstruction programmes were directed towards the lesser tsunami-affected south, people easily interpreted this divergence along pre-existing conflicting lines.[21] At the time the government announced that the tsunami aid had been expended, hundreds of families were still living in the so-called welfare camps in the east without a house, while in the southern Hambantota more houses than needed were being built (Hyndman 2009; Shaw et al. n.d.). In addition, the southern coastal areas that have not been affected by the war have also been popular destinations for international tourists, and rebuilding the affected tourism industry was a high priority for the Sri Lankan government (Shaw et al. n.d.).

Furthermore, inconsistencies in the government's tsunami recovery policies also became apparent when buffer zones were imposed near the shore. These zones aimed at preventing the rebuilding of houses within a given distance from the water (CPA 2005; Hyndman 2009). However, until the exact rules were determined, the policy created a lot of confusion, and questions were raised about the motives behind the rules. Because of the drawn-out process of coming to terms with the rules and conditions for the buffer zone policy, it took several months before people could start rebuilding their houses (Goodhand and Klem 2005). Furthermore, the buffer zone policy hit particularly hard on tenants and people living unauthorised within the zones (CPA 2005).[22] While many did receive some monetary relief, the structural problem with land ownership and housing for these people was left

[21] An approximate 20 % of the population was affected in the southern districts of Galle, Matara and Hambantota. However, it should be noted that within these southern districts, there were also pockets where the damages caused by the tsunami were very severe (Asian Development Bank et al. 2005).

[22] In a more general sense, people who prior to the tsunami had been occupying unauthorised land, either inside or outside the buffer zones, were particularly vulnerable groups in the post-tsunami Sri Lanka. The housing programmes generally had no mechanisms for dealing with the illegal squatters (Bastian 2009).

unaddressed in the tsunami recovery process (Bastian 2009).[23] It could also be noted that in line with the government's priority to rebuild the affected tourism industry, hotels were exempted from the ban and given the opportunity to rebuild.

The uneven distribution of wealth between different geographical areas in Sri Lanka and the urgent need for economic development in the war-torn northeast had been a core conflict issue during the war. It was accordingly also a central component of the peace process that was initiated a few years prior to the tsunami in the early 2000s (Åkebo 2013). The signing of a ceasefire agreement between the main belligerent parties in 2002 had enabled measures being taken for economic development in the war-torn areas. As a result, many of the war victims had been given the chance to experience at least some initial steps of improvements (ibid.). However, the peace process soon reached an impasse, and as a consequence the progress on the ground also halted. Thus, despite some initial progress, the needs for economic development and reconstruction of the war-affected areas were still urgent when the tsunami struck. In this context when vast amounts of funding was pledged for tsunami relief and reconstruction, as Orjuela (2011) argues, these reconstruction efforts 'came to overshadow' (p. 129) the attempts that had been made during the ceasefire period to improve the living conditions for the war victims.[24]

Furthermore, divergences in the distribution of aid between tsunami victims and war victims also accentuated cleavages and tensions along prevailing conflict lines. Before the tsunami hit Sri Lanka's coastlines, displacements had predominantly been a result of the protracted armed conflict. However, this changed in late 2004 as Sri Lanka saw a new group of tsunami refugees. Similar to the Aceh case, in Sri Lanka too, most of the aid being pledged in the wake of the tsunami was earmarked for tsunami victims. Indeed, many people displaced by the tsunami had also been displaced by the civil war, sometimes more than once. However, the people who were exclusively war refugees received none or only small amounts of aid (Hyndman 2009). For example, there were large groups in the western parts of the island, in particular in Puttalam and near Anuradhapura, who had been internally displaced since the early 1990s. When the tsunami struck, these people were still living in temporary houses, while at the same time people being displaced by the tsunami received new houses (Hyndman 2009). While it can be argued that the immediate focus on the tsunami victims indeed was necessary, it nonetheless also resulted in frustration and additional conflicts over inequalities in the distribution of aid (ibid.).

At the elite level, within weeks after the tsunami, there were disputes over reconstruction policies and priorities. A central issue of contention concerned how the tsunami aid was going to be distributed between different territories, groups and

---

[23] Households within the buffer zones were not required to demonstrate ownership of land in order to be provided by a house outside the buffer zone, while households outside the buffer zone on the other hand were required to do so (CPA 2005).

[24] According to Hyndman (2009), in the district of Ampara in the east, which was hit hardest by the tsunami and where 18,800 people had lost their homes, 3,136 houses were being built. At the same time, in the southern town of Hambantota, almost 4,478 new houses were built although it had been estimated that only 2,445 houses were needed.

sectors and in particular who was going to be in charge of the distribution (Orjuela 2011). To a large extent, the tsunami recovery in Sri Lanka became another arena for the two main warring parties in the battle to gain power, recognition and legitimacy. The government strived to strengthen the state's control over the relief and recovery process and over the aid distribution. When the UN Secretary General, Kofi Annan, was visiting Sri Lanka a few days after the tsunami, the government even refused him of entering LTTE-controlled areas (Goodhand and Klem 2005). At the same time, in the LTTE-controlled areas, the organisation's aid body—the Tamil Rehabilitation Organisation (TRO)—demanded that all aid coming to the area be channelled through the TRO. This way, the organisation tried to maintain its monopoly of power and control (ICG 2006). However, in parallel to these developments, negotiations between the government and the LTTE began quite soon after the disaster. The negotiations were conducted behind closed doors and focused on working out a structure to deal with the distribution of tsunami aid. The negotiations proceeded in secret because the parties 'did not want it to leak out and cause any kind of unnecessary publicity'.[25] However, as this procedure implies, the dynamics surrounding the negotiations of a post-tsunami management structure reveal that the politics of the conflict in Sri Lanka was more complicated than finding a common ground between two monolithic actors. Since March 2004, the LTTE had experienced internal challenges and disruption as the eastern LTTE leader Colonel Karuna decided to break away from the organisation and establish a separate fraction in the east. The so-called Karuna group[26] not only became a rivalry to the LTTE but in addition sided with the government (Goodhand et al. 2011). At the same time, the government had also been experiencing internal conflicts. In January 2004, President Kumaratunga dissolved parliament and formed a coalition which included the Sinhala nationalist People's Liberation Front (Janatha Vimukthi Peramuna, JVP) as a key coalition partner. The JVP made a strong argument not to legitimise the LTTE by putting the organisation in control over finance or disbursement of finance. Nevertheless, after several months of negotiations, in June 2005, the parties finally agreed on a Post-Tsunami Operational Management Structure (P-TOMS). The P-TOMS agreement was aiming at facilitating disbursement of aid and to speed up the process of relief and recovery in the six tsunami-affected districts in the northeast (P-TOMS 2005). However, for many it was unclear whether the agreement was part of the peace process between the government and the LTTE or solely a mechanism set up to administer humanitarian aid (Rainford and Satkunanathan 2009).[27]

[25] Interview with Jayantha Dhanapala, former Secretary General of the government's Secretariat for Coordinating the Peace Process (SCOPP) 2004–2005 and former political advisor to the president, Colombo, 24 January 2012

[26] The Tamil People's Liberation Tigers (Tamil Makkal Viduthalai Pulikal, TMVP) has commonly been referred to as the Karuna group.

[27] Furthermore, both parties tried to downplay the political weight of the P-TOMS agreement. They kept a low profile during the negotiations, and when the agreement was to be signed, they wanted people without significant political power to sign it. President Kumaratunga even publically stressed that the P-TOMS agreement was not to be considered as part of the peace process and that it was not a political body (Rainford and Satkunanathan 2009).

Moreover, in the beginning of June, the JVP critics left the government in protest of the forthcoming agreement. In addition, once the P-TOMS was signed, the JVP went to the Sri Lankan Supreme Court and challenged the agreement on the ground that it was seen as unconstitutional to give power over finance to a non-state actor (Goodhand and Klem 2005). The Supreme Court ruled in favour of the JVP and the P-TOMS collapsed in late 2005. This also illustrated the collapse of the last 'proper' negotiations ever to be held again between the government and the LTTE. By that time, the peace process had lingered on for several years without significant progress being made. In this context, the failure to implement agreed-upon structures, including the P-TOMS agreement to manage the tsunami recovery and relief, further deepened the distrust between the parties and contributed to the breakdown of the peace process (Orjuela 2011). As Goodhand and Klem (2005) conclude: 'In spite of initial hopes that the tsunami response would provide a space to re-energize peace negotiations, it had the opposite effect, deepening political fault lines' (p. 9). Thus, in the midst of a peace process without a roadmap and with competing actors striving to pursue their own interests and objectives, the P-TOMS disaster management structure can be seen as a form of power sharing being put to the test. Ultimately, various stakeholders in the Sri Lankan context failed to seize this opportunity. Thus, rather than contributing to unite different groups and fractions throughout the country, the tsunami response and recovery in Sri Lanka reflected and accentuated pre-existing tensions and competing systems of governance.

5.4 Comparative Insights from Aceh and Sri Lanka

Experiences from conflict zones around the world show that when major structural changes occur during ongoing violent conflicts—may it be regime shifts, economic crises or natural disasters—actions and events resulting from these external changes easily become intertwined with the dynamics of the conflicts. As the two case studies from Aceh and Sri Lanka also have shown, the impact of the tsunami and the dynamics of tsunami response and recovery can be related to prevailing structures and relationships that existed prior to the tsunami in these two conflict settings. Depending on ethnic identity, social position and geographical location, people seem to have experienced the natural disaster differently. There has also been divergence in the extent to which people have been able to take advantage of the extensive programmes for relief and reconstruction launched in both regions. In the context of ongoing conflicts and competing systems of governance, the processes of tsunami response and recovery have often become entangled into the dynamics of the violent conflicts, both reflecting ongoing conflicts and contributing to creating new ones.

A comparison of the tsunami impact, response and recovery in Aceh and Sri Lanka through the lenses of the protracted violent conflicts reveals both similarities and differences. Important in this respect is that the geography of tsunami and conflict has differed between the two cases. That is, the cases vary in the extent to which there has been an overlap between the tsunami-affected and the conflict-affected areas. In Sri Lanka, the impact of the tsunami was particularly severe in the same areas that also had suffered the most from the protracted war (i.e. northern and eastern parts). When the tsunami hit, these areas were already poor and in great need of economic development compared to other parts of the island. Tensions and cleavages resulting from the violent conflict, including land disputes, also exacerbated and gave rise to new conflicts in the wake of the tsunami. Furthermore, the disproportionate distribution of tsunami aid to the south and the failure to effectively rebuild the war-torn northeast strengthened perceptions of marginalisation and discrimination among minorities living in the war-affected areas. Furthermore, adding to the complexity, some of the territories affected by the tsunami were under control of the LTTE. When the tsunami struck, the LTTE was running de facto parallel state structures in the north and in parts of the northeast. Gaining control over the distribution of tsunami aid in these areas became a means of increasing the LTTE's power and influence over the territories which it claimed. Moreover, it was also in the LTTE's interest to speed up the reconstruction of the war-torn northeast as a way to gain legitimacy for its state project.

In comparison, Aceh displayed different geographical patterns in terms of tsunami- and conflict-affected areas. While the violent conflict in Aceh was concentrated to the inlands, the tsunami primarily affected the urbanised coast and many areas that historically had not been sites of the violent conflict. The urban centres along the coast were given a priority place in tsunami recovery programmes, both as a consequence of greater accessibility in these areas compared to the rural parts of the province and also as a result of the continuation of the violent conflict in the interior. This created divergences between urban and rural areas and resulted in tension between those who received tsunami aid and those who were left without. Important to note is that GAM did not have the same territorial control as the LTTE did in Sri Lanka, and they were not in the same position to challenge the government's control over the distribution and use of tsunami aid.

At elite-level negotiations, Aceh and Sri Lanka have proceeded along different trajectories. In order to understand these different outcomes, the timing of the natural disaster is deemed important. When the waves hit the coastlands of Aceh and Sri Lanka, the two regions were experiencing different situations in terms of peace and conflict. Importantly, ongoing peace initiatives were moving in different directions, the characteristics of international involvement differed between the two cases and they had different experiences from earlier engagements in efforts to finding peaceful solutions to their conflicts. These particular developments affected the dynamics of tsunami recovery in Aceh and Sri Lanka. In Aceh, new initiatives to engage in a peace process had taken place prior to the tsunami catastrophe, and it seem to have helped pushing the already initiated process further in the same direction (Åkebo 2013). As opposed to this, in Sri Lanka, the tsunami struck in a context of political stalemate. It has been argued that the country was on the brink of war when the tsunami hit in late 2004 and that it stopped the war from resuming. Nevertheless, the dynamics of tsunami recovery contributed to sustain and even aggravate ongoing conflicts. The large amounts of aid that was pledged for tsunami recovery soon became another matter of conflict, and the distribution of resources to particular towns, villages, neighbourhoods and families was to a large extent interpreted along existing conflict lines. These contentions became central also in the processes leading up to the failure of the P-TOMS agreement. As Goodhand and Klem (2005) puts it: ‘Each party saw the tsunami as an opportunity to strengthen their legitimacy through the control and distribution of resources’ (p. 58).

Moreover, the different trajectories in Aceh and Sri Lanka in the aftermath of the tsunami must also be understood in light of broader developments within Indonesian and Sri Lankan societies. In Aceh, the tsunami and the process of tsunami recovery took place in the context of Indonesia’s ongoing democratic transition. While this transition in many ways has been unstable, turbulent and complex, at the same time it nonetheless gradually generated space that enabled different viewpoints to come to the fore (Aspinall 2009; Miller 2009). In this context, both the government and GAM showed a willingness to transform and to reinterpret their interests, perspectives and goals. Thus, this enabled a shift in Aceh from a military approach to a political approach. In comparison, Sri Lanka has on the other hand been experiencing continuous political turmoil with key actors seemingly resistant to transform themselves and their relationships. The LTTE struggled to be recognised as a state actor and to a large extent continued striving for this objective during the tsunami recovery process. In this context, the LTTE interpreted the government’s relief and reconstruction work in the tsunami-affected northeast as attempts to undermine the organisation’s struggle for the Tamil nation (Orjuela 2011). At the same time, the government showed no intention to change its centralised nature and be willing to share power (Orjuela 2009). In the context of continuous political uncertainty, the patronage politics pervaded the tsunami recovery process in Sri Lanka. This context favoured nationalist groups in the south. In the context of continuous battles for power and recognition, the tsunami response and recovery were politicised and used by different actors in pursue of their own interests and agendas. As a result, these competing systems of governance exacerbated pre-existing inequalities between ethnic identity groups and different geographical localities.

5.5 Concluding Remarks

This chapter has focused on disaster governance in Aceh and Sri Lanka in the context of the 2004 Indian Ocean tsunami. In particular, it has emphasised the need for recognising pre-existing structures and relationships in order to understand the impact of the tsunami and the processes of tsunami response and recovery in each particular conflict setting. While a comparison between the two cases displayed both similarities and differences, the conflict context in both regions has nevertheless been important. Rather than treating the tsunami disaster as causal factor capable of automatically altering future developments, I have suggested that the tsunami can be better understood as a force triggering an already initiated process further along a trajectory. From this perspective, it can be proposed that ongoing dynamics and developments in Aceh and Sri Lanka were accelerated by the 2004 Indian Ocean tsunami. Furthermore, the chapter has also emphasised the importance of considering the *political* nature of disaster governance. By acknowledging existing power relations and competing interests and objectives, the analysis has stressed that the governance of natural disasters is not a 'neutral' enterprise. These insights underline the essential importance of recognising the political processes at play in order to understand the impact, response and recovery from environmental disasters in different societies, especially when dealing with disaster governance in the context of violent conflicts.

This edited volume particularly recognises the ongoing trends of urbanisation across Asia and how these processes could potentially affect experiences of environmental disasters and characteristic of disaster governance. In this chapter, I have highlighted how and to what extent the urbanisation of Aceh and Sri Lanka has influenced the people affected by the tsunami. I have also emphasised the primarily urban emphasis of the post-disaster recovery programmes and priorities in both Aceh and Sri Lanka. In the present study, I have often found the urban–rural divergences being more explicitly pronounced in accounts from the case of Aceh. One reason for this might be the comparatively low levels of urbanisation in Sri Lanka. While Sri Lanka is urbanising and in particular the coastal line is densely populated, many have noted that the extent and speed of urbanisation have been relatively slow.[28] In addition, the capital city of Colombo—which is the predominant urban hub in Sri Lanka—was only marginally affected by the tsunami. In comparison, in Aceh the larger cities are located along the coast, and they were nearly destroyed by the disaster, including the provincial capital city of Banda Aceh. However, considering the increasing urbanisation in Sri Lanka after the 2004 tsunami, it can be suggested that future disaster governance might be pervaded by new characteristics and dynamics. The Sri Lankan government has launched an urbanisation vision for 2030 as defined in the government's development policy framework—the Mahinda Chintana—and envisaged in the National Physical Planning Policy and Plan 2011–2030 (World Bank 2012). It is estimated that the urbanisation rate will increase from 0.3 % to 3–4 % a year and that 60 % of the population will be living in cities by 2020 (World Bank 2012, p. 8). The contemporary trends of urbanisation across Asia rest on dominant ideas about a strong relationship between urbanisation and economic development. In an era when economic development also becomes a salient component of contemporary peace processes, this raises interesting questions about the future of disaster governance in conflict-affected societies and to what extent measures to facilitate urbanisation will also be promoted as a means of managing intrastate conflicts.

[28] The available data on the level of urbanisation in Sri Lanka is however debated. The commonly held view is that the country is urbanising more quickly than the official data suggests, but more slowly than other countries in the region (World Bank 2012).

References

Åkebo, M. (2013). *The politics of ceasefires. On ceasefire agreements and peace processes in Aceh and Sri Lanka*. Dissertation. Umeå: Department of Political Science, Umeå University. http://umu.diva-portal.org/smash/get/diva2:640450/FULLTEXT01.pdf. Accessed 6 Nov 2014.

Asian Development Bank, Japan Bank for International Cooperation, & World Bank (2005, January 10–28). *Sri Lanka 2005 tsunami recovery program: preliminary damage and needs assessment*. http://www.jica.go.jp/english/news/jbic_archive/japanese/base/topics/050131/pdf/srilanka.pdf. Accessed 6 Nov 2014.

Aspinall, E. (2005). Indonesia after the tsunami. *Current History*, 104(680), 105.

Aspinall, E. (2009). *Islam and nation: Separatist rebellion in Aceh*. Stanford: Standford University Press.

Barron, P. (2008). Managing the resources for peace: Reconstruction and peace building in Aceh. In Aguswandi & J. Large (Eds.), *Reconfiguring politics: The Indonesia – Aceh peace process* (pp. 58–61). London: Conciliation Resources.

Barron, P., Clark, S., & Daud, M. (2005). *Conflict and recovery in Aceh: An assessment of conflict dynamics and options for supporting the peace process*. Washington, DC: World Bank.

Bastian, S. (2009). From research to policy: The case of tsunami rehabilitation in Sri Lanka. In M. de Alwis & E-L. Hedman (Eds.), *Tsunami in a time of war* (pp. 217–244). Colombo: International Centre for Ethnic Studies (ICES).

Choi, V. (2009). A safer Sri Lanka? Technology, security and preparedness in post-tsunami Sri Lanka. In M. de Alwis & E-L. Hedman (Eds.), *Tsunami in a time of war* (pp. 191–216). Colombo: International Centre for Ethnic Studies (ICES).

Center for Policy Alternatives (CPA) (2005). *Landlessness and land rights in post-tsunami Sri Lanka*. Colombo: Center for Policy Alternatives.

Center for Policy Alternatives (CPA). (2006, March). *Reflections on tsunami one year on: Lessons to be learnt*. Discussion paper. Colombo: Center for Policy Alternatives. http://www.cpalanka.org/wp-content/uploads/2007/8/RADA_Discussion_Paper.pdf. Accessed 16 Nov 2014.

de Silva, K.M. (1998). *Reaping the whirlwind: Ethnic conflict, ethnic politics in Sri Lanka*. India: Penguin Books.

Frerks, G., & Klem, B. (2005a, January). *Muddling the peace process: Post-tsunami rehabilitation in war-torn Sri Lanka* (CRU Policy Brief, 2). Clingendael Conflict Research Unit (CRU). The Hague: The Netherlands Institute of International Relations.

Frerks, G., & Klem, B. (2005b, March). *Tsunami response in Sri Lanka: Report on a field visit from 6–20 February 2005*. Clingendael Conflict Research Unit (CRU). Disaster Studies Wageningen University.

Goodhand, J., & Klem, B. (2005). *Aid, conflict, and peacebuilding in Sri Lanka 2000–2005*. Colombo: Asia Foundation.

Goodhand, J., Klem, B., & Sørbø, G.M. (2011) *Pawns of peace: Evaluation of Norwegian peace efforts in Sri Lanka, 1997–2009*. Report 5. Oslo: Norad.

Hyndman, J. (2009). The geopolitics of pre-tsunami and post tsunami aid to Sri Lanka. In M. de Alwis & E-L. Hedman (Eds.), *Tsunami in a time of war* (pp. 29–58). Colombo: International Centre for Ethnic Studies (ICES).

ICG (2006) *Sri Lanka: The Failure of the Peace Process*. Asia Report No 124, 28 November 2006.

ICG (2008, October 15). *Sri Lanka's Eastern province: Land, development, conflict*. Asia Report 159.

International Fund for Agricultural Development (IFAD) (2011). *Enabling poor rural people to overcome poverty in Sri Lanka*. Rome: International Fund for Agricultural Development. http://www.ifad.org/operations/projects/regions/pi/factsheets/srilanka.pdf. Accessed 6 Nov 2014.

Klem, B. (2006, April). *Sri Lanka: Possible futures and patterns of displacement*. Writenet Independent Analysis. http://www.bartklemresearch.nl/files/publications/displacement_scenarios_sri_lanka.pdf. Accessed 16 Nov 2014.

Mayer, M., & Salih, M. (2006). Sri Lanka: Business as an agent for peace. In *International Alert, Local business, local peace: The peacebuilding potential of the domestic private sector* (pp. 551–582). London: International Alert. http://www.international-alert.org/sites/default/files/publications/29_section_2_Sri_Lanka.pdf. Accessed 6 Nov 2014.
Mc Gibbon, R. (2006). Local leadership and the Aceh conflict. In A. Reid (Ed.), *Verandah of violence: The background to the Aceh problem* (pp. 315–359). Singapore: NUS Press.
Merikallo, K. (2006). *Making peace: Ahtisaari and Aceh*. Juva: WS Bookwell Oy.
Miller, M.A. (2006). What's special about special autonomy in Aceh? In A. Reid (Eds.), *Verandah of violence: The background to the Aceh problem* (pp. 292–314). Singapore: NUS Press.
Miller, M.A. (2009). *Rebellion and reform in Indonesia. Jakarta's security and autonomy policies in Aceh*. London/New York: Routledge.
Miller, M.A., & Bunnell, T. (2013). Urban–rural connections: Banda Aceh through conflict, tsunami, and decentralization. In T. Bunnell, D. Parthasarathy, & E.C. Thompson, (Eds.), *Cleavage, connection and conflict in rural, urban and contemporary Asia* (pp. 83–98). Singapore: Springer.
MoU (2005, August 15). *Memorandum of Understanding between the Government of the Republic of Indonesia and the Free Aceh Movement*.
Mulligan, M., & Shaw, J. (2007). What the world can learn from Sri Lanka's post-tsunami experiences. *International Journal of Asia Pacific Studies*, 3(2), 65–91.
Nazara, S., & Resosudarmo, B.P. (2007). *Aceh-Nias reconstruction and rehabilitation: Progress and challenges at the end of 2006*. ADB Institute Discussion Paper, 70. Tokyo: Asian Development Bank Institute. http://www.adbi.org/discussionpaper/2007/06/26/2288.acehnias.reconstruction.rehabilitation/. Accessed 6 Nov 2014.
Orjuela, C. (2004). *Civil society in civil war: Peace work and identity politics in Sri Lanka*. Gothenburg: Department of Peace and Development Research, Gothenburg University.
Orjuela, C. (2009). Domesticating tigers: The LTTE and peacemaking in Sri Lanka. In B. W. Dayton & L. Kriesberg (Eds.), *Conflict transformation and peacebuilding: Moving from violence to sustainable peace* (pp. 252–269). Oxon: Routledge.
Orjuela, C. (2011). Buying peace? Politics of reconstruction. In K. Stokke & J. Uyangoda (Eds.), *Liberal peace in question: Politics of state and market reform in Sri Lanka* (pp. 121–139). London/New York/Delhi: Anthem Press.
P-TOMS (2005, June 27). *Memorandum of Understanding (MoU) for the establishment of a post-tsunami operational management structure (P-TOMS)*.
Rainford, C., & Satkunanathan, A. (2009). *Mistaking politics for governance: The politics of interim arrangements in Sri Lanka 2002–2005*. Colombo: Centre for Ethnic Studies Publication.
Reid, A. (Ed.) (2006). *Verandah of violence: The background to the Aceh problem*. Singapore: NUS Press.
Sarvananthan, M., & Sanjeewanie, H.M.P. (2008). Recovering from the tsunami: People's experiences in Sri Lanka. *Contemporary South Asia*, 16(3), 339–351.
Shaw, J., Mulligan, M., Nadarajah, Y., Mercer, D., & Ahmed, I. (n.d.). *Lessons from tsunami recovery in Sri Lanka and India: Community, livelihoods, tourism and housing*. Australia: Monash University, RMIT University. http://artsonline.monash.edu.au/mai/files/2012/06/post-tsunami-1.pdf. Accessed 6 Nov 2014.
Schulze, K.E. (2005). *Between conflict and peace: Tsunami aid and reconstruction in Aceh*. London: London School of Economics.
Schulze, K.E. (2006). Insurgency and counter-insurgency: Strategy and the Aceh conflict, October 1976 – May 2004. In A. Reid (Ed.), *Verandah of violence: The background to the Aceh problem* (pp. 225–271). Singapore: NUS Press.
Stokke, K., & Uyangoda, J. (Eds.) (2011). *Liberal peace in question: Politics of state and market reform in Sri Lanka*. London/New York/Delhi: Anthem Press.
Thjin, C.S. (2005). Post tsunami reconstruction and peace building in Aceh: Political impacts and potential risks. In M. John & J. Illerhaus (Eds.), *A new dynamic for peace? Post-tsunami reconstruction and its impacts on conflict resolution* (23–30). Bonn: Friedrich Ebert Stiftung.

Uppsala Conflict Data Program (2008). *UCDP database: Indonesia*. http://www.pcr.uu.se. Accessed 22 Aug 2008.

Uppsala Conflict Data Program (2012). *Armed conflict dataset*. http://www.pcr.uu.se/research/ucdp/datasets. Accessed 19 Jan 2012.

World Bank Aceh (2005). *Aceh and Nias one year after the tsunami: The recovery effort and way forward*. A Joint Report of the BRR and International Partners. http://siteresources.worldbank.org/INTEASTASIAPACIFIC/Resources/1YR_tsunami_advance_release.pdf. Accessed 6 Nov 2014.

World Bank (2008). *Aceh poverty assessment 2008: The impact of the conflict, the tsunami and reconstruction on poverty in Aceh*. Jakarta: The World Bank Office Jakarta. http://www-wds.worldbank.org/servlet/WDSContentServer/IW3P/IB/2008/01/09/000020953_20080109160816/Rendered/PDF/421010Aceh0Pov1nt0P010437501PUBLIC1.pdf. Accessed 6 Nov 2014.

World Bank (2012). *Turning Sri Lanka's urban vision into policy and action*. Colombo: The World Bank. http://www-wds.worldbank.org/external/default/WDSContentServer/WDSP/IB/2012/10/12/000386194_20121012040706/Rendered/PDF/731820WP0P12800AL0UPN0Book005102012.pdf. Accessed 6 Nov 2014.

Chapter 6
Disaster Governance in Small Urban Places: Issues, Trends, and Concerns

Andrew Rumbach

Abstract A growing number of scholars have heeded calls for an urban research agenda grounded partly in the experiences of small urban places. This chapter makes a similar appeal to disaster researchers who have developed frameworks for thinking about urban disasters based largely on the experiences of larger cities. Disaster risk is also growing rapidly in small urban places, far from centres of trade, culture, and politics. Governing disaster risk in small cities will be an essential part of managing Asia's urban transition. I identify four unique characteristics of small cities important for disaster risk management. First, small cities are physically, politically, and culturally distant from capitals and other centres of power and influence. Second, small cities accumulate disaster risk in similar ways to larger urban centres, but without the attendant growth in infrastructure or governance capacity. Third, urbanization in smaller cities outpaces environmental learning. That is, the centres of environmental "memory" that are key to long-term risk governance may be absent. Finally, small cities often lack the redundancy that is the hallmark of resilient urban systems. I illustrate my arguments with evidence from three landslide-prone small cities in West Bengal, India. I conclude that more research is needed to understand what the growth of small cities means for disaster risk in rapidly urbanizing countries like India. Our conventional approaches to urban disaster risk management may be ineffective in smaller cities. The concept of disaster governance holds promise, however, because of its inclusion of a wider set of actors and institutions.

Keywords Disaster governance • West Bengal • India • Small cities • Urbanization • Landslides

A. Rumbach, PhD (✉)
Department of Planning & Design, University of Colorado Denver, Denver, CO, USA
e-mail: andrew.rumbach@ucdenver.edu

M.A. Miller, M. Douglass (eds.), *Disaster Governance in Urbanising Asia*,
DOI 10.1007/978-981-287-649-2_6

6.1 Introduction

Cities are increasingly central to our study of natural hazards and disasters. Today, for the first time in human history, more people live in urban areas than rural ones, a remarkable demographic shift that is reshaping human-environmental relationships. As larger shares of people and resources become concentrated in cities, international institutions, organizations and governments warn that urban disasters pose a significant threat to human health and safety and are a major barrier to poverty alleviation and sustainable economic development (Douglass 2013; World Bank 2011; UN-ISDR 2012).

As Pelling (2012) points out, however, city growth alone does not generate risk. Instead, urban risk is:

> …at its most basic level an outcome of competing values and visions of the city and the distribution of political, economic, and social power, which in turn determine expenditure on physical and social infrastructure and the application of technological innovation (146).

In other words, risk is not a natural outcome of urbanization but is socially, politically, and economically constructed (see also Cutter et al. 2009; Oliver-Smith 2010; Pelling and Wisner 2008b; Wamsler 2004; Wisner et al. 2003). As such, urban governance regimes play a central role in creating or reducing risk and determining geographies of risk within the city. Urban governance regimes have traditionally operated with an eye towards economic growth or livelihood generation, however, and rarely towards disaster risk management (Wamsler 2004, 11).

As urban disasters grow in size and impact, and with the rising threat of climate change, this may well be changing. The UN-ISDR's "Making Cities Resilient" campaign, for instance, counts over 1,500 participating cities and municipalities.[1] Many international institutions, development organizations, and foundations have made climate resilience a core theme of research and grant programming, with a strong focus on cities and city governments (e.g. see Rockefeller Foundation 2013; Swiss Re 2013; World Bank 2011). While perhaps not a paradigm shift, the introduction of resilience thinking into urban governance and development discourses is an important step towards safer and more sustainable cities and communities.

With a growing focus on cities and disasters, it is important to take stock of the basis for our understanding of urban risk. Even a cursory review of the urban disaster literature reveals that most of what we know comes from the experience of large cities, national and regional capitals, urban mega-regions, and other centres of power and influence. Small urban centres, those "faintly visible" cities that will account for the majority of urban growth in Asia in the coming decades, are largely absent (Davis 2004, 7).

In this chapter I seek to expand our understanding of urban disaster risk by focusing on smaller cities. I describe four characteristics of small cities that distinguish them from larger urban places in the context of disasters and disaster risk

[1] See http://www.unisdr.org/campaign/resilientcities/cities. Most recently accessed on 30 May 2014

management. First, small cities suffer from a tyranny of distance; their physical, cultural, and political separation from centres of power and influence makes disaster risk management policy, resources, and research less reflective of their needs and priorities. Second, there is a mismatch between risk and capacity. Small cities experience many of the same risks associated with urbanization without the attendant capacity for even basic planning or urban management. Third, the pace of urbanization in small cities often outpaces environmental learning. Finally, small cities lack redundancy in areas critical to disaster risk reduction, especially in infrastructure and in civil society.

I ground these arguments in a detailed case study of three small, landslide-affected cities in the Darjeeling District of West Bengal, India. The case study is based on data collected during five field visits to the Darjeeling District from 2009 to 2012.[2] The data includes 24 semi-structured and open-ended interviews conducted with (1) landslide-affected households and business owners; (2) current and former officials in the municipal, district, and state government; (3) representatives from non-governmental and community-based organizations (NGOs and CBOs); and (4) academics and researchers working on issues of environmental planning and urban development in West Bengal and India. Additionally, I gathered secondary and archival data on historic landslides in the region, reviewed local and district-level plans, and attended multiple workshops and conferences focused on disaster risk reduction in the Darjeeling District.

The chapter is divided into four parts. First, I briefly discuss the literature on urban risk and the need for an expanded research agenda that is partly rooted in the experiences of smaller cities. Next, I highlight the central importance of small cities in the urban disaster discourse by describing the pace and scale of small city growth in India over the past several decades. Third, I introduce my case study of the Darjeeling District and describe the four characteristics that distinguish small cities from larger ones. I conclude with some brief remarks on the challenges of small cities for researchers and policymakers interested in disaster resilience.

6.2 Cities and Disasters: Do We Need a "View from off the Map?"

> …the smaller cities are precisely where most of Asia will soon live. If megacities are the brightest stars in the urban firmament, three-quarters of the burden of future world population growth will be borne by faintly visible second-tier cities and smaller urban places: places where, as UN researchers emphasize, "there is little or no planning to accommodate these people or provide them with services" (Davis 2006, 7).

[2] Throughout my visits to the Darjeeling District, I have worked closely with Save the Hills (STH), a community-based organization headquartered in Kalimpong. STH is dedicated to reducing the risk of landslide disasters in the Darjeeling District through community-based education, research, and advocacy. Their insightful and regularly updated blog is located at www.savethehills.blogspot.com.

The literature on urban disaster risk and resilience is based largely on the experiences of large cities and national/regional capitals; the everyday realities of smaller cities are largely absent. Mark Pelling (2012), one of the foremost researchers in the field, argues that disaster researchers and policymakers "know relatively little about smaller cities" and that it is "tempting to project knowledge gained from larger cities" (150). Cross (2001), in one of the few papers that compare risk across different settlement types, argues that small cities struggle to gain recognition in disaster research and policy because of their limited human and economic impact relative to large-city events. Whereas disasters in a megacity or capital might affect hundreds of thousands of people and cause regional or global disruptions, small city disasters are usually felt locally. The bias towards larger cities is reflected in research. The most widely cited monographs on urban disasters, like Mitchell (1999), Pelling (2003), and Pelling and Wisner (2008a), largely rely on megacities and/or capitals as the primary basis for evidence. The bias gets reflected in the work of non-governmental organizations and foundations as well. The Rockefeller Foundation's "100 Resilient Cities" Campaign, for instance, defines a city as having 50,000 or more inhabitants, excluding many small cities from the scope of its work.[3] Of the first 33 cities selected, only a few have less than 250,000 residents. Initiatives like "Earthquakes and Megacities" do excellent work to advance disaster risk reduction, but equivalent programs do not exist for smaller places.

Despite our fixation on megacities, the majority of the world's urban population lives in small urban centres. The United Nations estimates that in 2011, more than half of the world's urban population (3.6 billion) lived in cities with populations of 500,000 or less (a total of 1.85 billion). By contrast, the number of people living in megacities, cities with 5 million people or more, was 642 million, or 17.8 % of the global urban population (UN 2012, 17–18). Pelling (2012) argues that towns and cities with fewer than 500,000 inhabitants "have been and will continue to be the type of urban settlement in which the largest share of the world's urban population resides" and that the population of those places exposed to environmental risk likely exceeds the total at-risk populations in megacities.

India is representative of these broader trends. The country is undergoing a massive demographic shift from rural to urban; in 1951, just 17 % of Indians lived in cities, compared to 31 % in 2011 (see Table 6.1).[4] This represents urban population growth of over 330 million, from 62 million in 1951 to nearly 400 million in 2011.[5] Much of this growth was captured in small cities. In 2011, the population of cities less than 100,000 people totalled 133 million, around 33 % of the total urban popu-

[3] See http://100resilientcities.rockefellerfoundation.org/pages/about-the-challenge.

[4] The World Bank (2013) argues that India's relatively low urbanization is due, in part, to an outdated and stringent definition of urban. Using the Agglomeration Index, the bank estimates that India is as much as 52 % urbanized (see page 24).

[5] The Census of India defines "urban" as a place that satisfies three criteria simultaneously: a minimum population of 5,000, at least 75 % of male working population engaged in nonagricultural pursuits, and a population density of at least 400 persons per sq. km.

Table 6.1 Distribution of India's population by settlement size[a]

	1951	1961	1971	1981	1991	2001	2011
Hamlets (%)	43	37	31	25	19	15	12
Small villages (%)	20	21	21	20	19	18	16
Medium villages (%)	17	18	20	21	22	23	23
Large villages (%)	5	8	10	12	14	16	15
Class IV–VI cities (%)	5	4	3	3	3	3	4
Class III cities (%)	3	3	3	3	3	3	4
Class II cities (%)	2	2	2	3	3	3	3
Class I (.1–1 m) (%)	1	2	4	6	7	8	10
Million-plus cities (%)	3	4	5	6	9	11	12

Source: Adapted from IIHS (India Urban Conference 2011, Figure 2)
[a]Following are the population cut-offs for each settlement type: hamlet and smaller (<1,000), small village (1–2,000), medium-sized village (2–5,000), large and very large villages (>5,000, but not classified as city), class IV–VI city (<20,000), class III city (20–50,000), class II city (50–100,000), other class I city (.1–1 m), million-plus city (>1 m). See India Urban Conference 2011

lation. Cities with populations between 100,000 and 1 million housed an additional 30 % (India Urban Conference 2011).[6]

Over the past decade, critics of mainstream urban studies have argued for a broader approach, one that reflects the experience of a wider range of cities. Robinson (2002) argues than the field's predominant focuses on global and world cities "impose substantial limitations on imagining or planning the future of cities around the world" (531). In this chapter I would like to advance a similar call for a more inclusive approach to urban disaster studies, one that is rooted, in part, in the experience of smaller urban centres. Such an approach would reflect actual trends in urbanization in most of Asia and Africa, where small cities will account for a substantial part of urban growth in the coming decades. It would also broaden the discourse on disaster risk, disaster governance, and resilience; our current body of knowledge about urban disasters, focused primarily on the experience of larger cities, limits our discourse—what can be said, or even imagined, about the problem of urban disasters, their origins and features, and our potential interventions.

6.3 Case Study: The Darjeeling District, West Bengal

Kalimpong, Kurseong, and Darjeeling are small cities in the Darjeeling District of West Bengal. Originally built as hill stations during British rule, they are now famous for their tea exports and are popular tourism destinations during the summer months. Darjeeling is the headquarters and largest city in the district with 132,016

[6]Additionally, large villages (those with 5,000 population or more) housed 200 million people in 2011 (India Urban Conference 2011). These villages share many of the same features as small cities, like higher population density and commercial activity (World Bank 2013).

Table 6.2 West Bengal hill station populations 1991–2011

City	1991	2001	2011	Net change	% change
Darjeeling	73,062	107,197	120,414	47,352	+64.8 %
Kalimpong	38,832	42,998	67,680[a]	28,848	+74.3 %
Kurseong	26,759	40,019	42,346	15,587	+58.2 %

Source: Census of India 1991, 2001, 2011

[a]This number is "estimated" in the 2011 census

inhabitants (Government of India 2001). Kalimpong and Kurseong are smaller, with 2011 populations of 67,680 and 42,346 respectively. Similar to small cities across India, these hill cities have grown quickly over the past 20 years (see Table 6.2).

The Darjeeling District is located in the foothills of the southern arc of the Himalayas, the most landslide-affected region on earth (Petley 2012). The average elevation ranges from 1,250 to 2,000 m (4,100–6,550 ft) with terrain characterized by steep slopes and deep river canyons, a perfect environment for landslides (the movement of rock, debris, or earth down a slope).[7] The Himalayas are, geologically speaking, "young" mountains with soft soils (Rahman et al. 2011). Landslides occur primarily during the monsoon season, from July to September, when heavy rainfall saturates the soil. Besides the landslide risks associated with regular monsoon rains, there are infrequent but severe rainfall events that have a more widespread impact. In 1968, for instance, torrential rainfall triggered more than 20,000 landslides across the region, killing thousands of people and crippling regional infrastructure (see Ives 1993).[8] Similar events in 1899, 1950, and 1958 caused devastation across the district (Sankrityayana 2009).

Another regional trigger of landslides are earthquakes. The Darjeeling District is located in a Zone IV seismic region, putting it at "high" risk of earthquake hazards. The region is long overdue for a major earthquake. In September 2011, a 6.8-magnitude earthquake centred in the state of Sikkim, which borders Darjeeling to the north, rattled the district, injuring thousands and causing damage to many homes and buildings. A strong earthquake with an epicentre closer to the hill cities is projected to cause significant damage and loss of life (e.g. Sinha 2011).

Global climate change will have a significant impact on the hazard profile of the Darjeeling District. Three trends are particularly worrying for mountain cities. First, climate change is expected to increase the frequency of heavy precipitation events, one of the main triggers of landslides (see IPCC 2012). Second, changes in heat wave activity, glaciers, and permafrost will all likely affect slope stability and the patterns of slides. Third, the intensity of tropical cyclones in the Bay of Bengal is expected to increase, increasing the likelihood that cyclonic storms will reach far enough inland to cause heavy precipitation events in the Himalayan foothills (IPCC 2012, finding 3.4.4).

Besides the natural drivers of landslide hazards like slope gradient, precipitation patterns, and lithology, there are important anthropogenic drivers as well, nearly all

[7] See http://www.ga.gov.au/hazards/landslide/landslide-basics/what.html.

[8] The total rainfall across the 3-day period was measured at 500–1,000 mm (1.6–3.3 ft).

of which are associated with urbanization and urban growth. Like most cities exposed to natural hazards, cities in Darjeeling have historically had a spatial development pattern that reflects the geography of hazards in the region. Land that is most stable and relatively safe from landslides, like that along ridgelines, has largely been exhausted, and development has now spread into more hazardous areas. In general, developable land is at a premium in mountainous terrains, and so many developers have resorted to "cutting" hillsides to create pads for construction, significantly destabilizing the slope. The increased runoff caused by the growth of impermeable surfaces has further affected soils and slopes downhill from cities.

Besides the direct effects of urbanization on landslide risk, there are several other important indirect effects. Widespread deforestation in the region, in part driven by the demand for construction materials, has increased soil erosion and decreased slope stability, especially during the monsoon (e.g. see Froehlich and Starkel 1993). Roadways, which in mountainous terrain require deep cuts into hillsides, contribute to slope instability and are commonly the initial points of slope failure (e.g. Kampa et al. 2008). A related driver of landslide hazards is the increased movement of vehicles along mountainside roads, especially heavy trucks, lorries, and buses, all of which put additional pressure on slopes.

Together these direct and indirect effects of urbanization have increased the incidence of both landslide hazards (the geological phenomenon) and landslide disasters (when landslides significantly impact human populations). Landslides are a classic case of compound disasters; individual events tend to create the condition for future slides, and human adjustments to the hazards (like repairing roadways through additional slope cutting) often exacerbate future hazard conditions (see Douglass, Chapter 2, in this volume; Sankrityayana 2009).

6.4 Disaster Risk in Small Cities

What distinguishes disaster risk in cities like Darjeeling, Kalimpong, and Kurseong, as opposed to larger cities like Kolkata or Kathmandu? In the remainder of the chapter, I describe four characteristics of small city risk based on a detailed case study of landslide risk management in the Darjeeling District.

6.4.1 *The Tyranny of Distance*

One of the central themes that emerges in interviews about landslide risk management in the Darjeeling District is *distance*.[9] Darjeeling is nearly 400 miles from the capital city, Kolkata, the epicentre of political and economic power in West Bengal

[9] Much of the discussion of distance relates to the scale at which disaster management decisions are made. Disaster governance in India is primarily a function of the states; the Disaster Management Act of 2005 created a disaster management authority at the central government level, but most disaster management roles and responsibilities are delegated to state authorities.

and home to the vast majority of agencies and institutions that play a role in disaster risk management in the region. While some state officials may visit the hill cities for a particular workshop, or function, they do not have the day-to-day interactions with landslide-affected communities they do with constituents in the capital or nearby districts. As a result, the opportunities to draw attention to landslide risk or to influence policymakers on disaster management priorities are extremely limited.

Distance is a useful metaphor for describing other features of small city risk and risk governance. There is certainly a *political* distance that limits the ability of the hill cities to put landslide risk governance on the state's agenda. The Darjeeling District has a population of over 1.8 million people, but that accounts for just over 2 % of the total population of West Bengal. Political representatives from the hill districts are a small minority in the Legislative Assembly and have little clout in state government (Census 2011, Provisional Population Totals). For state officials and the disaster management bureaucracy, the primary and overriding concerns are flooding and tropical cyclones, which predominately threaten high-population districts in the southern parts of West Bengal.

There is also an *experiential* distance between state officials and hill cities when it comes to natural hazards. As one interviewee described it: "they [state emergency management officials] do not care about landslides because they do not understand landslides!" For officials headquartered in Kolkata, a city regularly affected by floods and cyclones, landslides are a distant and abstract concern. The nature of landslides compounds experiential distance; because landslides tend to be geographically localized and the impact of any single event is quite limited, they do not have the capacity to cause the overwhelming damage or loss of life necessary to "shock" the consciousness of distant officials and spur them to action. Whereas a single earthquake causing widespread damage might focus attention on seismic hazards, hundreds of individual landslides with similar cumulative impacts have not.

Lastly, there is a distinct *cultural* distance between landslide-affected cities and those in power in the state.[10] In part, this is due to the ethnic make-up of the state of West Bengal; while the Darjeeling District is predominantly Nepali, the state as a whole has a large Bengali majority. A telling example of the cultural distance between state officials and community leaders from the hills occurred at the first West Bengal State Disaster Risk Reduction Congress in 2010. Congress organizers decided to make Bengali the primary language of the meeting, even though English is the lingua franca of the bureaucracy. Most hill residents speak Nepali and English and were effectively excluded from participating in panels or discussions. Several community-based organizations in attendance staged a walkout to voice their displeasure and remarked on the continued exclusion of the hills from state policy discussion. As one prominent official from the hills explained: "they [the Bengalis] run everything…in many ways, we still feel colonized."

---

[10] In many ways the metaphor of distance is synonymous with power; experiences of physical, experiential, and cultural distance reflect vast inequalities in power across the state and between the Bengali majority and minority ethnic groups.

Cultural distance extends beyond ethnicity. There is also a cultural distance between officials from the "big city" and constituents in smaller places. There is a pervasive feeling expressed amongst landslide activists in the hills that their local knowledge and experience is consistently undervalued by state and centre officials in favour of technical knowledge emanating from the capital. In one memorable workshop on risk reduction in 2010, slickly dressed officials travelled from Delhi to Kalimpong to give a presentation to an enthusiastic audience of local landslide experts and organizers who were hoping to secure central government support for a risk management agenda. The Delhi officials proceeded to lecture on the most basic and generic of preparedness principles ("during heavy rains, the risk of landslides rises"), discussed some general causes of slope failures without any reference to the region or its particular challenges, and promptly returned home. Later that day, one community activist summed up the general feeling that the officials had talked down to their audience: "they come from there [Delhi] and they lecture us like children! But we already know these things!" Locals were crestfallen but not terribly surprised; as another audience member put it: "they [outside officials] always talk, but never do they listen".

6.4.2 Urbanity Without the Capacity for Urban Governance

A second and perhaps defining feature of disaster governance in small cities is urbanity without the capacity for urban governance. Small cities generate many of the same drivers of disaster risk that are typical of larger urban centres, yet they do so without an attendant growth in the governance capacity that is the foundation of effective disaster risk management.

As described earlier, disaster risk has increased dramatically in the Darjeeling District due to rapid urbanization over the past three decades. The dynamics of urban risk are similar in Darjeeling as they are in larger cities: greater number of people living on hazardous land, a growing reliance on complex but fragile infrastructures, environmental degradation and intensification of development activities that produce or exacerbate hazards (e.g. Hardoy and Satterthwaite 1991; Graham 2010; Rumbach 2014b; Shaw 2008).

How is urban risk typically governed? Large cities rely on a broad and complex set of actors, institutions, regulation, governmental planning, laws, private sector entities, physical resources, and private sector and civil society entities (Tierney 2012). Examples include infrastructural investments like water, sewerage, and storm water management; environmental laws and regulations that restrict or manage growth in hazardous areas; hazard-sensitive building codes and design standards; dedicated disaster management agencies or organizations; robust non-profit and civil society engagement with disaster preparedness and recovery; and so on. Certainly disaster governance capacity varies from large city to large city, as does the degree to which that capacity is leveraged for reducing risk. Generally speaking, however, the key "ingredients" that have the potential to create effective disaster governance are more likely found in larger cities (especially capitals) than in smaller ones, even more so when comparing cities within the same country.

Nearly all of the literature on small cities highlights the reduced capacity for governance in such places (e.g. Montgomery 2008; Bell and Jayne 2009; Cross 2001). The same argument extends to disaster governance. The Darjeeling District is a case in point; while cataloguing every shortfall of governance in the region is beyond the scope of this chapter, several examples from planning and emergency management are illustrative of the broader argument. Despite historic and ongoing concerns over landslides across the district, there are no plans governing local or regional drainage, an infrastructural system crucial to reducing the impact of flood and landslide hazards. Similarly, in Kalimpong and Kurseong, there is no current land-use plan in effect, and in Darjeeling, the land-use plan is badly outdated and widely ignored. Similarly, there are no effective building codes in the district; while there are, strictly speaking, rules governing hillside development, they are rarely enforced and easily avoided. Despite the widespread and pervasive risks posed by landslides and earthquakes, there is no local emergency management or hazard mitigation office that focuses on the three cities. There are subdistrict headquarters of the state disaster management authority located in the district, but their jurisdiction extends only to villages and not urban areas.

Another key difference between large and small cities is the capacity for knowledge production. Larger cities have universities, policy research organizations, development organizations, and other knowledge-producing institutions that generate data useful for disaster governance at the local level. These institutions are particularly important for cities, where drivers of disaster risk are complex, dynamic, and interconnected; plans and policies require timely and context-relevant information that is best produced by institutions rooted in the community themselves. Without these institutions, small cities must rely on outside experts who often have, at best, a cursory understanding of the local social, cultural, and economic realities. One of the most durable findings in disaster research over the past several decades is that scientific knowledge about hazards is not nearly enough; human vulnerability to those hazards is primarily social, economic, and political in origin (Pelling 2003). That is, disaster risk is extremely *local* and without local information extremely difficult to manage.

6.4.3 Urbanization Outstrips Environmental Learning

Another characteristic of small urban centres is that urban growth tends to outstrip environmental learning. Urbanization and environmental hazards typically occur in very different timescales. While a torrential rainfall, large earthquake, or cyclonic storm may occur only once in 50 or 100 years, small cities often grow in a matter of years or decades and often do so in between major hazard events. This is especially true in rapidly urbanizing Asian countries like India, where the number of population of small cities has grown exponentially in a matter of just a few decades.

For larger cities with longer histories, natural hazards tend to play a more central role in the shape and character of urban growth over time. In those places there exists an environmental learning; past experiences with hazards and disasters shape the physical and spatial fabric of the city through land-use patterns, choices of housing styles and materials, the design of infrastructural systems, and the laws and policies relevant to disaster risk management.

Young and rapidly growing cities often do not adjust to risk because they have not experienced the trauma of disaster in their short lifetimes. There is also something of an economic exuberance that accompanies urbanization, especially in formerly rural places. The city brings with it the promise of economic prosperity and previously unheard-of material wealth. Once the city takes root, the headlong rush towards growth is rarely tempered by such worries as the distant and abstract possibility of disaster. Like the young person who feels invincible, small cities lack the wisdom to take adequate precautions against harm.

A telling example from the cities in Darjeeling District is the rapid increase in multistorey buildings. Urbanization is being driven partly by growth in the tourism sector, and many business owners are building massive new buildings to take advantage of economic opportunities (see Fig. 6.1). Owners of older, 2–3-storey buildings are adding extra floors. These new and updated buildings are constructed largely from concrete, rebar, and other weighty materials, wholly inappropriate for a region prone to slope failures and landslide disasters.[11] In mountainous terrain where flat land is at a premium, these buildings are largely sited on steep slopes and hang precariously from the hillside. The sheer weight of new and modified buildings, along with the thousands of people living, working, or staying in them, represents a sharp escalation in risk to earthquake and landslide hazards.

Fig. 6.1 Vernacular and modern architecture in Kalimpong, West Bengal (*Source*: Author's photographs)

[11]The vernacular architecture utilizes a "wattle and daub" construction, with lightweight walls made of reed matting, wire mesh, and mud plaster (see Fig. 6.2, left photograph. For more on Himalayan Architecture, see Bernier 1997). This lightweight style of construction is far better suited for landslide-prone areas, because it puts less weight on the slope and, in the event of a collapse, is less likely to crush its inhabitants.

From the perspective of disaster governance, there is a profound short-sightedness in this type of growth, knowing what we do that a single earthquake or heavy rain event could trigger the collapse of dozens or even hundreds of buildings. In larger cities, the exuberance of growth is often tempered by the wisdom of past disasters, encoded in laws, regulations, cultural norms, institutions, organizations, and people who carry forward the collective trauma experienced in such events. Cities typically learn from disasters and are often repaired or restored in ways that make them slightly safer than before (Olshansky 2009).[12] This incremental progress, the collective wisdom from past disasters, is a key ingredient for building urban resilience.

The hill cities are too young for this type of wisdom; in the past 20 years, they have seen massive growth but few major hazards. As Mukul (2011) argues, the lack of major earthquakes prior to September 2011 "led to a sense of complacency in the hills" and an attendant growth in unsafe building practices. Ironically, some locals feel that the 2011 Sikkim earthquake validates their choice of building design and materials; the fact that their building was not damaged or destroyed is proof that it is safe from future events. In the case of one five-story concrete hotel in Kalimpong that was largely undamaged, the owner has since decided to build two additional floors, since the building had "received high marks" from the earthquake (even though the quake was relatively minor compared to what is forecast for the region). In one local engineer's sober assessment, it will take a major disaster to change people's perception of risk. "The only way they will learn", he sombrely noted, "is when everything falls down".

6.4.4 The Redundancy of Urban Systems

A final characteristic that distinguishes small cities from large ones is a lack of redundancy. Redundancy, or a system's properties that allow for "alternate options, choices, and substitutions under stress", is a core feature of resilience (O'Rourke 2007, 25). The same is true for resilient cities, where redundancy ensures that shocks to one part of the urban system will not mean the collapse of the whole.

Infrastructure is a classic example. Urbanization typically implies a system that is increasingly reliant on interdependent and complex infrastructures like electrical grids, sewerage and waste disposal, transportation networks, and electronic communications. Without sufficient infrastructure to perform basic metabolic functions, the density that characterizes urbanity would be difficult to achieve.

Graham (2010) argues that complex infrastructure "permeates the everyday life of every urbanite", and as we become ever more reliant on such systems, the loss of

[12]This is not always the case, of course. In many larger cities also experiencing rapid growth, we see the same phenomenon of "youth" on the periphery, where fast-growing suburbs flout the environmental history of the place and are contributing to a massive increase in hazard risk through questionable land-use and planning decisions (Rumbach 2014a).

infrastructure during a disaster plunges modern cities into a desperate state (111). In larger cities, especially those in the industrial west, redundancy in infrastructural systems means that during a disaster, the failure of one part of the system does not spell collapse. During the 1989 Loma Prieta earthquake, for instance, a span of the San Francisco-Oakland Bay Bridge failed, severing a vital part of the regional transportation network. Because of the redundancy of transportation infrastructure in the Bay area, however, the impacts of the bridge collapse on the larger urban system were relatively minor.

In smaller cities, however, infrastructural systems are much less likely to be redundant, with serious implications during times of disaster. In the Darjeeling District, infrastructural systems like electrical lines, water delivery, solid waste disposal, drainage, and roadways are barely adequate for their intended purposes and often quite precarious.

Kalimpong, for instance, relies on just two major roadways for all of its imported food and supplies, including a great deal of drinking water that is trucked to the city. Earthquakes and monsoon rains trigger regular landslides along the roadways, sometimes in multiple locations (Fig. 6.2). After the 1968 flood, which triggered thousands of slides, hill cities were effectively cut off from the surrounding region for six weeks. The hill cities have grown exponentially since, and a similar event today would strand hundreds of thousands of city dwellers without sufficient access to food, water, or medicine.

Other infrastructures are similarly fragile; a heavy rain event or earthquake might conceivably knock out electricity or cellular phone connectivity. The water infrastructure is particularly at risk, due to the ad hoc nature of the water lines that supply drinking water for many city residents (Fig. 6.3).

Fig. 6.2 Landslide-affected roadway in Darjeeling District (*Source*: Save the Hills)

Fig. 6.3 Above-ground pipes and hoses for water delivery in Kalimpong (*Source*: Author's photograph (2012))

Redundancy extends to civil society as well. Whereas larger cities tend to have multiple, overlapping (and sometimes competing) organizations and institutions committed to issues like disaster risk management, smaller places may have few, if any, such groups. In the Darjeeling District, only a handful of organizations are committed to disaster governance, Save the Hills chief amongst them. Like many small advocacy organizations, STH is the product of a passionate individual who cares deeply about the landslide issue. His advocacy has brought together local leaders and environmental activists on behalf of the disaster risk reduction, and his work has elevated landslide preparedness as a priority in the eyes of state and centre politicians. The organization has achieved remarkable things with very few resources, but without redundant organizations doing similar or complimentary work, it begs the question: what happens when his leadership ends? Or as one young advocate worried, "when uncle is finished, will we be finished?"[13]

6.5 Conclusion

In this chapter I have argued for an expanded view of disaster risk, one that takes into account the experience of small urban places. As our world becomes increasingly urban, much of the growth will take place in small cities. Basing our approach to urban disaster risk management on the experiences of megacities, state and national capitals, and other notable places that are currently "on the map" for

[13] "Uncle" is an honorific used by young people to show respect for older men and does not imply blood relationship.

disaster, researchers risk putting forward any number of policy frameworks and interventions that will be ineffective in smaller places (Robinson 2002).

Clearly more work is needed to understand how city size affects disaster risk and disaster governance capacity. Here I have suggested four characteristics that distinguish small cities from large ones in the context of disasters, but it is a far from complete list. In this closing section, I would like to suggest a few additional ways we might think about urban disasters in a more inclusive manner.

The relatively new focus on disaster governance is a positive step towards addressing small city risk, because it takes into account a broader range of actors and institutions than our traditional focus on disaster risk management. As Tierney (2012) describes, disaster governance is a relatively new and uncommon term in the hazard and disaster literature. Governance is a "more inclusive concept" than disaster management in that it recognizes that many functions that were formerly the responsibility of the state are now "dispersed among diverse sets of actors that include not only government institutions but also private-sector and civil society entities" (342).

For small cities, locating disaster risk management within the state may be counterproductive because it simultaneously erodes community capacity for effective risk management while also concentrating much of the power, expertise, and resources related to risk reduction to bigger cities, typically administrative capitals. Even with disaster governance's expanded view of the role of non-state actors, however, the government still plays a central role and is likely the only entity that will bear the costs of many of the large-scale investments necessary to reduce risk in small cities. In Darjeeling District, for example, the drainage systems for all three cities require significant restoration and upgrading, a scale of investment that is beyond the capacity of any private or non-governmental actor.

This leads to an interesting question about urban disaster governance: given the central importance of both state and non-state actors, at what scale should our disaster governance institutions operate? Small cities present something of a paradox for disaster governance. Their size, diversity, and distance from capitals make effective governance from afar a difficult proposition, and yet decentralized approaches have proven ineffective because of the low capacity for planning and management at the local level. Small urban centres certainly have some distinct advantages over larger cities, especially a "flatter" governance structure that puts local leaders in closer contact with their constituents and civil servants (UN-Habitat 2006). Still, lack of local capacity precludes a purely decentralized or community-based model; such efforts would quickly find the limits of local actors and institutions.

The Darjeeling District is illustrative of this argument; Kalimpong, Kurseong, and Darjeeling lack the resources and capacity to mitigate the environmental risks associated with their rapid growth or to manage the resources necessary to mitigate risk, and thus are not well suited to bottom-up strategies, yet are too "distant" from centres of power and scholarship to be well served by top-down approaches.

The key may be to find a disaster management policy framework and funding structure that recognizes the uniqueness of smaller places, values local knowledge, builds local capacity, but recognizes the limitations of smaller places and works to fill the gaps. This is easier said than done, of course. One thing is certain—over the next 30 years, India's urban population will grow by over 300 million people, with much of that growth happening in small, hazard-prone cities. Looking forward, we need to take seriously the growth of small cities and what they mean for disasters, in India and elsewhere in Asia.

References

Bell, D., & Jayne, M. (2009). Small cities? Towards a research agenda. *International Journal of Urban and Regional Research*, 33(3), 683–689.

Bernier, R. M. (1997). *Himalayan architecture*. Madison: Farleigh Dickinson University Press.

Cross, J. (2001). Megacities and small towns: Different perspectives on hazard vulnerability. *Global Environmental Change*, 3(2), 63–80.

Cutter, S., Emrich, C. T., Webb, J. J., Morath, D. (2009). *Social vulnerability to climate variability hazards: A review of the literature. Final report to Oxfam America*. Columbia: Hazards & Vulnerability Institute, University of South Carolina.

Davis, M. (2004). Planet of slums. *New Left Review*, 26. http://newleftreview.org/II/26/mike-davis-planet-of-slums. Accessed 20 June 2014.

Davis, M. (2006). *Planet of slums*. New York: Verso.

Douglass, M. (2013). *The urban transition of environmental disaster governance in Asia* (Asia Research Institute Working Paper Series, No. 210). Singapore: Asia Research Institute.

Froehlich, W., & Starkel, L. (1993). The effects of deforestation on slope and channel evolution in the tectonically active Darjeeling Himalaya. *Earth Surface Processes and Landforms*, 18(3), 285–290.

Government of India (2001). *Census of India*. http://www.censusindia.net/. Accessed 13 July 2013.

Graham, S. (2010). Disruption by design: Urban infrastructure and political violence. In S. Graham (Ed.), *Disrupted cities: When infrastructure fails* (pp. 111–130). New York: Routledge.

Hardoy, J. E., & Satterthwaite, D. (1991). Environmental problems of Third World cities: A global issue ignored? *Public Administration and Development*, 11, 341–361.

India Urban Conference. (2011). *Urban India 2011: Evidence*. https://www.citiesalliance.org/sites/citiesalliance.org/files/IUC%20Booklet%20on%20Indian%20cities.pdf. Accessed 22 Oct 2013.

Intergovernmental Panel on Climate Change (IPCC). (2012). *Managing the risks of extreme events and disasters to advance climate change adaptation* (Special report of the Intergovernmental Panel on Climate Change). Cambridge: Cambridge University Press.

Ives, J. D. (1993). Mapping of mountain hazards in Nepal. In N. J. R. Allan, G. W. Knapp, & C. Stadel (Eds.), *Human impact on mountains* (pp. 154–164). London: Rowman & Littlefield Publishers.

Kampa, U., Growley, B. J., Khattak, G. A., & Owen, L. A. (2008). GIS-landslide susceptibility mapping for the 2005 Kashmir earthquake region. *Geomorphology*, 101(4), 631–642.

Mitchell, J. K. (1999). *Crucibles of hazard: Mega-cities and disasters in transition*. Tokyo: United Nations University Press.

Montgomery, M. (2008). The demography of the urban transition: What we know and don't know. In G. Martine, G. McGranahan, M. Montgomery, & R. Fernandez-Castilla (Eds.), *The new global frontier: Urbanization, poverty and the environment in the 21st century* (pp. 17–36). London: Earthscan.

Mukul, M. (2011, October 4). The earth shook and all hell broke loose. *The Bengal Post*, p. 8.

O'Rourke, T. D. (2007). Critical infrastructure, interdependencies, and resilience. *The Bridge*, 37(1), 22–29.

Oliver-Smith, A. (2010). Haiti and the historical construction of disasters. *NACLA Report on the Americas*, 43(5), 32–36.

Olshansky, R. B. (2009). The challenges of planning for post-disaster recovery. In U. F. Paleo (Ed.), *Building safer settlements. Governance, planning, and responses to natural hazards* (NATO Science for Peace and Security Series, Vol. 58, pp. 175–181). Amsterdam: IOS Press.

Pelling, M. (2003). *The vulnerability of cities: Natural disasters and social resilience*. London: Earthscan Publications.

Pelling, M. (2012). Hazards, risk and urbanisation. In B. Wisner, Gaillard, J.C., & Kelman, I. (Eds.), *Routledge handbook of hazards and disaster risk reduction* (pp. 145–155). New York: Routledge.

Pelling, M., & Wisner, B. (Eds.) (2008a). *Disaster risk reduction: Cases from urban Africa*. New York: Routledge.

Pelling, M., & Wisner, B. (2008b). Urbanization, human security and disaster risk in Africa. In M. Pelling & B. Wisner (Eds.), *Disaster risk reduction: Cases from urban Africa* (pp. 3–16). New York: Routledge.

Petley, D. (2012). Global patterns of loss of life from landslides. *Geology*, 40(10), 927–930.

Rahman, A., Khan, A. N., Collins, A. E., & Qazi, F. (2011). Causes and extent of environmental impacts of landslide hazard in the Himalayan region: A case study of Murree, Pakistan. *Natural Hazards*, 57(2), 413–434.

Robinson, J. (2002). Global and world cities: A view from off the map. *International Journal of Urban and Regional Research*, 26(3), 531–554.

Rockefeller Foundation. (2013). *100 resilient cities centennial challenge*. http://100resilientcities.rockefellerfoundation.org/. Accessed 1 Nov 2013.

Rumbach, A. (2014a). Do new towns increase disaster risk? Evidence from Kolkata, India. *Habitat International*, 43, 117–124.

Rumbach, A. (2014b, in press). Editor's introduction: Cities at risk. In F. Miraftab & N. Kudva (Eds.), *Cities of the global south reader*. London/New York: Routledge.

Sankrityayana, J. (2009, November 20). *Land use patterns and landslide risks in Darjeeling Hills*. Paper presented at the Save the Hills—2nd Landslide Hazard Workshop, Kalimpong, West Bengal, India.

Shaw, R. (2008). An overview of urban risk of South Asia: Issues, approaches and thoughts. *Journal of South Asia Disaster Studies*, 1(1), 69–84.

Sinha, R. (2011, September 25). After the quake. *The Bengal Post*, p. 16.

Swiss Re (2013). *Mind the risk: A global ranking of cities under threat from natural disasters*. http://media.swissre.com/documents/Swiss_Re_Mind_the_risk.pdf. Accessed 22 Nov 2013.

Tierney, K. (2012). Disaster governance: Social, political, and economic dimensions. *Annual Review of Environment and Resources*, 37, 341–363.

UN-Habitat. (2006). *Meeting development goals in small urban centres: Water and sanitation in the world's cities*. London: Earthscan.

United Nations (UN). (2012). *World urbanization prospects: The 2011 revision*. http://esa.un.org/unup/. Accessed 21 July 2013.

United Nations International Strategy for Disaster Reduction (UN-ISDR). (2012). *How to make cities more resilient: A handbook for local government leaders*. Geneva: United Nations.

Wamsler, C. (2004). Managing urban risk: Perceptions of housing and planning as a tool for reducing disaster risk. *Global Built Environment Review*, 4(2), 11–28.

Wisner, B., P. Blaikie, T. Cannon, & Davis, I. (2003). *At risk: Natural hazards, people's vulnerability and disasters*. New York: Routledge.

World Bank. (2011). *Climate change, disaster risk and the urban poor*. Washington: World Bank.

World Bank (2013). *Urbanization beyond municipal boundaries: Nurturing metropolitan economies and connecting peri-urban areas in India*. Washington: The World Bank.

Chapter 7
'Good' Time for Disaster: The Importance of Temporality in Governance Thinking

Zuzana Hrdličková

Abstract In this chapter, I look at why certain Indian states have well-functioning disaster preparedness, whereas others do not. I analyse the 'success story' of disaster governance in the Indian Union Territory of Puducherry. I look at how two recent disasters—the Indian Ocean Tsunami (2004) and cyclone Thane (2011)—shaped its priorities and approaches to disaster risk and facilitated the development of various tools of disaster preparedness among state as well as non-state actors. Furthermore, the chapter explores the opportunities that were presented through these events. Was it a matter of the 'good' timing of disaster? To answer this question, I explore the proliferation of technologies as well as the developments of the various tools of disaster preparedness that have been implemented on the ground. How is Puducherry different from other Indian states, which have been much slower in implementing new disaster management policies prescribed by New Delhi? In search of the answer, I take a closer look at the common and widely shared narrative about the reactive Indian population. The findings are based on my field research in 2012 and 2013, as well as the analysis on anthropology of disasters and science and technology studies.

Keywords Puducherry, India • Disaster governance • Tsunami • Cyclone • Empowerment • Social inclusion

7.1 Introduction

As numerous statistics and publications including the chapters in this volume show, disasters disrupt life and pose multiplicity of challenges to human society. On the other hand, some scholars (Carr 1932; Gunewardena and Schuller 2008; Homer-Dixon 2006; Prince 1920) have noted that catastrophic events also offer a range of opportunities for change. For example, following an earthquake, town layouts can be redesigned, quality of construction can be improved and additional infrastructure

Z. Hrdličková (✉)
Department of Sociology of Goldsmiths College, University of London, London, UK
e-mail: zuzahrd@gmail.com

M.A. Miller, M. Douglass (eds.), *Disaster Governance in Urbanising Asia*,
DOI 10.1007/978-981-287-649-2_7

can be developed. However, the window of opportunity is not eternal. It is limited by notions of priorities, which are undoubtedly connected to temporality. Certain time periods are characterised by varied perceptions of what needs to be done. In the liminal period of post-disaster recovery and rehabilitation, human societies become more perceptive to suggestions of change. Disasters thus seem to have the ability to generate political will, funds and dedication to allow alterations which had been unthinkable before. However, this momentum can fade away with time.

In this chapter, I emphasise that good timing is key to successful introduction of policies. I look at the case of Indian federal disaster management policies and their problematic implementation on the ground. I interpret the 'success story' of disaster governance in the Indian Union Territory of Puducherry, and I show how two recent disasters— the Indian Ocean Tsunami (2004) and cyclone Thane (2011) —shaped its priorities and approaches to disaster risk and facilitated development of the various tools of disaster preparedness among state as well as non-state actors. In order to make sense of the diversity of ways in which administrations of various Indian states implement (or fail to implement) disaster governance policies, I explore the 'reactive Indian' narrative that is common among disaster experts. The lens of temporality and time-bound priorities prove to be useful in the attempt to understand the paradoxes of public policies, disaster governance and narratives of epistemic communities of expert disaster practitioners. My findings are based on my field research in 2012 and 2013, and my analysis draws on anthropology of disasters as well as science and technology studies.

7.2 Disaster Governance in the Federal Republic of India

The Federal Parliamentary Constitutional Republic of India, the seventh largest country in the world, is vulnerable to many natural hazards. The Indian Vulnerability Atlas (published since the late 1990s) makes it clear that *all* of the 28 Indian states and 7 union territories face varied levels of natural hazards. Most areas are subjected to floods. Although earthquakes are also present all over the country, the mountainous regions are at high risk. According to seismologists, a powerful seismic event in the Himalayan region is long overdue (Revi 2010). Mountains are also prone to dangerous landslides. Coastal areas not only experience regular cyclones but are vulnerable to tsunamis too. Furthermore, inland regions are prone to droughts. The horror of recent natural catastrophes like the Latur earthquake (1993), Orissa supercyclone (1999), Gujarat earthquake (2001), Indian Ocean Tsunami (2004), the Kashmir earthquake (2005) and the recent Kedarnath flood (2013) is still in the living memories of contemporary Indians. The country has also experienced a number of technological and industrial disasters. Among these, the Bhopal gas leak (1984) is perhaps the most significant event. However, other 'more mundane disasters' have been affecting the Indian population in our time. According to the EM-DAT database, in the last 30 years alone, 651 million people in the country have been affected by drought, 326 million by floods, 61 million by tropical cyclones, 28 million by seismic activities, 11 million by storm surges and 96,000 by epidemics and hundreds of thousands by technological disasters (EM-DAT n.d.). Hence, it is no wonder that improving disaster governance has become one of the priorities of India.

The way the Indian administration has been dealing with disasters has been gradually evolving according to how disasters have been understood in relation to governance structures (Sinha et al. 2010). Until the 1980s the mode was mainly reactive, and individual disasters were within the competence of different departments. Government offices of the individual states and territories each had different guidelines for disaster response.

In the 1990s, India joined the United Nations' International Decade for Natural Disaster Reduction (IDNDR, 1990–1999), and this led to the country adopting a much more holistic structure to tackle the problem of disasters. The initiative was aimed to reduce the loss of life, property damage and social and economic disruption caused by natural disasters especially in developing countries. The 1990s was thus marked by much research and data collection in the global South. The participating countries were to improve accessibility to warning systems, introduce disaster mitigation plans and mainstream assessed natural hazards into development plans. Most importantly, the IDNDR has led to a significant shift in the way disasters are *understood*—from events that could only be responded to after their occurrence to events for which the impact could be mitigated through adequate preparedness. The transformed conceptualisation of disasters has had large-scale consequences: it has rendered disasters to be manageable and governable, leading to the establishment of institutions specifically working on disaster management in developing countries including India (Ministry of Home Affairs 2011; Sharma 2001; Sinha 1999). As mentioned by Mike Douglass in Chap. 2 of this volume, one could argue that disasters were discovered as a platform for governance involving a society-wide engagement with disasters, including governmental as well as non-governmental institutions.

While non-governmental organisations have always been involved in disaster-related work, in India, the process of holistic governing of disasters by the state began in 1999 with the establishment of a High Powered Committee on Disaster Management. This resulted in the passing of the Indian Disaster Management Act (DMA) in 2005. Subsequently, a number of state institutions have been created such as the National Disaster Management Authority (NDMA), National Institute of Disaster Management (NIDM) and the National Disaster Response Force (NDRF). The Act introduces a number of novel concepts and suggestions. It defines disasters as events that overwhelm the coping capacity of a community, and it promotes the building of disaster resilience through preparedness at all levels—from village to the state. The main message of the Act is that everyone has to plan for disaster and everyone should be involved in such planning. The Act suggests that Disaster Management Authorities will be set up on district and state levels. Furthermore, it suggests that disaster management plans will be put in place and that they will be created in a participatory manner—involving all stakeholders, including vulnerable segments of the population and non-state actors. The Act and guidelines associated with it thus provide quite a detailed step-by-step manual on how to build disaster preparedness in each state.

Nonetheless, it is up to the individual states to decide how to implement the Act. Hence, the formation of institutions in different states and union territories happens at different paces and with varying levels of enthusiasm. In some places, the State Disaster Management Authorities (SDMAs) function very well, while in others they are barely set up or do not exist at all. For example, the recent disastrous floods in the pilgrimage town of Kedarnath, Uttarakhand State (July 2013), have highlighted the issue of 'paper SDMAs'. Although the Uttarakhand State Disaster Management Authority has existed on paper since 2007, its executive committee has not met even once in 6 years (Tripathi 2013). According to my informants, Uttarakhand is not the only 'paper SDMA' in India. On the other hand, in some states like Gujarat and Odisha (formerly known as Orissa), disaster governance is given much priority, and local institutions pride themselves for using some of the most advanced disaster preparedness available, including both technical and management solutions. What then decides whether or not disaster risk reduction is viewed as a priority by the state and district administration? Why a catastrophic event is easily managed in one state, while it is a tragedy in another? The case of disaster governance in Puducherry will provide some answers for the aforementioned questions.

7.3 Governing Disasters in Puducherry: Time-Bound Priorities

The Union Territory of Puducherry is interesting and unique for several reasons, including its history, size and administration. Previously called Pondicherry (until 2006), it covers the area of former French India. It has cosmopolitan character, with inhabitants from various parts of India and abroad, speaking Tamil, English, Malayalam, Telugu, French, Hindi and other languages. Its tiny 492 km^2 is divided into four geographically separate coastal regions in South India: the two biggest—Puducherry and Karaikal—are both surrounded by the Indian state of Tamil Nadu; the Mahe region is neighbouring with Kerala; and Yanam is surrounded by Andhra Pradesh. Puducherry is a union territory (formed in 1963) with special privileges. As such, its administration falls within the jurisdiction of the central government in New Delhi. However, due to special constitutional amendments, Puducherry has its own cabinet of ministers and an elected legislative assembly, making it possible to pass legislation, which has to be ratified by the Centre. Administratively, the Union Territory is divided into two districts—Karaikal and Puducherry—which is further divided into three administrative regions: Puducherry, Mahe and Yanam.[1] In 2011, its total population was 1,244,464 with more than half of the Union Territory's inhabitants (654,392) living in Puducherry town (DRDMP 2013). Therefore, it can

[1] This chapter is not concerned with the latter two. It is based on research in Puducherry district administration covering mainly Puducherry town and surrounding settlements.

be argued that disaster governance in Puducherry is relatively equally divided between urban and rural. Similar to other Indian towns and cities, the medium-sized Puducherry town suffers extensive levels of environmental pollution due to industrial development and motor vehicular traffic (Krafft et al. 2004), as well as decreasing water table and serious water pollution. Some segments of the population live in slums with little access to basic facilities.[2] Among the rural population, agriculture and fishing are important sources of livelihood.

As a tropical coastal region, Puducherry is prone to heavy rain, floods, cyclones and to a smaller extent earthquakes. In recent history, it has experienced two major, yet completely different, catastrophic events: the Indian Ocean Tsunami in December 2004 and the severe cyclone Thane in December 2011. These disasters were distinct in terms of the scale of their impact on people's lives and living standards as well as in terms of relationships with governance structures. Crucially, they were experienced in two different administrative regimes in Puducherry: one *before* and one *after* the implementation of the Disaster Management Act.[3]

7.3.1 The Indian Ocean Tsunami

The most striking aspect of the Indian Ocean Tsunami which hit the coast on Sunday morning, 26 December 2004, was its unexpectedness. Nobody saw it coming. Although India has had some historical experience with tsunamis, it is fair to say that the living memory of contemporary population of the coast was oblivious to this hazard. It affected India badly, with heavy losses of life reported from the Andaman and Nicobar Islands, Andhra Pradesh, Kerala, Tamil Nadu and Pondicherry. The state of Tamil Nadu suffered heavy casualties with nearly 8,000 being dead,[4] several hundreds being missing and tens of thousands being homeless.[5] The Union Territory of Pondicherry too was affected, with 601 being dead (DRDMP 2013) and 30,000 being homeless (mapsofindia1 2011). However, the vast majority of deaths (over 450) appeared in one of its enclaves—the urban region of Karaikal (about 129 km away from Pondicherry town)—while the Pondicherry district lost over 100 people. Although the tsunami affected the inland population too (through affecting soil quality), it was mostly the coastal communities (17 villages in Karaikal

[2] According to estimates from the early 2000s, around a fifth of urban population in India was not covered by a sewer system: 26 % had no access to toilet, 31 % had no access to tap water and 14 % had no access to electricity (Bandyopadhyay 2013, p. 38).

[3] Note also that, at the time of the tsunami, the territory bore its old name—Pondicherry. Hence, I have used 'Puducherry' for the period after 2006 and 'Pondicherry' for the period before 2006.

[4] Exact figures vary depending on sources, with numbers of death being between 7,793 (mapsofindia1 2011) and 7,997 (Guleria 2009, p. 147).

[5] Exact figures vary depending on sources, with numbers of homeless ranging from 44,000 (mapsofindia1 2011) to 300,000 as having lost lives and property (Guleria 2009, p. 147) and 490,000 families being placed in sheltered camps (Guleria 2009, p. 147).

and 16 in Pondicherry districts), namely, fishermen, that bore the brunt of the disaster (Tsunami, n.d.). The city itself was affected less than the rural areas.

Upon realising the scale of the disaster, the government of Puducherry opened control rooms, and the Revenue Departments of Pondicherry and Karaikal were commissioned to organise search, recovery and relief measures through District Collectors with the help of police, rescue and fire departments. The administration was faced with multiple challenges including communication, coordination and management issues with over 40 relief camps set up in the Union Territory, alone. They had to find creative ways within the boundaries of the law regarding prescribed procedures of dealing with dead bodies.[6] The process of paying out ex gratia payments to the victims' families was complex, including lengthy recovery of bodies, identification, finding the right heirs, etc. Participatory governance was key to successful navigation and resolution of these issues with local village *panchayat* (municipal authority) leaders playing crucial roles.

The nature of the tsunami was so unexpected, and its impact was so vast and shocking that it generated much attention and unprecedented amounts of funding[7] from government and private donors in India and around the world:

> The financial response to the tsunami, the media and private response that gave rise to it, was quantitatively and qualitatively exceptional: 40 per cent (US$5.5 billion) of international resources for the tsunami came from the general public. The usual figure is nearer 15 per cent. It was the private response that meant that the international response was, for once, sufficient (together with substantial local resources) to cover both relief and reconstruction adequately. And it was the private response that made NGOs and the Red Cross Movement such important (and numerous) actors. (Flint and Goyder 2006, p. 8)

Though the disaster brought much suffering, it became an opportunity for development of coastal areas and human resources. Moreover, natural disasters became extremely present in the minds of the general public, non-state actors and administrators, with preparedness being seen as a priority. Therefore, it can be argued that the tsunami enabled new disaster governance to be formed.[8] The

[6] For example, according to the law, a post-mortem had to be carried out on those who died unnatural deaths. But this would be impossible with the sheer number of bodies. Hence, the act of post-mortem was substituted with the act of identification. Another concern was when bodies could not be found and the person was proclaimed missing and their families missed out on compensation payments.

[7] The exact amount of funding is unclear and varies across sources. For example, the Asian Development Bank donated a $205 million loan and grant assistance package, which was aimed at restoring livelihoods and rehabilitating and reconstructing damaged infrastructure in Tamil Nadu and Kerala (Indian Ocean Tsunami Response 2014).

[8] It would be a mistake to assume that Pondicherry did not have any form of disaster preparedness before the Disaster Management Act was passed in 2005. For example, some villages had communication infrastructure built by large with the assistance of NGOs. In Veerampattinam, one of Puducherry's coastal villages with a population of 6,200 people, hundreds of lives have been saved from the impact of the tsunami by the public address system (Subramanian 2005). Cyclone shelters built by the Red Cross in the 1980s dotted the coast, and some of them had been used as refuge (McKerrow 2010; see also Deville et al. 2014), and the elevated seafront protected the city from the wave.

aftermath of the tsunami saw unprecedented multiplicity of efforts undertaken by both the Puducherry state and non-governmental organisations (NGOs) which played an important role in recovery and disaster resilience building.

The post-tsunami period saw massive increase in funding of non-governmental organisations, which were experiencing a kind of unusual bonanza which lasted for several years. International NGOs which received funding from their home countries would mostly operate through locally established organisations, at first providing emergency relief and later shifting to livelihood recovery. The coordination of humanitarian and development activities was often extremely challenging (c.f. Regnier et al. 2008). Proportion of the post-tsunami funding was dedicated to building of disaster preparedness in material as well as in soft forms. This led to construction of more multi-hazard-resistant structures including housing and shelters as well as capacity building of local population. In addition, various agencies, including the UN and other NGOs, trained local communities and officers in disaster preparedness. Moreover, the coastal population benefitted from other courses aimed at more general welfare including nutrition and health. 'Everybody has been trained in everything in the coastal villages', one NGO worker told me. However, gradually towards the end of the 2000s, the post-tsunami funding ran out, and the number of local NGOs declined.

Arguably, the role of the state was crucial as it was overseeing the NGO activities through its system of regulations and permits. It was also active in post-tsunami recovery. The Puducherry Government set up the Project Implementation Agency (PIA), which was using the funds of the state and the World Bank to build new housing outside the coastal regulation zone. Officials of the Puducherry District Department of Revenue and Disaster Management perceived this disaster as an important catalyst that turned and fed governance priorities. In their view, disaster governance became a priority only after the Indian Ocean Tsunami, making it imperative for the Union Territory administration to follow the recommendations and guidelines of the DMA. The tsunami generated momentum of administrative will for action, and most importantly it made huge funds accessible. Therefore, in 2005, Pondicherry Union Territory started thorough implementation of the DMA. This included the formation of its State Disaster Management Authority, renaming the Collector's office as the Department of Revenue and Disaster Management (DRDM), and gradual estab lishment of the new crisis management framework.[9] Since 2009, there have appeared Disaster Management Action Plans, which detail departmental responsibilities and modes of cooperation during emergencies.[10] Furthermore, in 2011, the State Emergency Operations Centre (SEOC)

[9]The Relief and Rehabilitation Commissioner (RRC) is the incident commander for the entire Union Territory, and District Collectors are incident commanders for their respective regions (Puducherry and Karaikal). In addition, Regional Administrators in Mahe and Yanam (who are also sub-/deputy collectors) are incident commanders for their regions. The DRDM has three emergency support functions within the crisis management framework: (1) emergency public information, helpline and warning, (2) evacuation and (3) damage assessment.

[10]The plan effectively gives people within the public administration a clear idea on what to do and whom to contact in a crisis. As such the plan includes a large telephone directory of mobile

started its service too.[11] This hi-tech offsite facility is a source of much pride. During real-life events, the SEOC becomes the decision-making centre, where information is collected, analysed and disseminated. Through situation reports and information provided by teams on the ground and by the public (via the toll-free number 1077), the SEOC monitors arising situations. Emergency support officials who are present at the centre are allowed to make on-the-spot decisions so as to speed up the mobilisation of the department's resources while avoiding parallel or contradictory instructions.

On a daily basis, SEOC staff on duty constantly monitors information provided by global and Indian institutions (India Meteorological Department, Indian National Centre for Ocean Information Services) for potential threats, including storms and tsunamis. In case of a credible warning, they use the Puducherry Decision Support System (PDSS) modelling software,[12] which, depending on the type and magnitude of the expected event, can predict the location and extent of an impact, as well as generate information about the availability of shelters. Although the PDSS modelling software provides only estimates, it provides the administrators a rough idea about likely impact and required actions.[13] The system sends warning text messages to the mobile phones of key community people, emergency response teams and other relevant stakeholders. It is not without interest that the SEOC relies heavily on the ability to send warning text messages to cell phones. This technology became widely available only in the second half of the 2000s. In 2005, only 8 % of the entire Indian population had mobile phones, whereas in 2010 it increased to 70 % (Agar 2013, p. 102–107). By 2012, India had over 900 million telephone subscribers; and 96 % of them had mobile phones (Jeffrey and Doron 2013, p. 3). The successful proliferation of cell phones is attributed to extremely cheap handsets and service rates as well as its overall appeal and usefulness even for people with minimal literacy.[14] Given the mass expansion of mobile phones, the state and non-state actors find it crucial in disaster governance.

The people within the administration prepare for disasters through capacity building courses and mock-drills. For example, in November 2011, there was a

---

numbers. It is arguably this part that needs to be updated the most, as transfers are very common and people populating the administrative apparatus move *with* their phone numbers.

[11] The SEOC is a result of joint efforts of global, national and local disaster policies. It was jointly funded by the UNDP and the Puducherry territorial administration following the guidelines and design of the National Disaster Management Authority (Nair 2011).

[12] It is based on spatial database platform, and it was developed by Risk Mapping Software International (Delhi).

[13] For example, as a result of a test simulation of a cyclone, the PDSS generated a report of 102 pages, detailing the whereabouts of 83 safe shelters (lifeline buildings including schools, temples, churches, community halls and purpose-built shelters).

[14] Mobiles in India are popular for multiple reasons including socialisation, business and access to information. Given that it is a relatively simple device, even people with minimal literacy find it easy to use (Jeffrey and Doron 2013).

mock-drill[15] for the administration to test-run the standard operating procedures (SOP) for a case of tsunami. While emergency response teams—formed by members of various departments—were practising mutual cooperation in the whole cycle of the simulated disaster, top officials based in the SEOC aimed to avoid panic by carefully translating the information into commands.[16]

Clear decision-making systems and technology would be able to save more lives. Most importantly, disaster governance tools such as mock-drills, action plans and technology are able to decompose disasters into smaller more manageable units. Action plans make the command structure clear. The SEOC enables the administration to take single command and communicate warnings to the public simultaneously. The importance of good timing for successful implementation of such disaster governance structure cannot be overemphasised. The funding generated by the tsunami, the new policies and proliferation of technology made this framework relevant and possible.

Coastal communities, in particular, benefitted immensely from activities undertaken during the post-tsunami period. They got new modern housing,[17] they were supported to restore their livelihoods and their capacities were built through a variety of training, including disaster preparedness. Community members thus benefitted from the assemblage of state and non-state actors' activities, often without being aware of which actor exactly provided what (Hastrup 2011; interviews December 2013). Take, for instance, the fishing village of Veerampattinam, a few kilometres from Puducherry town, where the local NGO-run[18] Village Knowledge Centre (VKC) plays a central role in information dissemination and facilitating contacts between the village and the outside world. In networking with local officials, its public address system helped to save lives during the tsunami. In its aftermath, the village experienced a major boom—having new housing and modern cyclone shelter (community hall) built by the government-led Project Implementation Agency (PIA) and receiving assistance and training by a range of organisations. The village is prepared for disasters in multiple ways, and people can be warned of impending

[15] The scenario was that there was an earthquake in one of the South East Asian countries and that a tsunami was hurtling towards India. All coastal states and territories were put on first alert while receiving information from the sensors in the Bay of Bengal about the possible height and direction of the approaching wave. Based on this data and past experience, the modelling software kept providing an estimated picture of the impact on the Indian coast. The SEOC received messages every 10–15 min about the emerging situation.

[16] An interviewee in Puducherry (December 2013) said 'we were asked to successfully send messages down and what time it was taking. That time was to be noted and reviewed after the exercise. You know, easily you can say 'I sent a message to the collector in Karaikal.' But what time did it reach him? We had to get confirmation of the message receipt. I sent a fax, he said 'I sent an SMS message'; 'I sent an e- mail.' Only when he sees the e-mail he can respond. Fax message would have gone lying in the fax machine and no one would have taken it to the person who can take a decision. So, it's when he (Collector of Karaikal) was aware of the communication that was also in that drill for our learning system'.

[17] Around 7,500 *pukka* houses in Puducherry Union Territory were built for fishermen further inland. Yet they could retain the use of their old houses as work sheds.

[18] MS Swaminathan Foundation.

disaster through a variety of channels involving both state and non-state actors. Besides media, the people can learn about it through the announcement of the VKC, which receives information from the state-run Indian National Centre for Ocean Information Services (INCOIS) or from the elected *panchayat* leaders who receive text messages from the state-run SEOC. This example clearly suggests that collaboration of multiple stakeholders has the ability to improve disaster preparedness of communities.

With the benefit of some years' hindsight, the immediate years in the post-tsunami period seemed like a time full of frenzy (not only) in Puducherry, with disaster recovery activities taking place everywhere. The post-tsunami period saw much improvement in the lives of coastal communities, growing interest of the state in disaster management and high levels of NGO activity. However, towards the end of the 2000s, the outcomes varied for NGOs and the state. While the state-based disaster preparedness infrastructure and capacity have grown considerably, the NGOs gradually ran out of tsunami funding, and their activities gradually came to a halt. Could another disaster revive NGOs similar to the post-tsunami?

7.3.2 Thane

In December 2011, satellites showed a tropical disturbance to the west of Indonesia. It moved in northwest direction across the Bay of Bengal, developing into a depression (25th December) and later being declared as Cyclonic Storm Thane (26th December). The following day, its development slowed down, and on 28th December it became a very severe cyclonic storm, before reaching the landmass of India on 30 December 2011.

For days prior to its landfall, cyclone Thane had been monitored by the India Meteorological Department, which predicted the areas it may impact to be the coast of Andhra Pradesh, Tamil Nadu and Puducherry. Multiple types of warnings were issued through extensive media coverage as well as official channels.[19] Gradually, it became clear that the cyclone would make a landfall in northern Tamil Nadu and Puducherry. 'Great danger' signals were hoisted in numerous ports (Punithavathi et al. 2012), and fishermen were advised not to venture into the sea. People were told to stay in safe structures as there was danger of extensive damage to huts and houses with thatched roofs. They were also warned about possible fallen trees and electric cable disruptions. It was expected that tidal waves of over 1 m could inundate low-lying areas.

As the cyclone was approaching, the Puducherry district activated its various emergency support function teams (the multi-departmental teams defined by the

[19] 'The Area Cyclone Warning Centre, Chennai of IMD issued various user specific bulletins including fishermen warning, port warning, coastal weather bulletin, sea area bulletin, four stage warning for state and districts disaster management officials' (India Meteorological Department 2012).

Disaster Management Action Plan). The electricity department cut off the electric power supply to the town and villages in the evening before the landfall[20] to prevent electrocution. The Department of Revenue and Disaster Management issued warnings through press releases, radio, cable network and public announcement systems telling the public what to expect and how to protect themselves.[21] Further preparations were made, while search and rescue, evacuation and transport teams were on standby. Several teams of the National Disaster Response Force were deployed to the coast, and people were moved to safer areas and buildings including shelters. For example, in the previously mentioned coastal village of Veerampattinam, the people were warned through the media and public announcements by the District Collector[22] and by the Village Knowledge Centre. Villagers took shelter in the local Hindu temple, some stayed in their *pukka* (concrete) houses built in the post-tsunami period and some families used the multi-hazard-resistant shelter built by the PIA. The villagers' preparedness thus drew on the multiple resources provided by state and non-state actors.

Cyclone Thane made landfall after 6.30 am on 30 December 2011 with a wind speed of 120–140 km/h, affecting mainly Puducherry and Cuddalore (district of Tamil Nadu State). Other affected areas included Villupuram, Kanchipuram, Thiruvallur, Chennai, Nagapattinam, Thiruvarur and Thanjavur. During the landfall of the cyclone, the Puducherry SEOC room was manned by around 15 people. They received and recorded about 2,000 phone calls mainly by people trapped on the roads due to fallen trees, poles and wires. The SEOC passed the messages on to relevant departments so that their teams on the ground could assist the callers. Although the electricity was cut off, communication was disrupted only occasionally.

The damage caused by the cyclone was extensive affecting infrastructure and livelihoods of large segments of the population both on the coast and inland. Twelve people in Puducherry lost their lives, and the aftermath of the cyclone saw uprooted trees[23] and blocked roads, destroyed crops,[24] killed livestock, damaged houses[25] with missing roofs and fallen electrical posts, cables and transformers.[26] Life in urban areas was disrupted for as long as the trees and other debris were on the streets, affecting mobility of people and thus services and businesses. Therefore, in the immediate days after Thane, chainsaws were the main weapon of the Public

[20] December 29 at 9.30 pm.

[21] '[We informed the public] what are the do's and don'ts, and to put latches on the doors and stay inside' (Interview with DRDM official, July 2012).

[22] The head of the Department of Revenue and Disaster Management.

[23] In Puducherry town alone, 75 % of trees were damaged (Subramanian 2012).

[24] Around 17,000 ha of paddy is affected in Puducherry and Karaikal (Subramanian 2012).

[25] The cyclone fully or partially damaged about 350,000 huts and tiled houses in Cuddalore and Villupuram districts.

[26] 'In Cuddalore district alone, paddy on 58,000 ha, cashew on 23,500 ha, sugarcane on 5,752 ha and banana, pulses, oilseeds and cotton cultivated on 3,600 ha were damaged. About 27 high-tension electricity pylons, 4,500 distribution transformers, 36,000 electricity poles and cables running to 350 km were damaged,....' (Subramanian 2012).

Works Department as well as of the National Disaster Response Force teams. The next challenge was to restore electric power supply. In some areas, it took weeks, as majority of the cables were above the ground. However, Thane's effects were most severely felt by rural areas inland, which had neither the infrastructure nor the human capacity to prepare for the disaster. Moreover, they depended on agriculture for living. The cyclone damaged crops and thus affected the entire economy. In the countryside, it took several months for life to come back to normal.

The cyclone exposed the asymmetry of the post-tsunami disaster resilience building which had focused on the coast, but minimum activities had been undertaken inland. Having seen the amount of resources targeting the coast after the tsunami, it is no wonder that the inland communities had sense of injustice after Thane, when funding allocated to the rehabilitation came nowhere near the post-tsunami period. While non-state and state actors were fairly equal partners in the post-tsunami period, the rehabilitation after Thane was by large in the hands of local authorities,[27] and the profile of local NGOs has not expanded massively.

However, the cyclone has also shown that building disaster resilience pays off. According to my informants, had it not been for tsunami and all the subsequent efforts that have gone into building disaster preparedness in Puducherry, Thane would have claimed many more lives. They implied that governing structure and priorities were the decisive factors for the successful management of the event. The officials felt that the new disaster management structure and associated activities made them truly prepared and competent in dealing with the situation. They especially valued clarity of knowing what is to be done and by whom. This was provided and enabled by the incident command system, action plans and the newly built SEOC. Cyclone Thane was SEOC's first real-life disaster encounter, and it made their work efficient as it prevented parallel contradicting decision-making. Moreover, the cooperation of cross-departmental teams was fairly smooth, thanks to already existing personal relationships between various officials. This familiarity with people and process was by large mediated through the mock-drill which had taken place just a month before the cyclone:

> All these teams were formed and the teams gelled because of the mock drill. I know most of the members physically; one-to-one I know most of them. So if there were any issues during Thane I would call them directly. They knew who was calling and they responded… (Interview, DRDM official, Puducherry, July 2012)

The cyclone impacted on disaster governance in Puducherry: it rendered all the new disaster preparedness tools implemented after the tsunami to be useful. Thane came at a 'good time'—it convinced the government of Puducherry that they were on the right track and that the disaster management structure should grow further. Thus, recruitment of new officers for exclusive disaster management roles was speedily approved,[28] and the process of building up Puducherry's own State Disaster

[27] The Tamil Nadu Government pledged an 8.5 billion rupee package to provide relief to those affected by the cyclone in various districts in Tamil Nadu (Subramanian 2012).

[28] 'It happened after Thane only. The severity and you know … they [the Puducherry government] understood the importance of that and they understood that some work has gone into it. And this

Response Force commenced. The cyclone also showed that the informed public can cope better. The NGO sector has long been involved in building people's capacities in multiple directions. The cyclone has thus shown the benefits of raising disaster preparedness through activities of multiple stakeholders. However, this has not translated into increased funding for NGOs.

Given the positive outcomes of elaborate state disaster preparedness, one may ask: why are there so many Indian state governments which straddle behind Puducherry in implementing the Disaster Management Act? My research suggests the answer is to be found in how administrations define their governing priorities. This entails time-bound priority thinking, ideas concerning responsibilities of the state and attitudes of the general public as well as people populating administrative structures.

7.4 Reactive Indians and 'Good' Time for Disaster

> Here, in India, it is more of an attitude 'you let it come, I will face it'. The culture of prevention is not there. We will react, let it come. (Interview, Puducherry government official, December 2013)

> ...our mindset is reactive. A disaster happens, and immediately we react. A week later we are back to normal, we forget about it. We [Indians] believe in destiny. So, why safeguard myself by retrofitting this building? If earthquake happens and I have to die, I have to die. [...]So implementation of guidelines is not as strong, because mind-set at various levels, in government departments, in private sector, everywhere, the mind-set to give it [disaster preparedness] a priority is still not there. (Interview, NDMA official, New Delhi, March 2012)

As the quotes of Indian state officials above show, in the discourse of disaster management experts, Indian population, including those populating administrative structures, has particular culture of reactive, even fatalistic, risk thinking. 'People do not see the point in preparing for something that may or may not happen. They will react to it only when it comes. People believe in destiny'. This trope keeps reappearing throughout my interviews with officials in Puducherry, New Delhi as well as other places.However, some see the problem rather as lack of awareness and availability of options:

> I think this is utter nonsense, people will accept anything that is given to them... I think everyone would like an earthquake resistant house. It is totally unacceptable to say that people are fatalistic and all that. (Interview, NGO activist, Puducherry, October 2013)

This NGO activist perceives population as pragmatic. Perhaps it is true that in a certain sense of risk thinking, even superstitious individuals take precautions to prevent risks if they perceive these risks to be real and the precautionary measures

---

thing I'm going to tell, it's not boasting, [but compare] the impact in Cuddalore and Pondicherry... We had only 12 casualties...only now the government has approved new space office for disaster management...Posts are there, right, so we can recruit people now, we can employ people now. New people who will be in-charge of that work alone' (Interview, DRDM official, Puducherry, July 2012).

to be working. For example, given the widespread religiosity and dangerous traffic, it is a common sight in India to see professional drivers stopping by road shrines to pray to deities before driving on particularly dangerous sections of the road. Their prayers are their precautionary measures. Although these precautionary actions may be seen as superstitious, they show that the public are in fact willing to be protected from possible harm beforehand. Likewise, if one understands that there is serious possibility of harm and that there are means one could use to protect themselves, it is likely that they will take such precautionary actions.

Therefore, if one asks why the states have not been quick enough in implementing new disaster management structure in India, I would rather search for an answer within the realms of perceptions, limited resources and priority settings. As such, the ideas about the role of state administration and the imagery surrounding disasters (Guggenheim 2014) become very important. The Indian state on both federal and local levels has for decades been inclined towards strong welfare policies.[29] Innumerable schemes have been introduced throughout India to fight poverty, unemployment and homelessness. The state sees itself as a strong protector of the population. Further, according to the state's ideology, the administration is there to fight real, urgent problems. Therefore, implementing the suggestions of the Disaster Management Act in states with little living memory of major disasters seems like an unaffordable luxury to many administrators:

> It is just like asking for a cake when you don't have bread. You are asking him to eat the cake when he is not having the bread. It is a question of necessity and luxury. (Interview, Puducherry government official, December 2013)

They see other issues, such as poverty and unemployment, as more pressing. Disaster is not being imagined as a real possibility. The DMA with all its expensive and time-consuming work of creating disaster management action plans, setting up command structure and building of costly infrastructure seems unnecessary. In the view of the disaster expert community, usually real catastrophic events serve as eye openers for both the administration and the general public:

> Big lesson is when you lose something and that will be the motivation for you to be prepared. Apart from tsunami, in Pondicherry we have not had a major disaster. People are now much more aware compared to earlier. They were not serious about disaster management - prevention and preparedness, mitigation and all other concerns [of disaster management]. (Interview, Puducherry government official, December 2013)

When it strikes, disasters become a real possibility for which one should prepare. Moreover, it becomes a priority, as it causes acute damage to the population, temporarily more acute than poverty and unemployment. Hence, to fulfil its role of a protector, the state feels compelled to implement recommendations of the DMA. As such disaster preparedness is seen not as a luxury but as a top priority enabling the

---

[29] 'Here, it's a totalitarian welfare state' (Interview, Puducherry government official, December 2013).

continuation of the relationship between the protector and the protected.[30] This logic is illustrated by the examples of Odisha (formerly known as Orissa)[31] and Gujarat[32], both of which experienced major disasters and subsequently dedicated significant resources to build preparedness infrastructure. So, in areas which were often covered only by the non-governmental sector, the state has started being more present. Disasters begin to be planned for:

> I think tsunami had made a big change in the planning exercise as well as where the state is concerned. Normally, the state has a tendency of knee-jerk reaction. Whenever something happens they ask the central government or other groups for money and normally it is not channelised. There is no planning. But, I think things have changed after tsunami. Now, disasters are better managed and I think we appreciate the role of the state in this as well as of various organizations. (Interview, Puducherry NGO activist, October 2013)

7.5 Conclusion

This chapter has examined the case of disaster governance in the Indian Union Territory of Puducherry, which has experienced two recent disasters—the Indian Ocean Tsunami (2004) and cyclone Thane (2011). Both disasters have exposed different vulnerabilities of the territory, showing that the rural and urban poor get affected the most. However, both disasters have also become opportunities for positive developments. In the view of local administration, the experience of tsunami has prompted new attitude towards risk and has enabled disaster management to become a significant part of the local administrative structure. The tsunami came in the 'right' time for Puducherry administration to realise the importance of implementing the recommendations of the Disaster Management Act (2005). Moreover, the suffering caused by the disaster generated much sympathy and funding which enabled the disaster management framework to grow and render disasters manageable through a set of tools including mock-drills, action plans, improvements to

[30] This idea is reinforced also by stories about how the Disaster Management Act itself was passed: the Act was ready since 2002, and dust was falling on it with little political will to push it through the Lok Sabha (lower chamber of federal parliament). At that point India has already experienced two major disasters: the 2001 Bhuj earthquake in Gujarat and the Orissa supercyclone in 1999. In this narrative, the Indian Ocean Tsunami was the final straw in 2004 that prompted the passing of the Act.

[31] In 1999, 10,000 people lost their lives in Orissa in a supercyclone. Since then, much effort and resources by governmental and non-governmental actors have been used to develop their disaster preparedness infrastructure and systems. In October 2013, these systems were tested by the Phailin cyclone (with wind speed of about 200 km/h), which resulted in minimal loss of life (*BBC News India*2013).

[32] In January 2001, a major earthquake of 7.7 magnitude struck Bhuj in the state of Gujarat leaving around 20,000 dead. Since then, the state of Gujarat has developed their disaster preparedness to very high standards.

infrastructure and technology. In 2011, another disaster—cyclone Thane—came in 'good' time, too. It allowed for the state-led disaster management structure to grow further.

The Union Territory is perceived as 'advanced' compared to other states in India.[33] In general, high-standard disaster preparedness is being associated with 'developed' status. Such associations are unhelpful, as they render disaster preparedness an unnecessary luxury and contribute to the lack of commitment of many other states in India to sincerely implement the new disaster management legislation prescribed by the central government. Moreover, the trope about 'reactive Indian attitudes' usually renders disaster preparedness as difficult to achieve in the discourse of disaster experts. However, looking at it rather through the lens of temporality and time-bound priorities, pragmatism plays an important role. Disasters then serve as events that change priorities, shift political wills and allow space for new developments. Disasters can happen in the 'right' time, and I argue that good timing is key to successful introduction of policies. In this case, it was the timing of disasters in South India that helped to promote and implement new disaster governance policies while maintaining the important role of non-governmental sector involvement at community level.

Acknowledgements The data presented in this chapter was collected during fieldwork for the ERC-funded 'Organizing Disaster: Civil Protection and the Population' project (263731). I would like to extend my gratitude to my informants as well as all people and institutions that have supported my work, especially the National Institute of Disaster Management, the Pondicherry University and the French Institute of Pondicherry. I would also like to thank the editors of this book for their patient guidance which provided me with a sense of direction. Moreover, I would like to express my appreciation of the insightful comments of my colleagues Dr. Michael Guggenheim and Dr. Joe Deville.

References

Agar, J. (2013). *Constant touch: A global history of the mobile phone*. London: Icon Books Ltd.

Bandyopadhyay, N.C. (2013). *Training module on urban risk mitigation*. New Delhi: National Institute for Disaster Management.

BBC News India. (2013, 13 October). *India's cyclone Phailin leaves trail of destruction*. http://www.bbc.co.uk/news/world-asia-india-24510464. Accessed 13 May 2014.

Carr, L.J. (1932). Disaster and the sequence-pattern concept of social change. *The American Journal of Sociology*, 38(2), 207–218.

Department of Revenue and Disaster Management Puducherry (DRDMP). (2013). *Disaster management action plan 2013*. Puducherry: Government of Puducherry.

[33] 'Still, I won't say we are complete, we have just started... we have a better start compared to other agencies in the country. Other agencies they don't have this framework, this tier system is not there, they don't have an Emergency Operations Centre. ...We are advanced state, we are going to get the state accord in emergency operations, very high quality emergency operations...' (Mr. K., DRDM, Puducherry, July 2012).

Deville, J., Guggenheim, M., & Hrdlickova, Z. (2014). Concrete governmentality: Shelters and the transformations of preparedness. In M. Tironi, I. Rodriguez-Giralt & M. Guggenheim (Eds.), *Disasters and politics. Materials, preparedness and governance* (pp. 183–210). Sociological Review Monograph. London: Wiley.

EM-DAT: The OFDA/CRED International Disaster Database. (n.d.). *Database*. www.emdat.be. Accessed 25 Sept 2014.

Flint, M., & Goyder, H. (2006). *Funding the tsunami response. A synthesis of findings*. London: Tsunami Evaluation Coalition (TEC).

Guggenheim, M. (2014). Introduction: Disasters as politics – Politics as disasters. In M. Tironi, I. Rodriguez-Giralt & M. Guggenheim (Eds.), *Disasters and politics. Materials, experiments, preparedness* (pp. 1–16). The Sociological Review Monograph. Chichester: Wiley-Blackwell.

Guleria, S. (2009). *Integrated coastal zone management plan for tsunami affected coastal areas in Cuddalore, Nagapattinam and Kanyakumari Districts, Tamil Nadu, Southern India*. PhD dissertation. Tuticorin: Suganthi Devadason Marine Research Institute.

Gunewardena, N., & Schuller, M. (Eds.) (2008). *Capitalizing on catastrophe: Neoliberal strategies in disaster reconstruction*. Lanham: AltaMira Press.

Hastrup, F. (2011). *Weathering the world: Recovery in the wake of the tsunami in a Tamil fishing village*. New York: Berghahn Books.

Homer-Dixon, T. (2006). *The upside of down: Catastrophe, creativity, and the renewal of civilization*. London: Island Press.

India Meteorological Department (2012). *Very severe cyclonic storm "THANE" over the Bay of Bengal (25–31 December, 2011): A report*. New Delhi: India Meteorological Department. Cyclone Warning Division. http://www.imd.gov.in/section/nhac/dynamic/THANE.pdf. Accessed 8 May 2014.

Indian Ocean Tsunami Response. (2014). *Asian Development Bank*. http://www.adb.org/themes/climate-change/disaster-risk-management/indian-ocean-tsunami-response. Accessed 6 May 2014.

Jeffrey, R., & Doron, A. (2013). *The great Indian phone book: How cheap mobile phones change business, politics and daily life*. London: Hurst.

Krafft, T., Kremer, A, & Schraeder, S. (2004). Environmental degradation and health: Urban growth and health risks in Pondicherry. In A.K. Dutt, A.G. Noble, G. Venugopal, & S. Subbiah (Eds.), *Challenges to Asian urbanization in the 21st century* (pp. 175–188). The GeoJournal Library 75. Springer Netherlands. http://link.springer.com/chapter/10.1007/1-4020-2531-9_11. Accessed 16 Apr 2014.

Mapsofindia1. (2011). *Tsunami: A disaster to be remembered for ever!*http://blog.mapsofindia.com/india-map/tsunami-a-disaster-to-be-remembered-for-ever/. Accessed 1 May 2014.

McKerrow, B. (2010). You smell like a fish and drink like one too! *Bob McKerrow – Wayfarer mountains, travel, humanitarian work and opinion*. http://bobmckerrow.blogspot.co.uk/2010/04/you-smell-like-fish-and-drink-like-one.html. Accessed 1 Oct 2012.

Ministry of Home Affairs. (2011). *Disaster management in India*. New Delhi: Government of India.

Nair, R.B. (2011, April 30). Emergency operation centre to be operational hub. *The Hindu*. http://www.hindu.com/2011/04/30/stories/2011043051570300.htm. Accessed 2 May 2014.

Prince, S. (1920). *Catastrophe and social change, based upon a sociological study of the Halifax disaster*. New York: Columbia University.

Punithavathi, J., Tamilenthi, S., & Baskaran, R. (2012). A study of Thane Cyclone and its impacts in Tamil Nadu, India using geographic information system. *Archives of Applied Science Research*, 4(1), 685–695.

Regnier, P., Neri, B., Scuteri, S, & Miniati, S. (2008). From emergency relief to livelihood recovery: Lessons learned from post-tsunami experiences in Indonesia and India. *Disaster Prevention and Management*, 17(3), 410–429.

Revi, A.K. (2010). Afterword. In A. Revi & S.B. Patel (Eds.), *Recovering from earthquakes. Response, reconstruction and impact mitigation in India* (pp. 384–404). New Delhi: Routledge.

Sharma, V. (2001). *Disaster management*. New Delhi: National Centre for Disaster Management, Indian Institute of Public Administration.

Sinha, A. (1999). *Culture of prevention: Natural disaster management, India*. New Delhi: National Centre for Disaster Management, Indian Institute of Public Administration.

Sinha, A.K., Satendra, & Srivastava, S. (2010). Disaster management in India. Policy interventions and recent initiatives. In A. Revi & S.B. Patel (Eds.), *Recovering from earthquakes. Response, reconstruction and impact mitigation in India* (pp. 238–254). New Delhi: Routledge.

Subramanian, T.S. (15 January 2005). Their own warning systems. *Frontline*. http://www.frontline.in/navigation/?type=static&page=archive. Accessed 2 May 2014.

Subramanian, T.S. (14 January 2012). Cyclone fury. *Frontline*, http://www.frontline.in/navigation/?type=static&page=archive. Accessed 12 May 2014.

The Times of India. (2004, December 29). *Tsunami toll in TN, Pondy touches 7,000*. http://timesofindia.indiatimes.com/india/Tsunami-toll-in-TN-Pondy-touches-7000/articleshow/974891.cms. Accessed 1 May 2014.

Tripathi, P.S. (2013, July 26). State of paralysis. *Frontline*. http://www.frontline.in/the-nation/state-of-paralysis/article4894175.ece. Accessed 24 Sept 2014.

Tsunami. (n.d.). *Chapter VII*. http://pandr.puducherry.gov.in/Plan%20Formulation/Plan%20Document/DAP%202006-07/pdffiles/int/C-7.pdf. Accessed 12 May 2014.

Chapter 8
Intergenerational Transmission of Local Knowledge Towards River Flooding Risk Reduction and Adaptation: The Experience of Dagupan City, Philippines

Fatima Gay J. Molina

Abstract Lying approximately 1 m above sea level, Dagupan in the province of Pangasinan is one of the cities in the Philippines perennially vulnerable to river flooding. It serves as the catch basin of waters that drain into *Lingayen* Gulf and two river deltas, the *Agno* and *Toboy-Tolong*, and is subject to the onset of high tidal movement. The excess waters would find their way into low-lying areas resulting in disruption of people's access to social services and livelihood activities such as milkfish farming, as well as loss of life. Interviews and discussions revealed that *Dagupeños* cultivated local knowledge to cope with the situation over time. In looking at the said knowledge, its intergenerational transfer will be traced to the following generations: before 1990, 1991–2000 and 2001 to present. Through the integration of intergenerational local knowledge with urban disaster governance, there will be a better understanding on variations of risk perception and making risk reduction more sensitive to different sectors in urban settlements. Furthermore, determination of impacts from one generation to another would enable people to have a more holistic approach to disaster governance. Finally, identification of local knowledge that can address thematic concerns on disaster risk reduction such as preparedness, prevention and mitigation, response and rehabilitation and recovery would promote decentralised governance in vulnerable and urban poor communities.

Keywords Capacity • Dagupan • Governance • Intergenerational • Local knowledge • Resilience

---

F.G.J. Molina (✉)
Center for Disaster Preparedness, Quezon City, Philippines
e-mail: fatimagaymolina@gmail.com

M.A. Miller, M. Douglass (eds.), *Disaster Governance in Urbanising Asia*,
DOI 10.1007/978-981-287-649-2_8

8.1 Introduction

About 38 % (close to two billion people) of the global population inhabits flood-prone areas (Baker 2012). Flooding is a consequence of cyclones and storm surges that affect 2 % of the world's land area, which comprises 13 % of the world's urban populace (McGranahan et al. 2007). It also poses a threat to about 24 % (1.4 billion people) of the world's inhabitants that occupy densely populated coastal areas including the Philippines, Central America, the Caribbean, Bay of Bengal and China (Baker 2012). Flooding and the location—along the sea or waterways—of most of the major cities threaten a total of 616 urban areas in the world, which are located in metropolitan regions. Among these, 334 are in Asia and Oceania, 90 in Europe, 83 in North America, 60 in Africa and 49 in South America and are inhabited by a total of 379 million people (Sundermann et al. 2013). The population in large cities exposed to cyclones is projected to increase from 310 million in 2000 to 680 million by 2050 (Baker 2012). So far, from 1970 to 2010, the number of people affected by flooding in Asia has doubled from 30 million to 64 million (Miller and Douglass 2014).

The Philippines is considered to be one of the high-risk countries to flooding in the world. The country's vulnerability is a result of its location, being situated in the Pacific Ring of Fire and Western Pacific Basin, where tropical cyclones are formed. Its disasters range from typhoons, floods, landslides, fire, earthquake, drought, volcanic eruption and other hazards and was ranked 3rd out of 173 countries identified as the most disaster-prone nations in the world (CRED 2013). It was also identified as the most disaster-affected country with 2,360 deaths due to natural disasters in 2012 (Beck et al. 2012). An average of 20 typhoons affects the country annually, and as a result, densely populated coastal cities are exposed to greater risk. Alongside this, the role of sociopolitical concerns such as poverty, waste management, urban planning and governance must be taken into account when assessing disaster impacts and planning risk reduction.

The Philippines serves as home to 98.4 million people as of 2013 (United Nations Population Fund 2013). As of 13 September 2014, the Philippines is now composed of 144 cities classified as highly urbanised, except for Isabela City (Component City) in Region IX and Cotabato City (Independent Component City) in Region XII, whose provinces (Basilan and Maguindanao) are under the regional jurisdiction of the Autonomous Regional of Muslim Mindanao (National Statistics Office 2014). Since 2010, 49 % of the county's total population inhabits urban settlements (United Nations Population Fund 2011). In Metro Manila alone, it was noted by the Housing and Urban Development Coordinating Council (HUDCC) that as of July 2010 there were about 581,059 informal settlers. In the whole of Philippines, informal settler families reached an estimated 550,771 households as of 1 August 2007 (Cruz 2010). Given the growing population and high concentration of people in urban settlements, tight competition on space utilisation and the struggle for access to social services exacerbate the impacts of disaster. In times of disaster, people's concerns on acquisition of safe spaces and appropriate shelter need to be addressed by disaster governance. The differentiating capacity to acquire space and shelter given the varying social classes should also be considered. People who do not own land and are living below the poverty line are more vulnerable and at risk. With these repercussions of urbanisation driven by the growth of population, the people in cities surrounded with bodies of water are exposed and more vulnerable to flooding incidents.

One of the cities in the Philippines that often experiences flooding is Dagupan, located in Pangasinan province in Region I. The city suffers from regular flood inundation as it is close to the *Lingayen* Gulf and surrounded by seven main river tributaries and an upstream meandering river and has poor elevation. Floods that occur in the city are categorised into different types, such as tidal rise flood [for areas that are 0.9–1 m above sea level (MASL)], rain-induced flood, river swelling flood and coastal flood (for areas affected by storm surge) and floods due to release/spilling of water from dams surrounding the city (Dagupan City Government 2014). Among these, river swelling causes the greatest impact, resulting in inundation of 23 *barangays*[1] out of the 31 in the city. Once combined with other factors contributing to the situation such as dam spilling, torrential rains and tidal rise, the entire city becomes swamped with floodwater.

With this exposure, the *Dagupeños* developed and cultivated local knowledge as a mechanism to address the impacts. This chapter discusses the role of intergenerational local knowledge transmission as an approach towards risk reduction, resilience and adaptation to river flooding. The focus will be on river flooding as it causes immense damage to the city. Through the integration of intergenerational local knowledge with urban disaster governance, a better understanding on the variations of risk perception among different generations would make risk reduction more sensitive to different sectors in urban settlements. Recognition of the different experiences and needs of different generations would pave the way for a better comprehension of strategies to be applied in the varying contexts of risk from one generation to another. Due to this, representation and identification of the risks and vulnerabilities among different age groups, such as children, middle age and elderly, would lead to more inclusive disaster governance.

Furthermore, the understanding of risks and impact from one generation to another would enable people to prepare a more holistic approach to disaster governance. Being guided by the diverse experiences of different generations, the plans and actions to be implemented will be in concurrence with past and present interventions. This would stimulate advance mechanisms in order to address the changing context of disasters. Also, with the knowledge of past and present context of disaster impacts, a departure from the habitual formula in disaster governance will lead to a more advanced disaster perception and imagination. The worst-case scenario will be taken into account in projecting the possible impacts. Thus, people will not prepare just based on what they have experienced, but by thinking beyond what they have gone through.

[1] The smallest political unit of local government system in the Philippines, a village

Identification of local knowledge that is more conscious of the thematic concerns on disaster risk reduction such as preparedness, prevention and mitigation, response and rehabilitation and recovery would promote decentralised governance among vulnerable and urban poor communities. With the acknowledgement of the differences of interventions in every thematic concern, people from different sectors will be given a chance to partake in the process that recognises their capacities. Local knowledge itself is one of the capacities that they could capitalise in disaster governance. The varying knowledge of each sector and age group could be considered in developing actions for disaster governance.

With the integration of local knowledge to disaster risk reduction, the governance process will also be devolved, giving an enabling environment for the people to define their own action. Thus, the application of local knowledge in flood risk reduction has also endowed "agency" and "voice" to the people in the communities. It not only provides them an avenue to influence their leaders in disaster risk reduction and management councils but also a chance to apply and promote their own disaster governance agenda. They would have a voice in the decision-making process and have the autonomy to present a new perspective that complements what is being practiced and advocated by the government. Further, they would be in a position to exert influence in and have better participation in disaster governance. This is evident in the usage of local knowledge in the early warning system in which they were able to participate and define the protocols for communicating risks and preparedness mechanisms to their fellow community members.

This integration also provided the local citizens a "sense of ownership" and a "sense of identity" because their process of governance gave them an avenue to extend their *Dagupeño*-self by employing the practices that they have learnt from their ancestors since. Therefore, one's process of urban disaster governance becomes an extension of one's self, and this makes leadership more conscious to the needs, context and nature of their people. This also empowers people since it enables them to define the process of governance according to their standards. This makes the process of disaster governance more flexible and sustainable since it is according to their management perspective and experiences. Thus, their sense of governance is more people-centred, highlighting the totality of people's vulnerability.

The application of new ways to disaster governance recognises that the process is not a "one size fits all" approach. Given the city's changing coastal urban landscape and evolving nature of disasters over time, experiences from the following three generations will be traced: before 1990, from 1991 to 2000 and from 2001 up to present. Taking into account the experiences of these generations is essential to validate whether the integration of the local knowledge to disaster governance enabled people to adapt to changing nature of hazards and varying contexts and develop knowledge. This would also allow people to assess what is still applicable and develop new ways and practices to adapt to current disaster experiences.

In this chapter, the term "local knowledge" will be used to refer to the form of knowledge inherent or acquired by local people based on their experiences, society-nature relationships and community practices and institutions and/or from previous generations (Fernando 2003; Sillitoe 1998). It also originates within the community, maintained through nonformal means of dissemination, collectively owned, and has developed over several generations and subject to adaptation and embedded in a community's way of life as a means of survival (Shaw and Baumwoll 2008). However, this knowledge is not static; it can change as a result of experiences and interactions with external knowledge (Wisner 2010). Further, it is also based on local needs and reflects a clear foundational understanding of communities' adaptation strategies (Shaw and Sharma 2008). The value of such knowledge has been of interest to scholars who study environmental hazards and disasters (Cronin et al. 2004a, 2004b; Mercer et al. 2010). It can also be classified into two dimensions: technical, which encompasses informal and hard-to-pin-down skills captured in the term "know-how", and cognitive, which comprises of schemata, mental models, beliefs and perceptions (Rahman 2000).

Through intergenerational transfer, the enrichment of local knowledge on one's preparedness initiative towards the mitigation of impacts, such as those caused by flooding as a result of typhoons, enabled risk reduction and adaptation of people. Their observations and experiences over the years made them conscious of the need to be constantly aware of nature's varying behaviour. In the case of *Dagupeños*, they became aware that the torrential rains would result in the river overflowing, which in turn leads to flooding of the low-lying areas in the city and later on the entire city as well. Thus, this knowledge of nature's behaviour aided them in defining their strategy for risk reduction and the next step, which is geared towards adaptation. In this case, they would know when to move to a safer place. As this chapter progresses, more examples of local knowledge and practices relevant to river flooding risk reduction and adaptation will be discussed.

In looking at local knowledge, the circle of capacities framework will be used to argue that communities can reduce risks and adapt to the changing climate through the integration of local knowledge in disaster governance. With the usage of data from an ongoing fieldwork in northern Luzon, experiences from the abovementioned three generations would be presented to see how perceptions and experience of environmental risks could vary across different stakeholders given the changing and heterogeneous population of an urban settlement and the transformation of the landscape over time.

The harnessing of the local knowledge transferred intergenerationally resulted in the reduction of risks brought about by flooding. As we all know, disaster governance is not just limited to technical approaches that are passed on by the experts or scholars in disaster studies. It encourages people to capitalise on their existing capacities and mechanisms in their communities. Therefore, risk reduction and disaster governance must be led and practiced in partnership with all sectors of the community for these people serve as the frontliners in disaster risk reduction and management. Thus, the process of flood governance becomes more inclusive and comprehensive where multi-sector, multidisciplinary and multi-stakeholder engagement is encouraged.

8.2 Methodology

With the practice of "anthropology at home" (Davies, 1999; Jacobs-Huey 2002; Peirano, 1998; Sarmiento 2011), a more grounded perspective on the process of urban disaster governance in the area of study is highlighted in this chapter. With the author being a local of *Mangaldan*, the municipality located to the east of the city of Dagupan, the embodiment of disaster impacts experienced will be reflected more consciously as these areas are geographically close to each other. The people of these localities have encountered the same disasters, although the impacts are significantly different to each other. The discussion of urban disaster governance will be focused on the city of Dagupan and its people's experiences.

The past disaster experiences of the author are common to the people of Dagupan. The earliest devastating disaster faced by the author was the 1990 earthquake, which registered a 6.8 magnitude on the Richter scale and greatly affected Dagupan and nearby Baguio City among others. Despite being only 3 years old at that time and not being totally aware of what was happening, the strong ground movement is still remembered. This experience points to the need for an intergenerational approach on the transmission of information on disasters and practices in order to prepare the younger generation and promote proper urban governance. This must be done for all hazards experienced by the people in the area because apart from geophysical hazards, they have also lived through the impacts of hydro-meteorological hazards like floods. In 1998, the author experienced the first inundation in their house due to Typhoon *Gading* (International name: Vicki). Back then, the family had to evacuate to the paternal grandmother's house to save themselves before the floodwater continued to rise inside the house. There was no form of disaster preparedness in the *barangay* because the households were not informed by the local leaders on the rising floodwaters. There was no clear evacuation system or evacuation centres identified in the *barangay*. People were given information about the possible occurrence of flooding only during the event itself—thus, disaster governance was highly "reactive" at that time. The government acts upon the situation once the hazard is already being experienced. After three years, Typhoon *Feria* (International name: Utor) inundated the same areas. However, that time people were more prepared since some members of the *barangay* have practiced mitigation practices that involved the renovation of their houses and building another shelter to ensure that they have an alternative safe space. This postulates another dimension when speaking of the capacity to prepare in times of disasters. Not all people have the same level of preparedness since the capacity is not equal among different stakeholders. The government as duty bearers must provide an enabling environment and have mechanisms to ensure "equilibrium" in the process of risk reduction and management and governance among its people.

In 2004, the migration to the capital of the Philippines exposed the author to a greater scale of urban disaster impacts. Among the typhoons that have struck the megacity of Metro Manila over the years, the worst was in 2009 after Typhoon *Ondoy* (International name: Ketsana) made its way to the area. The turn of events has resulted in advances in disaster risk reduction and management and climate change policies. In the same year, Republic Act (RA) 9729 also known as the *Climate Change Act* was enacted, and in the following year, RA 10121 also known as the Philippine Disaster Risk Reduction and Management Act was ratified that repealed Presidential Decree 1566 implemented in 1978 by the Former President Ferdinand Marcos. In spite of these actions, people were not spared from the great devastation of Typhoon Yolanda (International name: Haiyan) in 2013. It was identified as the most disastrous typhoon that ever landed in the Philippine area of responsibility. This suggests that it is not just enough to have policies; they should also be implemented and be responsive to high-risk disaster-prone areas. Its effects have to be felt by the people up to the *barangay* level since the communities are the ones immediately affected.

Among the different kinds of communities, urban coastal communities are the ones that suffer greatly from the effects of flooding. The city of Dagupan, with its geographical characteristics, heterogeneity of population and the complexity of impacts accompanied by its rich disaster experience is one of the cities that represents the context of high-risk urban coastal communities. Thus, it was selected as the area of study. With its vulnerability over the years, the people of the city have developed a way to adapt to impacts. They have areas to capitalise on apart from the mainstream scheme of disaster risk reduction that is generally prevailing in the country. This was learnt through the ethnography started in the scoping visit in October 2013 and completed in the succeeding field visits in March and August 2014. Apart from gathering secondary research materials on Dagupan, conduct of focus group discussions (FGDs) with fishermen and people living in high-risk villages from the following generations, before 1990, 1991–2000 and 2001 to present, was completed to understand intergenerational perspective. Such categorisations were made based on the existing chronology of disaster experience (see Table 8.1 for further information). In addition, the geographical landscape of the city has changed after the 1990 earthquake. This classification would help to identify whether the change of landscape

Table 8.1 Dagupan City timeline of flood experience as of 2011

Year	Disaster	Impacts
1935	The biggest flood in Pangasinan *(name of typhoon to be identified)*	Entire Dagupan City was flooded; Colegio de San Alberto Magno and Franklin Bridge at the village of Calmay were destroyed and the aforesaid structures sank in Calmay River
1972	Flood	Agricultural crops, fishponds and other properties were damaged
1998	Typhoon *Gading* (Int'l name: Vicki) and *Iliang* (Int'l name: Zeb)	Php 2.8 million damage on infrastructure
		Php 36.01 million damage on agriculture
		Php *3.3 million* damage on infrastructure
		Php 42.11 million total damage
1999	Monsoon rains	Php 7.5 million damage on infrastructure
2000	Typhoon *Ditang* (Int'l name: Kirogi) and *Edeng* (Int'l name: Kai-tak)	Php 24.4 million damage on agriculture

(continued)

Table 8.1 (continued)

Year	Disaster	Impacts
2001	Typhoon *Feria* (Int'l name: Utor)	Php 8.5 mdamage on infrastructure
		Php *17.215 million* damages on agriculture
		Php 25.715 million total damage
2002	Typhoon *Gloria* (Int'l name: Chataan)	Php 16.32 million damage on agriculture
2003	Typhoon *Chedeng* (Int'l name: Linfa)	Php 10.7 million damage on infrastructure
		Php *30.57 million* damage on agriculture
		Php 41.27 million total damage
2004	Typhoon *Marce* (Int'l name: Aere)	21 villages in the city were flooded
		17,139 households affected
		87,409 of the population affected
		5,505 business establishments affected
2007	Typhoon Mitag (Int'l name: Mina)	Php 13.8 million damage on aquaculture
2008	Typhoon *Cosme* (Int'l name: Halong)	31 villages affected by the very strong winds which occurred in only about four (4) hours
		18,383 houses were reported totally damaged with roofs blown off while other structures collapsed to the ground, affecting 24,973 families
		The city was isolated for about two (2) days
		All lifelines were down for about four (4) days
		4 persons injured and 4 casualties
		Php 552 million damage on fisheries and crops sector
		Php 29 million damage on infrastructure (schools and public facilities)
2009	Typhoon *Pepeng* (Int'l name: Parma)	Inundated all the villages in the city
		29,278 households affected
		141,000 population affected
		5,000 or more establishments were affected by the flooding caused by rainwaters and the occurrence of storm surge that inundated the city
		All fishponds and croplands were inundated
		Php 549.50 million total damage
2010	Typhoon *Juan* (Int'l name: Megi)	Php 22 million damage on the agriculture and aquaculture
2011	Typhoon *Quiel* (Int'l name: Nalgae)	Php 19 million damage on the agriculture and aquaculture

Source: Dagupan City Government (2014)

resulted in development of new local knowledge and practices to cope with disasters. In order to have a better understanding on the governance, key informant interviews (KIIs) were conducted with heads of the City Agriculture Office, City Assessor Office, City Disaster Risk Reduction and Management Council and the technical consultant of the council who used to be the former *barangay* captain (village chief) of the most high-risk community to flooding.

8.3 The Research Field Site

8.3.1 Geographical Profile

The city is located in the province of Pangasinan, approximately 1 m above sea level along the seacoast and at the eastern margin on the delta of Agno River (the 3rd largest river in the country) that is part of Region 1. It has a total land area of 4,446 km^2, of which 42 % is land and the remaining 58 % is covered by water. In the sixteenth century, it was known as *Nalimpit*, the seat of power of *"Luyag na Pangasinan"*, one of the two ancient kingdoms of the province of Pangasinan (Vistro and Reyes 2005). It was later called *Bagnotan* when claimed under the jurisdiction of the Spanish colonial government. After rebuilding due to fire damage in 1661, its name was changed to *Nandaragupan* and was later shortened to Dagupan in 1720 (Vistro and Reyes 2005). It became a chartered city under Republic Act 170 in 1947 and was formally inaugurated in 1 January 1948 (Cortes 1974). It is one of the areas in the Philippines perennially vulnerable to river flooding during the rainy season (Iglesias 2007; Victoria 2008). It serves as the catch basin of all the waters that drain into *Lingayen* Gulf and two river deltas, the *Agno* and *Toboy-Tolong*, and is subject to the onset of high tidal movement that poses greater threat to the urban settlement, whose inhabitants total 163,676 (Dagupan City Government 2014). The area is thickly populated, having a population density of almost 4,000 persons per square kilometre (Center for Disaster Preparedness, CDP 2009). The tidal backflow creates seven rivers (*Bayaoas, Calmay, Dawel, Magueragday, Pantal, Patogcawen* and *Tanap*), which in turn form islets out of the eastern villages of *Mamalingling, Bolosan, Tambac, Manguin* and *Salisay* (Dagupan City Government 2008). The geophysical setup of the city can be seen in Figs. 8.1 and 8.2, while Table 8.1 shows the timeline of flooding experience.

During the rainy season, rivers overflow affecting low-lying areas of the city which are populated by vulnerable informal settler families (as shown in Table 8.2) resulting in disruption of people's access to social services, devastation of livelihood activities such as *bangus* [milkfish (*Chanos chanos*)] farming and loss of life to some. Rivers presently have a high sediment load and do not drain easily into the sea, which further aggravates the flooding situation (Iglesias 2007). These sediment loads are from the unabated use of commercial feeds in the milkfish industry. The flooding situation is further worsened once two major dams (*Binga* and *San Roque*) located closely to the city reach the spilling level after a heavy rainfall (CDP 2009).

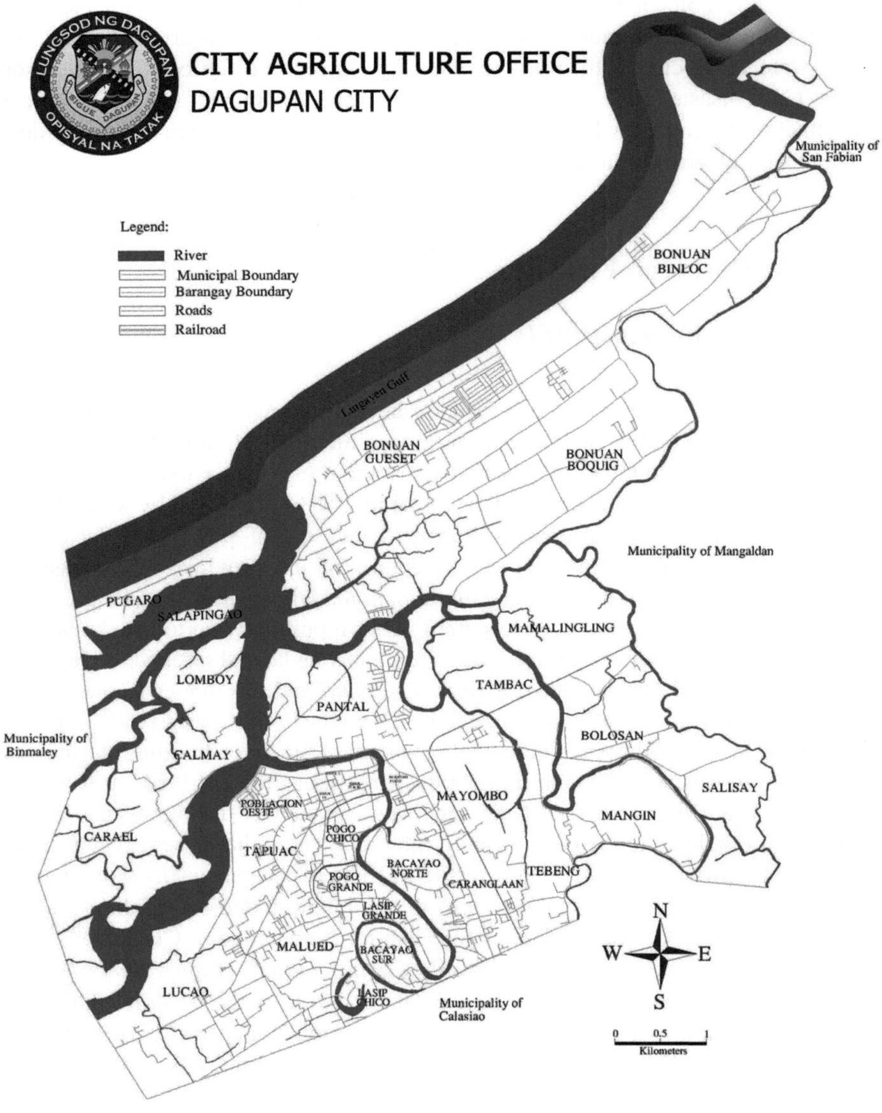

Fig. 8.1 Dagupan city map as of 2008 (*Source*: Dagupan City Government 2008). River deltas: (*A*) Agno, also known as Bonuan Sabangan, the mouth of Agno River, and (*B*) Toboy- Tolong, also known as Bued-Angalacan River; 7 rivers: (1) Mangueragday River, (2) Tanap River, (3) Dawel River, (4) Patogcaoen River, (5) Bayaoas River, (6) Calmay River and (7) Pantal River

The water from the dams ends up in the rivers, thereby significantly contributing to the city's inundation. With the larger impact of climate change, seasonal variability further exacerbates river flooding during the aforesaid season, which may now begin as early as February and extend even until December.

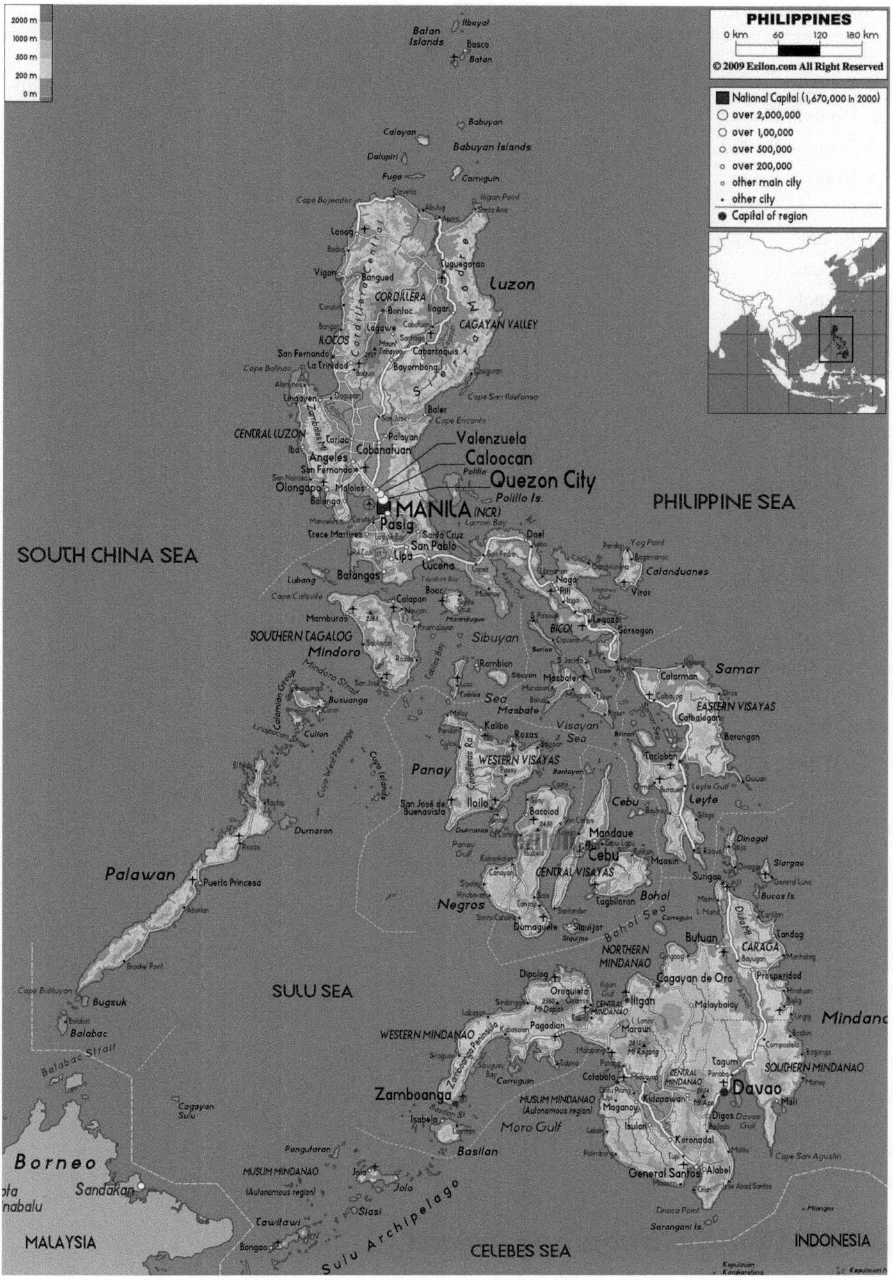

Fig. 8.2 Philippine political map (*Source*: Ezilon Maps, n.d.)

Table 8.2 Distribution of informal settlers in Dagupan City as of 2011

	Informal settlers			
Barangay	Private lands	Public lands	Total	Rank
1. Bonuan Gueset	220	740	960	2
2. Bonuan Binloc	150	260	410	4
3. Bonuan Boquig	280	58	338	6
4. Herrero-Perez	50	34	84	
5. Pantal	708	520	1,228	1
6. Tapuac	76	8	82	
7. Carael	68	53	121	
8. Calmay	120	110	230	9
9. Salapingao		52	52	
10. Pugaro	528	220	748	3
11. Lomboy		19	19	
12. Lucao	44	50	94	
13. Mayombo	64	302	366	5
14. Caranglaan	25	268	293	7
15. Poblacion Oeste	175	60	235	8
16. Barangay I	50	50	100	
17. Barangays II and III	70	70	140	
18. Barangay IV	20		20	
19. Mamalingling	6	11	17	
20. Salisay	36	23	59	
21. Manguin		15	15	
22. Tebeng	12	3	15	
23. Bolosan		50	50	
24. Tambac		96	96	
25. Bacayao Sur	102	25	127	
26. Bacayao Norte		35	35	
27. Pogo Chico	205	24	229	10
28. Pogo Grande				
29. Malued	20	30	50	
30. Lasip Grande		12	12	
31. Lasip Chico				
Total	*3029*	*3196*	*6225*	
	47 %	*53 %*		

Source: Dagupan City Government (2014)

8.3.2 *Disaster History and Impacts*

The city of Dagupan has been exposed to flooding incidence as early as 1935. The impacts have resulted in sizeable damages affecting critical sectors and facilities in the city such as agriculture, aquaculture, infrastructure, settlements, educational institutions, businesses and public services. The chronology of disaster events are listed as follows:

Other than hydro-meteorological hazards and disasters, the city has also experienced geological impacts such as the earthquake in 1990. The earthquake measuring magnitude 6.8 on the Richter scale rocked the city and other cities and municipalities in Northern Luzon (Pablo et al. 1999). This resulted in Dagupan City sinking 3 m as immense geophysical pressures caused the widespread liquefaction of compacted silt and black sand gushed up, making the landscape look like "a beach without a sea" (Bankoff 2003a; Coping 1990). The earthquake made a meandering pattern that caused the dynamic lateral shifting of *Pantal* River. The shift left numerous abandoned channels and created a low-lying flood-prone terrain made up of levees and back swamps (Iglesias 2007). Prior to the aforementioned earthquake, the city experienced one of the most disastrous earthquakes in the country in the year 1880 that also affected its neighbouring municipalities such as *Mangaldan* and *San Jacinto* (Cortes 1990). It was followed by the 1892 earthquake, which is believed to have the same intensity as the 1990 earthquake, while the earliest earthquake experience was in 1962 (Vistro and Reyes 2005).

8.4 Understanding Vulnerability and Elements at Risk

In order to comprehend the gravity of impacts of a particular disaster, the people's vulnerability and the elements at risk in the area must be identified. Vulnerability is described as a set of conditions and processes resulting from physical, social, economic and environmental factors, which increase the susceptibility of a community to the impact of hazards (Victoria 2003). According to the United Nations Office for Disaster Risk Reduction (UNISDR), there are many aspects of vulnerability, arising from the aforementioned factors. Examples include poor design and construction of buildings, inadequate protection of assets, lack of public information and awareness, limited official recognition of risks and preparedness measures and disregard for wise environmental management. Vulnerability substantially differs within a community and over time. Wisner and Luce (1993) analysed vulnerability in terms of the following variables: class, gender, age, ethnicity and disability. Understanding the vulnerability of the city therefore sets the parameters in identifying the elements at risk. Dagupan's vulnerability resonates around the following aspects: geographical location, geophysical set-up, the composition and condition of its population (e.g. class, gender, age, ethnicity and disability) and anthropogenic factors, livelihood activities (e.g. fishing, fish pen and fishpond construction) and many others. Floating communities as seen in Fig. 8.3 are at risk during disaster occurrences as stipulated by the Dagupan City Disaster Coordinating Council Manual in 2008.

Fig. 8.3 Elements at risk include floating communities inhabiting *Calmay* River situated at the village of Calmay (Photograph by F. Molina 2014)

8.4.1 **Dagupeños** ***at Risk***

As of 2010, the population at risk inhabiting 31 *barangays* reached 163,676 individuals (Dagupan City Government 2014). Ever since the city has experienced flooding, eight *barangays* have been classified as highly vulnerable: *Bacayao Sur, Bacayao Norte, Lasip Chico, Lasip Grande, Manguin, Pogo Grande, Salisay* and *Tebeng*. However, due to the ongoing construction of the Agno River Integrated Irrigation Project's (ARIIP's) programmed expansion for the Ambayaoan-Dipalo Irrigation System (ADRIS) and Lower Agno River Irrigation System (LARIS) that was later renamed to Agno River Irrigation System Extension Project (ARISEP), a change in the course of the rivers' flow caused the rechanneling of the water path, resulting in more communities to be at flood risk. *Malued, Pogo Chico, Herrero-Perez, Pantal and Barangays 1, 2 and 3* are now included among those identified as high risk, which used to just consist of the barangays of *Bacayao Norte, Bacayao Sur, Lasip Chico, Lasip Grande and Pogo Grande*. As of 2008, there are 3,165 households facing direct risks in the event of any disaster (Dagupan City Government 2008). Apart from the registered households, the informal settler families (ISFs) that inhabit the city are also at risk. As of 2011, about 47 % of ISF occupy private lands and 53 % inhabit public lands (Table 8.2). *Pantal*, the *barangay* with the highest concentration of ISFs, was also identified as one of the high-risk communities.

These settlers are a mix of poor indigenous *Dagupeños* and people who originated from other cities and municipalities in the province. At the sectoral level, 15,852 individuals are classified as highly vulnerable as of 2008. Among them are children, people with disabilities and women. Also categorised as highly vulnerable are the single or surrogate parents. About 1,060 senior citizens face the same degree of vulnerability. Only 20 % of the total population is considered safe during floods as they face very low vulnerability levels. There are also 107,594 persons exposed to medium vulnerability. Of these, 41 % are children and youth (aged between 1 and 17 years old), while 6 % are senior citizens (aged 60 years and above).

8.4.2 Vulnerability in Bangus *Farming*

Vulnerability in the city is greatly reflected in the *bangus* industry. Being considered as an industrial hub in Pangasinan, the city is categorised as urban in its entirety, though one-third of its total villages are still devoted to agricultural production while nine remain as fishing communities. In as early as 1575, people put fishponds in their backyard (Cortes 1990). At present, a total of 985 ha of the city is devoted to fishpond management (Sotelo 2012). In 2003, fishponds were used for brackish water culturing production for *bangus*, prawns, shrimps, crabs and mussels (Iglesias 2007). The province is currently the country's top producer of *bangus*. From 2001 to 2003, the city's *bangus* production has contributed 16.8 % to the total provincial production (Dagupan City Government 2008). On the other hand, the production of *bangus* was noted to have an impact in the event of flooding. According to Sotelo (2012), a study by the World Wide Fund for Nature (WWF) noted that the overproduction of *bangus* has contributed to land subsidence and saltwater intrusion into habitable lands in the city. When the swampy areas are converted to fishponds, the city's aquifer is depleted since the sea advances and intrudes the emptied aquifers. This could result in the sinking of the land because the coastal city has exceeded its carrying capacity.

One of the major disasters identified by the people in relation to the rivers not just revolves around natural hazards such as flooding but also the incidence of fishkill. Most of the informants revealed that *tangok* (fishkill) happens when there's a sudden change of water temperature as influenced by the change of weather. For instance, the abrupt change of water temperature from warm to cold (due to sudden rain during summer) affects the flow of oxygen within the river system. The number of fingerlings placed in a fish pen or fishpond also influences the incidence of *tangok*: too many fingerlings in the river or pond mean competition over oxygen. *Tangok* is more likely to occur in fish pens[2] than in fishponds[3] because the competition over

[2] Fishpens are artificial enclosure constructed within a body of water (i.e. in the city rivers) for culturing of fish and fishery resources. It is made up of bamboo poles closely arranged in an enclosure with either fine bamboo or wooden materials, screen or nylon netting to prevent the escape of the fish (City Ordinance No. 1768–2003: Dagupan City Coastal Fisheries Resources Management Ordinance of 2003).

[3] Fishponds are man-made enclosure constructed within dikes, lands and swampy areas for culturing of fish and fishery resources (City Ordinance No. 1768–2003: Dagupan City Coastal Fisheries Resources Management Ordinance of 2003).

space in the rivers is more prevalent compared to ponds since it is open access to public. Those who are unable to own or afford a land construct fish pens in the river instead of fishponds. But not all have access to the river. As stated in the Dagupan City Coastal Fisheries Resources Management Ordinance of 2003, only residents of Dagupan can have access to the city's public domains, such as the river. But, they have to secure the permit in order to operate under the approval of the city government. It is also more cost efficient to put up fish pens than fishponds since the resources needed are more affordable. Due to this, some people set up fish pens illegally. As shared by the City Agriculture Office (CAO) head, "over the years, it has been observed that even after a series of demolitions, there are still isolated cases of illegal fish pens engulfing the rivers". With this condition, the state control over public spaces creates conflict. The state becomes a hazard (Wisner and Luce 1993). This is also an example of entitlement. As Peluso and Watts (2001) noted:

> Entitlements by which differentiated individuals, households, and communities possess or gain access to resources within a structured political economy. It grants priority to how these entitlements are distributed, reproduced and fought over in the course of shaping, and being shaped by, patterns of accumulation. (p. 5)

Based on the interviews, due to siltation of the rivers, rechanneling of river pathways and sea-level rise, some rivers already occupy private lands. As shared by *Nanita*, a 74-year-old fish vendor and wife of a fish pen and fishpond caretaker: "Before, there was still a space for me to walk on whenever I have to feed the bangus in our fish pen. But now, the space is gone, it is already covered by water".[4]

According to some informants, this resulted for some private landowners to demand payment of rights to fish pen operators in the river that occupied a part of their land. With this, some disagreements have surfaced among the operators and landowners. The city government through the implementation of Article 457 of Philippine Civil Code addressed this concern where the Law on Accretion is stipulated.

River siltation due to the accumulation of commercial feeds in riverbeds has also contributed to the abrupt overflowing of river during the rainy season or typhoon in the communities. This condition intensifies flooding situations in the villages as revealed by Eddie, a 45-year-old caretaker of a fish pen: "At present, the rivers have become shallow due to the accumulation of feeds. Due to this situation, the rivers will easily overflow. With this condition, our village will experience flooding at a faster rate".[5]

This demonstrates the need for urban disaster governance to prioritise proper coastal resource management in order to have a more integrated approach in disaster risk reduction and climate change adaptation.

---

[4] *"Nensaman, wala ni akaran ko no manpakan ak ya bangus ed sa may fish pen mi. Balet natan, agko la naakaran ta wala lay danum"* (original quote).

[5] *"Natan, so ilog aliwan maralem ta dakel la so feeds ya umbebeba dyad ilog. Ed sayan sitwasyon, maples lan unatagey sa danum. Lapod tan, maples lan mandelap ed barangay mi"* (original quote).

8.5 Looking into Local Knowledge

8.5.1 Framework on Understanding Local Knowledge

In situating local knowledge in urban disaster governance, one relates to the circle of capacities framework and another to power and politics of knowledge as well as to what constitutes local knowledge in Dagupan City. An inquiry to look into intergenerational transmission of local knowledge towards river flooding risk reduction, adaptation and resilience led to the application of the circle of capacities framework (see Fig. 8.4). In identifying capacities in this chapter, the following definition will be adapted: "Capacities refer to the ability of a person or group to take actions in order to resist, cope with, and recover from disasters" (López-Marrero and Wisner 2012; Victoria 2003; Wisner et al. 2004; Wisner et al. 2012). These capacities are

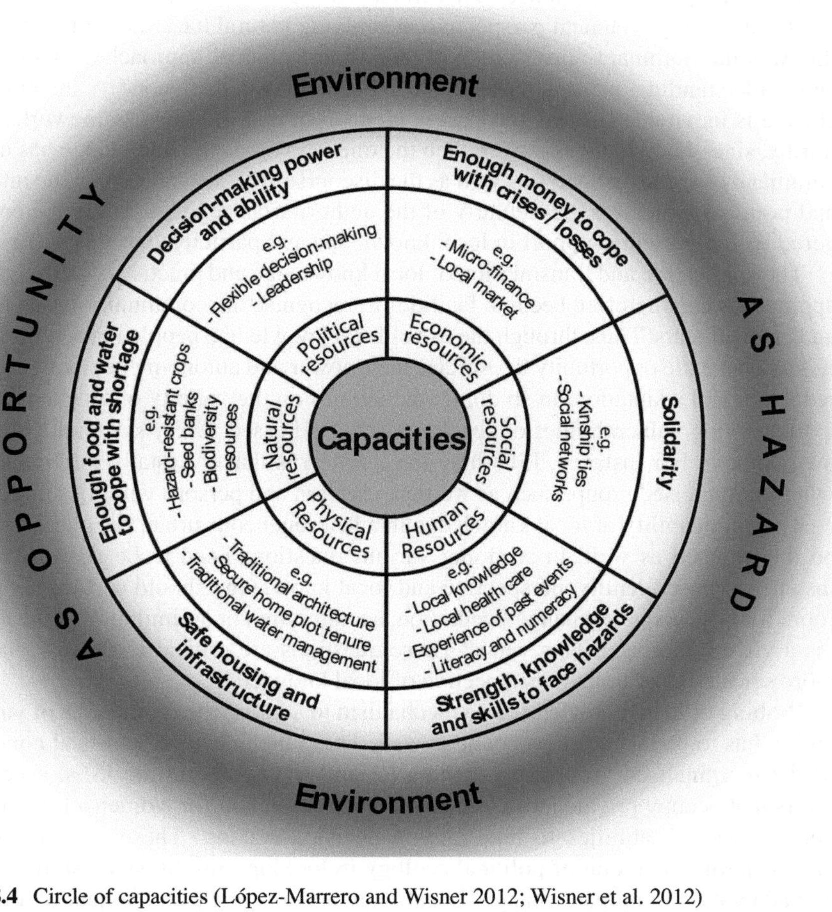

Fig. 8.4 Circle of capacities (López-Marrero and Wisner 2012; Wisner et al. 2012)

based on the availability of and access to various resources and assets, such as natural, physical, economic, human, social and political (López-Marrero and Wisner 2012; Wisner et al. 2004). One of the capacities underscored in the circle of capacities model to be utilised in dealing with the environment is human resource. This includes local knowledge together with local health care, experience of past events and literacy and numeracy (see Fig. 8.4 for further information). In perceiving local knowledge as capacity, it is also seen as a mechanism to promote resilience for it serves as a method to flood risk reduction and adaptation. Resilience in this discussion will be derived from this description where it was identified as the ability of a system, community or society exposed to hazards to resist, absorb, accommodate and recover from its effects in a timely and efficient manner, including through the preservation and restoration of its essential basic structures and functions (UNISDR 2009).

Conversely, Dekens (2007) argues that local knowledge is marginalised due to power relations and states this in three ways: firstly, local knowledge is a political issue, giving prime importance to it will change power relations. From 1960s until the 1970s, a very technical perspective in studying natural hazards proliferated and this was the dominant approach. With this, non-technical approaches in studying and understanding natural hazards have been relegated. Secondly, institutional change is inevitable if local knowledge is incorporated in policy frameworks; and thirdly, since knowledge is power, then the one who holds the knowledge has direct command over the resources as well as the discourse. She also noted that the interest and political structures, particularly of the authoritarian regimes, would be endangered by the growing support in local knowledge and participatory approaches.

The application and transmission of local knowledge and practice of participatory approaches can therefore become facilities to recognise the community-based needs and engagements. Thus, through the use of local knowledge, people in the communities can have the opportunity to exercise their power and autonomy in practicing risk reduction and adaptation. In so doing, the inquiry on the validity of this knowledge to the varying vulnerabilities of people must be addressed. Every stakeholder has its own context. For instance, ISFs may have concerns that are totally different from other marginalised groups such as women, children and persons with disabilities.

The applicability of local knowledge in a heterogeneous urban population has to be interrogated as well. In so doing various questions need to be answered. For instance, whose identity, perspective and local knowledge should surface? Is there room for knowledge exchange, mediation, acculturation or assimilation of the locals vis-à-vis migrants and vice versa? These questions are critical in order to have a more sensitive and holistic perspective of local knowledge.

Probing of local knowledge as a mechanism to address the root causes of vulnerability has to be taken into account as well. These includes sociopolitical concerns of the marginalised groups, e.g. access to land rights among the ISFs, access to rivers that occupy private lands and enabling environment for women, children and persons with disabilities to take part in decision-making. These concerns open channels for discussion of political ecology in looking into disaster risk reduction and adaptation to climate change. Local knowledge is just a component of a larger picture of ecosystem management. As I go through the discourse, the relationship between people's actions and the condition of the river would be highlighted.

8.5.2 Examples of Local Knowledge[6] in Dagupan City and Its Intergenerational Transmission

One of the prevalent local knowledge among the people interviewed was the usage of *kanungkong* to provide early warning advisory to the people whenever a disaster is about to happen. Victoria (2008) noted that through the use of *kanungkong*, Dagupan City has combined local knowledge with modern scientific knowledge and equipment for use in disaster risk reduction. The early warning instrument is made out of bamboo and was traditionally used to call community members to assemble at the village hall for meetings, alert people (i.e. occurrence of a crime, like robbery incidence at night) or to call children home (as seen in Fig. 8.5). When hit with a stick, it produces the sound, *"kung, kung, kung"*, that serves as the signal to people. As narrated by one of the informants, *kanungkong* comes from the word *mangkanungkong* that literally means to make sound. It functions as a local relay

Fig. 8.5 One of the barangay officials tests the *kanungkong* to get the attention of the people (Photo by F. Molina 2014)

[6] This study presents the local knowledge in the city based from the existing literature and preliminary interviews conducted.

communication medium within a zone, locally known as "purok", a smaller part of a *barangay*.

The corresponding signal for every alert level was agreed upon by the representatives of all the *barangays* through a multi-stakeholder consultation that was led by the city government in partnership with civil society groups. The number of strikes at designated time intervals to correspond to specific actions (as seen in Table 8.3) was also identified during the consultation.

The agreed warning codes are integrated with the early warning system that was implemented by all the villages in the city (Table 8.4). These codes are based from the flood markers placed near the rivers and flood-prone locations in the *barangays* being monitored by the people (Figs. 8.6 and 8.7). These markers are painted in secure structures in the community that are easily seen for monitoring and facilitation of early warning. The measurements of the flood markers were identified by the people and were scaled in feet. Thus, this early warning system gives the people the autonomy to decide for themselves. They are also guided by the flood markers prepared by the Philippine Atmospheric, Geophysical and Astronomical Services Administration (PAGASA) that are placed under the bridge (as seen in Fig. 8.8). With the leadership of the village chieftain and the *Barangay* Disaster Risk Reduction and Management Council as mandated by the RA 10121, the sharing of the information regarding the need to prepare for disasters or impacts of climate change is facilitated by the people in the communities. In doing so, they need not wait for the notice of the city government to practice the necessary protocols to put people in the safe zone, such as pre-emptive evacuation, in times of disasters. They are the ones who govern themselves towards disaster preparedness. This underscores that through the application of local knowledge in community-based governance in disaster risk reduction, the people in the communities become more empowered since they are able to exercise decision-making to promote resilience in their respective communities.

Thus, for the local knowledge to become more relevant to the context of disaster risk reduction and adaptation, it has to be integrated with the existing scientific mechanisms of early warning and governance.

Table 8.3 Warning codes adopted in Dagupan City

Colour	Alert level	Kanungkong warning signal
White (ready)	Normal	
Yellow (get set)	Alert (there is danger)	5 strikes at 20 min interval
Orange (go)	Prepare for evacuation or proceed to holding area (there is possibility of flooding)	10 strikes at 20 min interval
Red	Full evacuation is encouraged (forced evacuation if necessary)	Non-stop (15 strikes at 10 min interval)
		Non-stop (20 strikes at 5 min interval)
Green	Back to normal	

Source: Dagupan City Government (2008), Victoria (2008)

Table 8.4 City level early warning system

Warning level control	Status	Flood indicators	Barangay indicators	Depth indicators	CDCC/committee	Response
1. White "ready"	1. High tide forecast	Rain + High Tide	Lasip Chico	Hips (<2 ft.)	CDCC Deputised Officer	1. CDCC convened and conducts immediate briefing to CDCC members
	2. Heavy rain downpour (within 3 days)		Pogo Grande	Knee (<1 ft.)	CDCC Operation Team(warning)	1. Establish proper coordination and update report with Line Agencies, PAGASA, PHIVOLCS, RDCC, OCD, Agno Flood Control Forecasting System
	3. Flood alert forecast by Agno flood control		Pantal (Riverside and Arellano)	Knee (<1 ft.)	CDCC Staff Team	2. Establish proper coordination and update report with BDCC with the aid of early warning device installed/improvised
			Bacayao Sur	Hips (<2 ft.)	CDCC Staff Team (communication)	1. Media information dissemination thru press release as well serves as warning information
			Bonuan Gueset (Bagong Barrio)	Knee (<1 ft.)	CDCC Staff Team (supply/logistics)	1. Inventory of logistics and manpower resources
			Herrero-Perez (near creek)	Knee (<1 ft.)		
2. Yellow "get set"	1. High tide forecast	Rain + high tide + river overflow	Lasip Chico	Hips (<2.5 ft.)	CDCC	1. CDCC activated, EOC operational
	2. Heavy rain downpour (more than 3 days)		Pogo Grande	Hips (2 ft.)	CDCC Operation Team (warning)	1. Establish proper coordination and update report with PAGASA, PHIVOLCS, RDCC, OCD, Agno Flood Control Forecasting System, Red Cross, DepEd
	3. Flood alert forecast by Agno flood control		Bacayao Sur	Hips (2.5 ft.)		2. Update situational report to CDCC Deputized Officer and establish the possible scenario from the gathered reports

(continued)

Table 8.4 (continued)

Warning level control	Status	Flood indicators	Barangay indicators	Depth indicators	CDCC/committee	Response
			Bacayao Norte	Knee (<1 ft.)		3. Establish proper coordination and update report with BDCC with the aid of early warning device installed/ improvised
			Tebeng	Knee (<1 ft.)		4. Instruct the BDCC to conduct disseminate warning. If possible by house-to-house basis
			Salisay	Knee (<1 ft.)	CDCC Staff Team (communication)	1. Announcement of warning alert and information dissemination thru media, radio, press release
			Mangin	Knee (<1 ft.)	CDCC Operation Team (medical)	1. Medical, search and rescue teams on alert
			Lasip Grande	Knee (<2 ft.)		2. Medical and relief services deployed to community evacuation centres
			Pantal	Hips (2 ft.)		3. Close coordination with Red Cross, PNP, BFP, POSO
			Herrero-Perez	Knee (<1 ft.)	CDCC Operation Team (rescue)	1. Rescue Team activated
			Pogo Chico	Knee (<1 ft.)		2. Close coordination with Red Cross, CDCC Staff Team Transportation, Security and other line agencies
					CDCC Operation Team (evacuation/ relief)	1. Evacuation operation to the first priority areas with head counts and report submission
						2. Relief services deployed to evacuation centres and other temporary shelters
3. Orange "go"	1. High tide forecast		Lasip Chico	Waist (<3 ft.)	CDCC	1. CDCC on high alert
						2. Recommends declaration of state of calamity for release of calamity fund

	2. Heavy rain downpour (continuous)		Pogo Grande	Waist (<3 ft.)	CDCC operation Team (warning)	
	3. Dam release		Pantal	Hips (<2.5 ft.)	CDCC Operation Team (engineering/ damage control)	1. Clearing/monitoring operations for the route to the affected communities
	4. Neighbouring towns flooded		Bacayao Sur	Waist (<3 ft.)		2. Continuous on site flood monitoring on a 24-h basis and conduct damage, affected elements (people, livelihood, infrastructure) assessment
	5. Routes not passable by light vehicles		Bacayao Norte	Waist (<3 ft.)	CDCC Operation Team (medical/ relief)	1. Medical, search and rescue teams on alert
			Tebeng	Hips (<2.5 ft.)		2. Medical and relief services deployed to city and community evacuation centres
			Salisay	Hips (<2.5 ft.)	CDCC Operation Team (evacuation)	1. Force evacuation operation to all affected areas with head counts and update reports 2. Institute evacuation management at evacuation centres
			Mangin	Hips (<2.5 ft.)		
			Lasip Grande	Waist (<3 ft.)		
			Herrero-Perez	Hips (<2.5 ft.)		
			Pogo Chico	Hips (<2.5 ft.)		
			Mayombo	Hips (<2.5 ft.)		
			Caranglaan	Hips (<2.5 ft.)		
			Tapuac	Hips (<2.5 ft.)		
			Bolosan	Hips (<2.5 ft.)		
			Mamalingling	Hips (<2.5 ft.)		
			Tambac	Hips (<2.5 ft.)		
			Barangay IV	Knee (<1.5 ft.)		
			Poblacion Oeste	Knee (<1.5 ft.)		
			Lucao	Knee (<1.5 ft.)		

(continued)

Table 8.4 (continued)

Warning level control	Status	Flood indicators	Barangay indicators	Depth indicators	CDCC/committee	Response
4. Red	1. Super typhoon	Typhoon rain + dam release + high tide	City wide (31 barangays)		NDCC Takeover	1. Declares state of calamity
	2. High tide forecast				OCD	2. Implementation of emergency evacuation plan
	3. Heavy rain downpour (continuous)					3. Full evacuation operation to all affected areas with head counts and update reports
	4. Dam release					
	5. Neighbouring towns flooded					
Green Back to normal						

Source: Dagupan City Government (2008)

List of agencies: *BFP* Bureau of Fire Protection, *CDCC* City Disaster Coordinating Council, *DepEd* Department of Education, *EOC* Emergency Operations Center, *NDCC* National Disaster Coordinating Council, *OCD* Office of Civil Defense, *PAGASA* Philippine Atmospheric Geophysical and Astronomical Services Administration, *PHIVOLCS* Philippine Institute of Volcanology and Seismology, *PNP* Philippine National Police, *POSO* Public Order and Safety Office, *RDCC* Regional Disaster Coordinating Council

Fig. 8.6 Flood marker placed near the river (Photo by F. Molina 2014)

Fig. 8.7 Flood marker in a Daycare Center (Photo by F. Molina 2014)

Fig. 8.8 Flood maker under the bridge (Photo by F. Molina 2014)

The usage of *kanungkong* cuts across all sectors and age groups. Another example of local knowledge from the people of Dagupan that highlights the importance of intergenerational transfer towards disaster preparedness is evident from Fermin's narrative, as translated below:

> My grandfather advised me that when the wind comes from the place where the sunrise or the sunset is observed, the typhoon is greatly devastating. All houses must be fastened with a rope to withstand the strong winds. People also experience flooding frequently. If there's a typhoon, they no longer go to sleep in order to observe and to know when to evacuate.

Fermin is a fisherman from *barangay Calmay*—one of the vulnerable communities to river flooding in Dagupan, as it is surrounded by *Calmay* River. His life story implies that *Dagupeño*s have acquired local knowledge on disaster preparedness over a period of time.

Other examples of knowledge and practices that people do to prepare and protect themselves during river flooding are presented (Fig. 8.9) intergenerationally: before 1990, 1991–2000 and 2001 to present.

8.5.3 Coding of the Local Knowledge

1. Observation of the environment—people's perception of the nature's behaviour such as movement of the wind, colour of the river, celestial bodies (e.g. moon, stars, animal behaviour, flora and fauna) that serve as early warning signal for

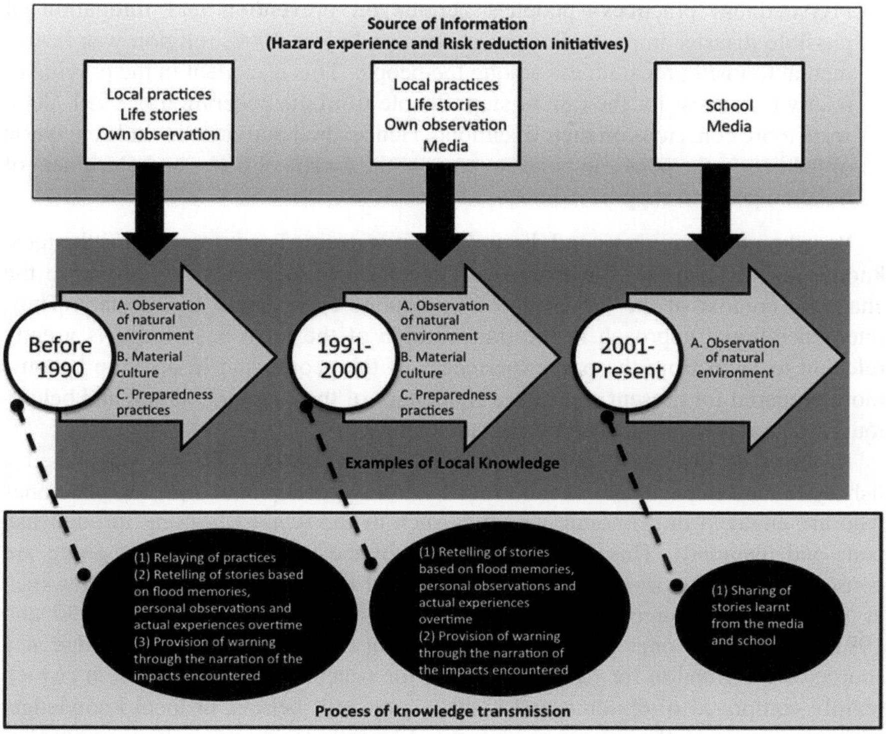

Fig. 8.9 Intergenerational local knowledge transmission—the relationship of the source of information on hazard and risk reduction initiatives and the transmission of the local knowledge from one generation to another (*Source*: Interviews with the people of Dagupan City)

occurrence of the change of season, a particular hazard or a disaster, e.g. torrential rains, typhoon occurrence. The first two categories of generations (i.e. before 1990 and 1991–2000) have shared the following: observation of *alimbusamos* (strong winds), observation of the change of river colour, observation of ants collecting food and migrating to higher locations, observation of chickens moving to the roof of the house, observation of flowering trees (golden shower starts to bloom during summer), observation of the wind (northeast/southwest monsoon), observation of the moon (large ring around it), observation of the sunset (colour of the horizon), absence of the stars and observation of the sky (change of colour). The last category of generation (i.e. 2001 up to present) has also observed a similar pattern of ants' behaviour as well as a rainbow after the rain.

2. Material culture—people's utilisation of local materials in order to prevent and mitigate for the occurrence of hazards and disasters. This includes but not limited to shelter design, usage of tools and books. The first two generations (i.e. before 1990 and 1991–2000) have shared the usage of *lunario* (booklet on the moon movement), usage of *kanungkong* as a warning tool for people to move to safer places and house design where the indigenous dwellings are elevated with stilts.

3. Preparedness practices—people's actions on prevention and mitigation of possible disaster impacts. For the generation before 1990, religion was instrumental to instil preparedness among the people. This is evident in the praying of rosary to request for the rain to stop. People from the generation of 1991–2000 were more conscious on their livelihood. Hence, their activities include elevation of fishnets in the pens and ponds to prevent the escape of fishes and placement of bamboo posts to support fishnets.

Based on the interviews and discussions, it is deemed as important for the local knowledge to be passed on from one generation to another so as to bridge the changing context of the urbanity and to respond accordingly to disasters. Thus, intergenerational approach in the transmission of the local knowledge is greatly relevant to learn from the past experiences of the people and in order to become more prepared for present and future challenges of the changing nature and behaviour of disasters as influenced by climate change.

In terms of disaster experience, the generation 1991–2000 has postulated a fishery-related impact such as the fishkill. This age group also represents the ones who are engaged in economic activities such as the *bangus* farming through fish pens and fishponds. This reflects that most of the informants of this group are conscious of the environmental repercussions of the anthropogenic activities such as the fishery resource management. Among the generations, before 1990 and 1991–2000 are the ones that demonstrated great reliance on local knowledge as a source of information for disaster risk. As for generation 2001 to present (which mainly comprised of children and youth), some still believe in local knowledge. However, most of them have very limited knowledge because they are highly influenced and are more dependent on the information they gather from media (television and radio) as well as the lessons learned in schools. They revealed that they no longer witness some of the local knowledge that they learn from their parents because most of them move closer to the city centre where their schools are located. Given these, it is therefore important to have an intergenerational transmission of knowledge in order to make the present generation aware of the indigenous indicators of possible environmental threats.

For the local knowledge on river flooding risk reduction and adaptation, it is notable that most of the actions fall under preparedness that comprise initiatives on early warning, prevention and mitigation. There is still a huge data gap on short-term response and long-term adaptation.

As for the river flooding experience, some of the informants perceived it as a normal condition in their lives—for them, water just come and go. Due to this, people do not evacuate to another place; but they prepare for possible evacuation. The flooding incidence that most of them remember and deem as the "strongest" was based on the impact that they had experienced. For example, most of them recall Typhoon Cosme (International name: Halong) due to its devastation on the shelters and infrastructures in the city.

The generation before 1990 is the only group that imparted faith-based action such as the praying of rosary. This speaks of the changing values and beliefs of different generations over time. Likewise, the generation 1991–2000 is the only group that shared about mitigation activities on fisheries management as elaborated before.

In the process of knowledge transfer, the generation before 1990 seems to be the most experienced among all, given their variety of knowledge transfer processes.

8.6 Conclusion

In order to understand the relevance of intergenerational transmission of local knowledge for disaster risk reduction, this study highlights three factors: firstly, the importance of an intergenerational approach in local knowledge transmission to promote decentralised governance linked with the physical and sociopolitical change of a settlement; secondly, the interplay of environment and society in the creation of risk; and lastly, the need for the integration of local knowledge with the existing and shifting scientific approaches and governance structures to promote resilience in river flooding risk reduction and adaptation.

The evolving nature of urban settlements (e.g. Dagupan) that includes migration of people, change of landscape (influenced by the disasters such as the 1990 earthquake and other typhoons) and the changing context of the nature of disasters entails intergenerational understanding to have better risk reduction and adaptation mechanisms. The inadequate historical perspective to understand the contexts and roots of disaster causality hampers the fuller appreciation of the nature of vulnerability (Bankoff 2003b; Lees and Bates 1984; Oliver-Smith 1986). The sequence of environmental events is of critical importance in addition to occurrence, frequency and intensity (Bankoff 2003b; Winterhaler 1980). Just as history reveals that vulnerability may take centuries in the making, so is the sense of its complex construction important to the way in which disaster is increasingly conceptualised (Bankoff 2003b). Part of the information necessary in the transfer of local knowledge is the idea of the kind of hazard and disaster it addresses. The practices related to disaster risk reduction and adaptation governance become alive and more relevant to the lives of people once it is linked with the past experiences. Therefore, understanding how a hazard is perceived over the years is pertinent in order to inform the present with a holistic understanding of vulnerability and to develop mechanisms for river flooding risk reduction and adaptation. Thus, intergenerational local knowledge transmission will help people to govern preparedness mechanisms for risk reduction and adaptation initiatives.

Apart from the transformation of the space's terrain, the sociopolitical shift from a rural settlement to an urban settlement also speaks of aggravated vulnerability and loss of potential capacities as highlighted by Miller and Douglass (2014): "The link between increasingly frequent floods and urbanisation in Asia has been exacerbated by the transition from rural to more urban societies, which has displaced traditional knowledge and social mechanisms to cope with environmental hazards" (p. 10). In this regard, the importance of intergenerational transmission of local knowledge is further needed in urban settlements since people in the cities are more at risk given the relegation of local knowledge and social mechanisms that gives greater exposure to its people to the impacts of hazards and disasters. With the integration of intergenerational transmission of local knowledge to urban disaster governance, there would be a better appreciation of local knowledge in risk reduction and adaptation. This would also give attention to harnessing and continuity of knowledge. Having this as the situation, the mainstreaming of local knowledge with the current scientific information on risk and disaster governance will also become a realisation and a priority for it will be further recognised by people. Given this, the sharing of local knowledge also become a cause and a vehicle to promote social integration and mechanisms for urban disaster governance.

Another point to consider is the relationship of environment and society in constructing the idea of risk and capacity. River flooding in the city of Dagupan is not just a product of its geography and behaviour of typhoons that have affected the area, but it is highly related to the actions made by the people over the years. Excessive practice of *bangus* farming and culture of other marine resources, competition for space in the rivers, overlapping ownerships of land and river and conflict of the state and people further intensify the incidence and impact of disasters. However, despite the fact that risk is built around the people's actions, it is also in people's local knowledge and practices where capacity to govern urban disasters can be anchored. With the intergenerational transmission of these local knowledge and practices on urban disaster governance, people will become more conscious of the actions that gear towards risk reduction and adaptation.

People's actions today are informed by the past initiatives and practices but not all actions are transmitted to the succeeding generations. Given that the location is urban, the exchange of practices among the heterogeneous group may result in loss, modification or mediation of knowledge and practices due to the contact of one identity to another. Another concern is the fact that the nature of disaster changes along with the changing climate. This condition may entail new approaches and initiatives. Thus, certain knowledge and practices may not address the changing context of disasters and impacts brought by climate change. Therefore, it is necessary to raise concerns on the need to integrate local knowledge with the existing and shifting scientific approaches and governance structures to promote resilience in river flooding risk reduction and adaptation.

References

Bankoff, G. (2003a). *Cultures of disaster society and natural hazard in the Philippines.* London: RoutledgeCurzon.

Bankoff, G. (2003b). Constructing vulnerability: The historical, natural and social generation of flooding in Metropolitan Manila. *Disasters*, 27(3), 95–109.

Baker, J. (Ed.). (2012). *Climate change, disaster risk and the urban poor cities building resilience for a changing world.* Washington, DC: The World Bank.

Beck, M., Shepard C., Birkmann, J., Rhyner, J., Welle, T., Witting, M., et al. (2012). *World risk report 2012*. Berlin: Bündnis Entwicklung Hilft (Alliance Development Works).
Coping. (1990). Coping as town sinks. *Asiaweek*, 16(31), 34.
Center for Disaster Preparedness. (2009). *Project PROMISE (Program for hydro- meteorological disaster mitigation in secondary cities in Asia) Terminal Report*.
Centre for Research on the Epidemiology of Disasters (CRED) Research Institute Health & Society (IRSS). (2013, March). Disaster data: A balanced perspective. *CRED Crunch*, 31, 1–2.
Cortes, R. M. (1974). *Pangasinan, 1572–1800 (1)*. Quezon City: New Day Publishers.
Cortes, R.M. (1990). *Pangasinan, 1801–1900 (2)*. Quezon City: New Day Publishers.
Cronin, S.J., Gaylord, D.R., Charley, D., Alloway, B.V., Wallez, S., & Esau, J.W. (2004a). Participatory methods of incorporating scientific with traditional knowledge for volcanic hazard management on Ambae Island, Vanuatu. *Bulletin of Volcanology*, 66(7), 652–668.
Cronin, S.J., Petterson, M.J., Taylor, M.W., & Biliki, R. (2004b). Maximising multi-stakeholder participation in government and community volcanic hazard management programs: A case study from Savo, Solomon Islands. *Natural Hazards*, 33(1), 105–136.
Cruz, J.E. (2010, October 4–5). *Estimating informal settlers in the Philippines*. A paper presentation for the 11th National Convention on Statistics (NCS) at EDSA Shangri-La Hotel, Philippines.
Dagupan City Government. (2008). *2008 Dagupan City disaster coordinating council manual*. Dagupan: Dagupan City Government.
Dagupan City Government. (2014). *2014 Dagupan City disaster risk reduction and management council manual*. Dagupan: Dagupan City Government.
Davies, C.A. (1999). *Reflexive ethnography*. London: Routledge.
Denkens, J. (2007). *Local knowledge for disaster preparedness: A literature review*. International Centre for Integrated Mountain Development (ICIMOD) and The European Commission Humanitarian Aid (ECHO).
Ezilon Maps. (n.d.). *Philippines*. http://www.ezilon.com/maps/images/asia/Philippines-physical-map.gif. Accessed 31 Oct 2014.
Fernando, J.L. (2003). NGOs and production of indigenous knowledge under the condition of postmodernity. *The Annals of the American Academy of Political and Social Science*, 590(1), 54–72.
Iglesias, G. (2007, April). Cooperation between local authority and communities reducing flood disaster risk in Dagupan City, Philippines. *Safer Cities*, 16, 1–8.
Jacobs-Huey, L. (2002). The natives are gazing and talking back: Reviewing the problematics of positionality, voice, and accountability among “native” anthropologists. *American Anthropologist*, 104(3), 791–804.
Lees, S. & Bates, D. (1984). Environmental events and the ecology of cumulative change. In E. Moran (Ed.), *The ecosystem concept in anthropology* (pp. 133–159). Boulder: Westview Press.
López-Marrero, T., & Wisner, B. (2012). Not in the same boat: Disasters and differential vulnerability in the insular caribbean. *Caribbean Studies*, 40(2), 129–168.
McGranahan, G., Balk, D., & Anderson, B. (2007). The rising tide: Assessing the risks of climate change and human settlements in low elevation coastal zones. *Environment and Urbanization*, 19(1), 17–37.
Mercer, J., Kelman, I., Taranis, L., & Suchet-Pearson, S. (2010). Framework for integrating indigenous and scientific knowledge for disaster risk reduction. *Disasters*, 3(1), 214–239. doi:10.1111/j.03613666.2009.01126.x.
Miller, M., & Douglass, M. (2014, forthcoming). Introduction: Governing flooding in Asia's urban transition. *Pacific Affairs Special Issue* (pp. 1–21).
National Statistics Office. (2014). *The Philippines in figures 2014*. Quezon City: National Statistics Office.
Oliver-Smith, A. (1986). Disaster context and causation: An overview of changing perspectives in disaster research. In A. Oliver-Smith (Ed.), *Natural disasters and cultural responses* (pp. 1–38). Williamsburg: College of William and Mary.

Pablo, C., Papa, A., Batino, C., Nazarino, R., Burgonio, T., Medina, R., et al. (1999, December 3). 3 Killed as quake Jolts Luzon; Power cut off anew. *Philippine Daily Inquirer* (Interactive).
Peirano, M. (1998). When anthropology is at home: The different contexts of a single discipline. *Annual Review of Anthropology*, 27, 105–128.
Peluso, N., & Watts, M. (Eds.). (2001). *Violent environments*. New York: Cornell University Press.
Rahman, A. (2000, October 30–November 1). *Development of an integrated traditional and scientific knowledge base: A mechanism for accessing, benefit-sharing and documenting traditional knowledge for sustainable socio- economic development and poverty alleviation*. Paper presented at the United Nations Conference on Trade and Development (UNCTAD) Expert Meeting on Systems and National Experiences for Protecting Traditional Knowledge, Innovation and Practices, Geneva.
Sarmiento, R. (2011). The Catandunganons—Making and seeing selves through the eye of the typhoon. *AghamTao*, 20, 68–86.
Shaw, R., & Baumwoll, J. (Eds.). (2008). *Indigenous knowledge for disaster risk reduction good practices and lessons learned from experiences in the Asia- Pacific region*. Bangkok: UNISDR.
Shaw, R., & Sharma, A. (2008). *Transferable indigenous knowledge (TIK): Progress and challenges*. A PowerPoint presentation for the Kyoto University's Graduate School of Global Economic Studies, Japan.
Sillitoe, P. (1998). The development of indigenous knowledge: A new applied anthropology 1. *Current Anthropology*, 39(2), 223–252.
Sotelo, Y. (2012, August 14). Bangus overproduction causes sinking- study. *Inquirer Northern Luzon* (Interactive).
Sundermann, L., Schelske, O., & Hausmann, P. (2013). *Mind the risk – A global ranking of cities under threat from natural disasters*. Zurich: Swiss Reinsurance Company Ltd.
United Nations International Strategy for Disaster Reduction. (2009). *2009 UNISDR terminology on disaster risk reduction*. http://www.unisdr.org/files/7817_UNISDRTerminologyEnglish.pdf. Accessed 1 June 2014.
United Nations Population Fund. (2011). *2011 State of the world population people and possibilities in a world of 7 billion*. http://foweb.unfpa.org/SWP2011/reports/EN-SWOP2011-FINAL.pdf. Accessed 1 Sep 2014.
United Nations Population Fund. (2013). *2013 State of the world population motherhood in childhood facing the challenge of adolescent pregnancy*. http://www.unfpa.org/webdav/site/global/shared/swp2013/EN-SWOP2013-final.pdf. Accessed 1 Sept 2014.
Victoria, L. (2003). *Kahandaan, Katatagan at Kaunlaran ng Komunidad: Gabay sa Pagsasanay sa Disaster Management*. Quezon City: Center for Disaster Preparedness.
Victoria, L. (2008). Dagupan City, Pangasinan, Philippines combining indigenous and scientific knowledge in the Dagupan City flood warning system. In R. Shaw & J. Baumwoll (Eds.), *Indigenous knowledge for disaster risk reduction good practices and lessons learned from experiences in the Asia- Pacific region* (pp. 52–54). Bangkok: UNISDR.
Vistro, J.V., & Reyes, C. (2005). *Nandaragupan: The story of a coastal city and Dagupan Bangus*. Dagupan: Dagupan City Heritage Commission, Office of the City Mayor.
Winterhaler, B. (1980). Environmental analysis in human evolution and adaptation research. *Human Ecology*, 8, 137–170.
Wisner, B. (2010). Climate change and cultural diversity. *International Social Science Journal*, 61, 131–140.
Wisner, B., & Luce, H. (1993). Disaster vulnerability: Scale, power and daily life. *GeoJournal*, 30(2), 127–140.
Wisner, B., Blaikie, P., Cannon, T., & Davis, I. (2004). *At risk: Natural hazards, people's vulnerability and disasters*. London: Routledge.
Wisner, B., Gaillard, J.C., & Kelman, I. (Eds.). (2012). *Handbook of hazards and disaster risk reduction*. London: Routledge.

Chapter 9
Securing the Safety of Informal Settler Families Along Waterways in Metro Manila, Philippines: Government-Civil Society Organisation Partnership

Benigno C. Balgos

Abstract This chapter examines how and to what extent social institutions such as the government (national, local and community levels) and civil society organisations contribute to disaster governance in Metro Manila, Philippines. Most importantly, it looks at collaborative responses of state and not-state actors addressing disaster vulnerability of informal settlers living near the waterways and the factors that assist and impede such engagement. It further examines the participation of informal settler communities in such initiatives. The chapter argues that civil society facilitates urban resiliency by providing resources and transferring technologies. However, the political will, commitment and social accountability of the state are imperative to accelerate the process of urban resiliency and sustain the gains of the collaborative governance.

Keywords Informal settler families • Civil society organisations • Urban resilience • Collaborative governance • Metro Manila

9.1 Introduction

The scale and the pace of urbanisation in today's world are unprecedented in human history (Davis 2006). Rural populations, particularly in developing countries, are declining, while migration to urban areas is typically increasing (Davis 2006). As a result, cities are expanding rapidly. At present, Asia leads the growing pace of urbanisation worldwide (ADB 2006 as cited in Shaw et al. 2009, p.77). The rise of urbanisation in Asia is exceptional. In fact, the World Urbanisation Prospects released by the United Nations projects that in 2015, there would be 550 megacities worldwide. Asian Development Bank (ADB) report (2012) indicates that in 2010,

B.C. Balgos (✉)
Independent Researcher, Manila, Philippines
e-mail: ninoybalgos@gmail.com

M.A. Miller, M. Douglass (eds.), *Disaster Governance in Urbanising Asia*,
DOI 10.1007/978-981-287-649-2_9

12 of the 23 worldwide megacities were in Asia; and these numbers are expected to increase to 21 of 37 worldwide megacities by 2025 (ADB 2012). The United Nations Global Assessment Report on Disaster Risk Reduction states that over a billion people have been added to Asian cities from 1980 to 2010, while it is estimated that another billion will be added to Asia's urban centres by 2040 (UN 2011).

Economic growth is considered to be the primary driver of such phenomenal urbanisation (Jha and Stanton-Geddes 2013; Shaw et al. 2009). Davis (2006) and Shaw et al. (2009) argue that the expansion of the Asian economy is largely driven by globalisation. The so-called exodus of rural workforce to megalopolis is propelled by agricultural deregulation and financial discipline policies imposed by the World Bank and the International Monetary Fund (Davis 2006). He further notes that increased flow and movement of people, goods, services and information, among others, have led to the rural-urban nexus. Quarantelli (as cited in Shaw et al. 2009, p. 82) mentions that urbanisation is irreversible. Thus, urban vulnerability has become a reality. Urbanisation will tremendously influence the future of disaster governance as many Asians choose to dwell in unsafe and vulnerable locations in urban areas (Shaw et al. 2009; UNISDR 2013).

Davis (2006) proposes two kinds of cities that describe this epochal transition. He notes that there are "megacities" (with population between 8 million and 20 million) and "hypercities" (with population of more than 20 million). According to the 2013 edition of the Demographia of World Urban Areas, there are ten cities that fall into the aforementioned "hypercities" category—one of which is Metro Manila, capital of the Philippines comprising of 16 cities and a municipality with approximately 21,241,000 residents.

With 94,013,200 people, in 2010, the Philippines ranks as the 12th most highly populated country in the world (NSCB 2012). Simultaneously, the urban population has reached 66.4 % (ADB 2010), indicating that 6 out of 10 Filipinos live in urban areas. ADB estimates that by 2030, the aforementioned figure would reach 67 %. Affirming the link between globalisation and economic growth to urbanisation in Asia, the Philippine Development Plan 2011–2016 attributes the lack of economic opportunities in rural areas as the main reason for the rise of urban population (NEDA 2011). The Philippine Government estimates that there are 104,000 informal settler families inhabiting disaster-prone areas in Metro Manila such as creeks, rivers and waterways, thus being at high risk. For instance, during the September 2009 massive floods[1] in Metro Manila, such families were the worst affected.

With this in mind, this chapter examines how and to what extent social institutions such as the government (national, local and community levels) and civil society organisations contribute to urban disaster governance in Metro Manila. More pointedly, it looks into collaborative responses of state and not-state actors to address disaster vulnerability of informal settlers living near the waterways. In the same way, the chapter examines the factors that assist and impede such collaborative engagement as well as how and to what extent informal settler communities are actively involved in the initiatives. The chapter argues that civil society facilitates

[1] These floods were a result of Tropical Storm Ketsana (local name: Ondoy).

change by providing resources and transferring technologies, but the strong commitment, political will and social accountability of the state are imperative to accelerate the process of change and sustain the gains of the collaborative governance.

The discussion and findings of this chapter are mainly drawn from: (1) semi-structured interviews with key informants (i.e. local government and informal settlers); (2) focus group discussions and community risk assessment workshops with the local government and informal settlers; (3) participation and observations in meetings between informal settler families and the national/local government units, civil society organisations and a donor organisation; and (4) review and content analysis of government strategic policies on informal settling in the Philippines.

The chapter is organised into three sections, which begins with a discussion of the exposure, vulnerability to hydrometeorological hazards (and to some extent geological hazards) and the context of informal settling in Metro Manila. In so doing, the case of Tropical Storm Ketsana (2009), a catastrophic event that submerged Metro Manila to floodwater, would be discussed. The second section provides the policy frameworks and programmes of the Philippine government to address disaster vulnerability and exposure, particularly of the informal settler families living in eight major waterways in Metro Manila. In relation to the policy frameworks, the section also looks into the collaborative engagement of the government, donor organisation [Department of Foreign Affairs and Trade (DFAT) – Australian Government], civil society organisations and communities in ensuring the safety of high-risk communities along the waterways. Finally, the challenges, implications and lessons learned on collaborative disaster governance in relation to urban resiliency would be presented.

9.2 The Tale of Typhoon Ketsana and the Informal Settling in Metro Manila

Metro Manila, located in the island of Luzon, is the Philippines' National Capital Region (NCR) comprising 16 cities and a municipality. As the seat of political and economic power, it plays a significant role in the country's growth and development. Over the decades, Metro Manila has experienced rapid urbanisation with unprecedented growth of population and infrastructure development. It houses state-of-art buildings, a three-line mass transit system, sprawling shopping malls, condominiums and many other attractions that encourage rural to urban migration. All of these including 21, 241,000 residents make up the urban Metro Manila. However, this very characteristic of Metro Manila makes it prone to geo-, pyro- and hydrometeorological hazards.

The high exposure and vulnerability of Metro Manila to disaster risk are a result of the ballooning number of urban dwellers, particularly living in danger areas such as creeks and rivers. Government figures reveal that there are approximately 104,000 informal settler families in Metro Manila alone (DILG 2013), of whom 60,000 are

Fig. 9.1 Informal settler houses (*Source*: Author (taken on 26 August 2014))

considered living in dangerous areas. Figure 9.1 shows typical houses of informal settlers in communities along a waterway.

When Metro Manila was submerged in the wake of Tropical Storm Ketsana that made a landfall on 26 September 2009, many of the informal settlers along the waterways had been affected. Arlene Dulog,[2] resident of *Barangay* (community) Dona Imelda in Quezon City, mentioned that during the disaster, the floodwater was so deep that it reached the second floor of their house. As a result, she and her family had not been able to carry out their daily tasks. Likewise, Elizabeth Camara,[3] an informal settler from *Barangay 894* in Manila, recounted that they were really scared because it was their first experience of a very deep flood, one which rose quickly and lasted long. She mentioned the floodwater rose to the second floor while they were transferring items from the first floor. In a focus group discussion (FGD) conducted in *Barangay 894*,[4] informal settlers shared the different impacts the storm Ketsana had in their community. These include food scarcity, scattered waste after the flood, conflicts between community leaders and residents due to uneven relief distributions, damage to houses and other community properties, loss of lives, respiratory and skin diseases and scarcity of potable drinking water.

According to the Philippine Atmospheric Geophysical and Astronomical Services Administrations (PAGASA), the country's weather bureau, the volume of water from the 6-h continuous rainfall brought by Tropical Storm Ketsana was equivalent to what the region normally receives in a month. Thus, Metro Manila experienced a deluge, which prompted the then administration of Gloria

[2] Interview on 26 August 2014.

[3] Interview on 20 August 2014.

[4] FGD on 21 August 2014.

Macapagal-Arroyo to declare a State of National Calamity. Never was there a catastrophe of this magnitude in recent decades.

According to the Final Report of the National Disaster Coordinating Council (NDCC; now the National Disaster Risk Reduction and Management Council or NDRRMC), 247,000 homes were destroyed, and 632,000 ha of crops were damaged from the storm. It further states that the storm displaced about 700,000 families (mainly informal settler families) and affected nine million people (NDCC 2009). The Post-Disaster Needs Assessment (PDNA) of the World Bank Office, Manila, reports that damages caused by the storm totalled US$ 4.38 billion (IPC 2010).

Tropical Storm Ketsana paralysed both the national government and the affected local government units of Metro Manila. As a matter of policy, the Philippine disaster management system has been predominantly reactive—the state responds only when disaster strikes. The aforementioned disaster revealed the unpreparedness of the Philippine Government in an event of catastrophe (Balgos 2013). In addition, it also revealed the ineffectiveness and inefficiency of the Philippine disaster management system. However, it made the government and the public realise the possibility of the re-occurrence of similar disasters in future. Therefore, in a way, it made the government recognise the need to relocate families living in unsafe, disaster-prone areas, such as creeks and waterways.

However, before undertaking necessary actions, the government had to carefully study the risks and vulnerabilities of Metro Manila to be able to come up with an informed decision and strategic measures to address the problem of flooding while at the same time considering the long-term security and safety of the informal settlers. Hence, in the aftermath of Tropical Storm Ketsana, the Philippine Government initiated the Metro Manila Post-Ketsana Recovery and Reconstruction Program. One of the programmes in the said undertaking is the Enhancing Risk Analysis Capacities for Flood, Typhoon, Severe Wind and Earthquake for Greater Metro Manila Area (GMMA), otherwise known as the Risk Analysis Project (RAP). The project, funded by the Department of Foreign Affairs and Trade (DFAT) of the Australian Government (formerly AusAid), had three primary components, namely: (1) Flood Risk Modelling which covers the Pasig-Marikina River Basin, (2) Tropical Cyclone Severe Wind Risk Modelling for the Greater Metro Manila Area and (3) earthquake vulnerability. The study revealed that in the event of catastrophic flooding in the Pasig-Marikina River Basin, damage will be experienced in four areas: the Marikina River near Tumana, along the banks of the Manggahan Floodway, along banks of San Juan River and at various locations along the Lakeshore and Taguig-Pateros regions (Badilla 2014). It is in these river systems where great numbers of informal settler families are concentrated. Additionally, the RAP study reveals that frequent intense cyclones are likely to persist as a result of the change in global temperature. Therefore, the Philippines will continue to be vulnerable to severe wind. Hilario and Abadilla (2014) argues that this is apparent in three things: (1) the frequency, duration and intensity of rainfall have changed over the years; (2) more extreme weather events are expected to occur in several parts of the country; and (3) more people are affected by hydrometeorological-related hazards. Also, the RAP results reiterated that Metro Manila is extremely vulnerable to severe winds, particularly the northeastern portion of the Greater Metro Manila Area (Hilario and Abadilla 2014). The types and ages of the establishments in Metro Manila exacerbate the exposure and vulnerability. As a result, retrofitting of buildings is necessary (Hilario and Abadilla 2014). In short, Metro Manila, including the informal settler families within its administrative boundaries, would continue to be vulnerable and exposed to hydrometeorological hazards.

The weather patterns in the Philippines from 1951 to 2009 have given meteorologists an idea of what could be expected in 2020–2050. A separate study entitled Climate Change in the Philippines conducted by PAGASA under the Millennium Development Goals Fund (MDGF) Joint Program entitled Strengthening the Philippines' Institutional Capacity to Adapt to Climate Change reveals that a regular southwest monsoon in June, July and August is expected to bring heavier rainfall, particularly in Luzon and in the Visayas. The 2012 massive flooding in Metro Manila, which also greatly affected informal settler families, was a result of the southwest monsoon. The report also highlighted that rainfall would be more frequent, while a dry spell could be expected during the hot seasons (PAGASA 2011). Although there is a belief that large-scale disaster events have a hundred-year return period, it took only three years after the onset of Tropical Storm Ketsana for a similar disaster to occur in Metro Manila. Learning from the experience of that storm, the southwest monsoon put to the test the systems and mechanisms that have since been put in place to improve the disaster management of the country. Notwithstanding, the 2012 southwest monsoon showed that Metro Manila including its growing informal settler families is still vulnerable to hydrometeorological hazards. Aside from flooding, these families are also prone to earthquakes based on the findings of the earlier Metro Manila Earthquake Impact Reduction Study (MMEIRS) in 2004 and the Risk Analysis Project studies. The findings of the MMEIRS and RAP underscore that Metro Manila is long overdue for a big earthquake, which is expected to result from a rupture in the West Valley Fault (formerly the Marikina Fault). According to Dr. Renato Solidum, Director of the Philippine Institute of Volcanology and Seismology (PHIVOLCS), the RAP results show that in the event of a 7.2 magnitude earthquake, the fatalities would be 31,228. Concurrently, the projected injuries are as follows: 14,000 very seriously injured, 112,000 seriously injured and 385,000 slightly injured (Solidum 2014). With this stark reality, Metro Manila was ranked second among the world's 10 riskiest cities in relation to earthquakes, hurricanes, cyclones and tsunamis (Michael 2014).

9.3 Zero Casualty: Government and Civil Society Partnership on Urban Resiliency

Birkland (2010) argues that a catastrophic event can provide windows of opportunities for policy change. He further claims that a disaster event can do what policy entrepreneurs, lobby groups and researchers, among many others may fail to do: to put forward Disaster Risk Reduction (DRR) in the policy-making agenda (Birkland

2010). The then President Gloria Macapagal-Arroyo's signing of the Republic Act 9729[5] or the Climate Change Act of the Philippines (GOP) after Tropical Storm Ketsana was a testament that the government was starting to consider the impacts of climate change to the country. As a result, the Philippines is now one of the few countries that have a law on climate change in place. Undoubtedly, Tropical Storm Ketsana created significant changes in the Philippine disaster management system. Not only did President Macapagal-Arroyo ratify the Climate Change Act, but in May 2011 she also signed the Republic Act (RA) 10121 or the Philippine Disaster Risk Reduction and Management Act (GOP 2010). The Republic Act 10121 replaced the three-decade-old President Decree (PD) 1566, which was ratified in 1978 by then President Marcos as a response to a series of massive floods that hit Metro Manila and its provinces in 1970, 1972, 1974, 1976 and 1978 (Warren 2013). This was a significant achievement for DRR in the country as it prompted the shift from a reactive to a proactive approach in risk reduction (Balgos 2013).

Prior to the enactment of the Philippines' existing legislations on disaster risk reduction and climate change adaptation, the country's disaster management system subscribed to the highly reactive approach as put forth in PD 1566. As a result of it, cities in Metro Manila aligned their disaster management plans to focus mainly on emergency response. The Disaster Risk Reduction Network Philippines (DRRNet Philippines), an alliance of civil society organisations in the Philippines working on disaster risk reduction, notes three main differences between PD 1566 and the RA 10121: (1) PD 1566 was top-down in nature, with highly centralised management by the National Disaster Coordinating Council (NDCC), while RA 10121 capitalises on a bottom-up approach and participatory disaster risk reduction; (2) PD 1566 viewed disaster as a form of physical hazard, while RA 10121 recognises disaster as a reflection of people's vulnerability; and (3) PD 1566 focused on disaster response and anticipation, while RA 10121 emphasises on an integrated approach to genuine risk reduction to sustainable development (DRRNet Philippines 2011).

As a consequence of the vulnerability and exposure of Metro Manila to various forms of hazards and as a commitment to RA 10121, the Metro Manila Disaster Risk Reduction and Management Council (MMDRRMC) was formed. In keeping with the Republic Act 7924,[6] the Chairman of the Metro Manila Development Authority acts as the Chairman of the council (GOP 1994). Concurrently, the Office of Civil Defense serves as the secretariat and coordinating institution of the council. The other members of the council that act as Vice-Chairpersons are: Department of Science and Technology (DOST),NCR, for prevention and mitigation; Department of Interior and Local Government (DILG),NCR, for preparedness; Department of Social Welfare and Development (DSWD),NCR, for response; and the National

[5] Government of the Philippines (GOP) (2009). An act mainstreaming climate change into government policy formulations, establishing the framework strategy and programme on climate change, creating for this purpose the Climate Change Commission.

[6] Government of the Philippines (GOP) (1994). An act creating the Metropolitan Manila Development Authority, defining its powers and functions, providing funding therefore and for other purposes.

Economic and Development Authority (NEDA),NCR, for recovery and rehabilitation. Apart from these, the executive department of the Metro Manila Development Authority and the regional and field stations operating in NCR are members of the council. All local governments in Metro Manila are likewise members of the council. The 17 local government units of Metro Manila form part of the MMDRRMC interim organisation. Concurrently, given that Metro Manila is also vulnerable to earthquake, the MMDRRMC organised the Task Force Rainbow specifically designed to respond to emergencies as a consequence of earthquake. Similarly, after a series of consultations, the Department of Public Works and Highways (DPWH) in partnership with various stakeholders came up with the Metro Manila Integrated Flood Risk Management Master Plan. The master plan was deemed imperative given the vulnerability and exposure of Metro Manila to flooding. According to DPWH Secretary Rogelio Singson, the three most important reasons for the master plan to be developed are: (1) lack of integrated plan and strategic programme to address perennial flooding and other water-related disasters such as landslides, lack of potable water and siltation; (2) massive urbanisation and lack of affected land use plans in the Greater Metro Manila Area (GMMA), which has exacerbated flooding; and (3) the illegal structures that have narrowed the waterways of Metro Manila causing massive floods (Singson 2013). He also indicates that due to the lack of a master plan, flooding in Metro Manila has annually caused US$160 million financial damage in addition to 640 cost of lives, 3.4 million displaced people and 71,000 damaged houses of Metro Manila residents (Singson 2013).

In the Philippine Development Plan 2011–2016 under the Aquino Administration, among the policy directions and programmes in relation to addressing recurrent flooding in Metro Manila are: (1) preparation of flood control master plan for major river basins; (2) prioritisation of the construction of flood control structures in high risk areas; (3) application of CCA and DRRM strategies in the planning and design of flood management; (4) increasing the local government and community participation in DRR and CCA initiatives; (5) creation of a master plan for flood management and clearing of waterways in Metro Manila; (6) initiation of a water convergence programme with the support of various government agencies such as the Department of Public Works and Highways (DPWH), Department of Agriculture (DA), Department of Agrarian Reform (DAR) and the Department of Environment and National Resources (DENR); and (7) fund allocations for various flood control projects and 11 small water impounding projects (SWIPS) (NEDA 2011).

Under the new integrated approach for Metro Manila, among the urgent structural and non-structural measures identified are: (1) flood modelling, forecasting and warning system through Project NOAH; (2) enforcing the easement requirements and clearing of priority rivers and waterways of informal settlers and obstructions; (3) creating resettlement action plans and provision of housing options; (4) upgrading of pumping stations; and (5) widening the waterway channel, dredging and construction of dikes and river walls (Singson 2013). The long-term measures include: construction of flood control dam upstream, use of natural flood plains near waterways, land use ordinances, strict enforcement of waterway easement laws and increase participation of local government and communities (Singson 2013).

According to the Official Gazette of the Republic of the Philippines, the DPWH under the flood master plan for Metro Manila would carry out 11 structural mitigation measures until 2035. These are: Pasig-Marikina River Improvement and Dam Construction, Meycauayan River Improvement, Malabon-Tullahan River Improvement, South Parañaque-Las Piñas River Improvement, East Manggahan Floodway (Cainta and Taytay River Improvement), West Laguna Lakeshore Land Raising, Land Raising for Small Cities around Laguna Lakeshore, Improvement of the Inflow Rivers to Laguna Lake, Manila Core Area Drainage Improvement, West Manggahan Area Drainage Improvement and the Valenzuela-Obando-Meycauayan (VOM) Improvement. The estimated project cost for all these initiatives is Php 351,718,000 billion (US$ 7,999,590).

During the FGDs and workshops, residents of high-risk communities mentioned that informal settler families lined up for hours just to get a bag of relief goods that includes 2 kg of rice, canned goods, toiletries and used clothes. In fact, many residents living in flood-prone areas in Metro Manila are now familiar with these kinds of events. Disasters are becoming a social event for them. Residents have claimed they prefer not to relocate according to local government orders because of the real or perceived dangers at the designated evacuation sites. Instead, they prefer to remain on the roofs of their houses and receive relief goods provided by the government using helicopters.

Scholars and development organisations define slums in various ways. Booth notes that informal settlers dwell in areas "characterised by amalgam of dilapidated housing, overcrowding, disease, poverty, and vice" (Davis 2006, p. 22–23). The Victorian Calumnies holds that these places are "characterised by overcrowding, poor or informal housing, inadequate access to safe water and sanitation, and insecurity of tenure" (*Ibid*). On the other hand, in the Nairobi meeting in October 2002, the United Nations highlighted that slums are "restricted to the physical and legal characteristics of the settlement and eschews the more difficult to measure social dimensions although it equates under most circumstances to economic and social marginality" (Davis 2006, p. 23).

Given the long history of informal settling in Metro Manila and the vulnerability of these families to both geological and hydrometeorological hazards as evident in the experience during Tropical Storm Ketsana, the government and civil society organisations undertake strategic measures to address this vulnerability and exposure, particularly of the residents near waterways. The following provides, by far, the most comprehensive programme and response of the government and civil society organisations for the informal settler families in Metro Manila.

9.3.1 Response: National Government

As a result of the severity of urban disasters in the country, President Benigno Aquino III has three programmes of action in relation to the problem of informal settler families. Firstly, Php 50 billion (US$ 1,136,363,636) was allocated for housing developments for the national government to carry out the provisions in the Urban Development House Act (UDHA), which was ratified in 1992 to solve the urban housing problem of the country. The goal of this law is to provide decent and affordable housing to urban dwellers. In order to facilitate the efficient release of the allocated funds, a National Technical Working Group (NTWG) on informal settler families was established. The NTWG, chaired by the Department of Interior and Local Government, comprises various government agencies authorised by the President to discuss and find solutions to the issue of informal settling as well as allocation and utilisation of the fund. In relation to the use of the aforementioned fund, a Joint Memorandum Circular (JMC) was drafted among member agencies. Of the 50 billion (US$ 1,136,363,636), 10 billion (US$ 227,272,727) in 2011, 7.5 billion (US$ 170,454,545) in 2012 and another 10 billion (US$ 227,272,727) in 2013 have so far been requested from Department of Budget and Management. However, it took some time for the JMC to be signed by all concerned authorities/agencies due to politics of the country.

Secondly, President Aquino urged to improve the government's preparedness efforts through the call for "zero casualties", particularly during the rainy months in the country. With the call to make Metro Manila both flood-resilient and inclusive, the DILG was asked to prioritise housing projects by relocating the informal settler families within the 3 m easement of waterways. In so doing, informal settler families in eight river systems traversing ten cities in Metro Manila were prioritised (Table 9.1).

Finally, these families were given the option to relocate to areas developed by the National Housing Authority or by the civil society organisations themselves through the creation of a series of "People's Plans". According to the JMC, a community would not be relocated without the appropriate People's Plan which highlights, among other things, the kind of housing and the amount that they are willing to pay for their monthly amortisation. Likewise, if they opt to voluntarily relocate and dismantle their houses, they would automatically receive a sum of Php 18,000 (US$

Table 9.1 Ten cities in Metro Manila and the traversing major waterways

Metro Manila cities	Waterways
Makati City	Pasig River and Tripa de Gallina
Mandaluyong City	San Juan River and Pasig River
City of Manila	San Juan River, Pasig River, Estero de Sunog Apog
Quezon City	San Juan River and Tullahan River
Pasig City	Pasig River and Manggahan Floodway
Pasay City	Tripa de Gallina and Maricaban Creek
Caloocan City	Tullahan River and Estero de Maypajo
Valenzuela City	Tullahan River
San Juan City	San Juan River
Malabon City	Tullahan River

Adopted from DILG (2013)

409) as financial assistance. Under this policy, informal settlers would not be evicted without a genuine consultation.

9.3.2 Response: Department of Foreign Trade and Affairs (DFAT), Australian Government

In relation to financial and technical assistance, the DFAT, Australian Government, is now the leading multilateral donor of the Philippines. Although Japan Overseas Development Aid previously held this position, their financial assistance declined after the 2011 Great East Japan Earthquake. Over the years, the DFAT increased its regional presence through its various development assistance activities in the Asia Pacific region. The two strategic objectives of the DFAT are: (1) to strengthen basic services for the poor and (2) to reduce vulnerabilities arising from climate change and conflict (DFAT 2013).

In the Securing the Safety of Informal Settler Families (SSISF) in Metro Manila project, DFAT assists the Philippine Government to address the informal settling issue in the country. This project is helmed by the DILG, which is mandated to supervise disaster preparedness of communities and local government units.

According to the DFAT, the core problem being addressed in this initiative is the high density of urban areas in Metro Manila, which make dwellers remain vulnerable to adverse impacts of natural hazards (DFAT 2013). Three significant reasons for this problem are identified as: (1) the significant number of informal settlers in river system, lakeshores and river banks; (2) weak and overburdened capacity for efficient and effective urban planning; and (3) lack of capacity of the local governments in relation to disaster risk reduction and management (DFAT, Australian Government, 2013).

As earlier mentioned, learning from the onslaught of Tropical Storm Ketsana, the Aquino administration has intended to clear the waterways of Metro Manila, so that there would not be any obstructions to the water flow during a flood. In fact, the clearing of waterways and the relocation of informal settler families are considered as one of the intended legacies of the current administration.

The joint initiative of the Philippine Government and DFAT, Australian Government, has three specific components that involve different civil society organisations. These are: Component 1, socioeconomic profiling of the informal settler families undertaken by the school of urban and regional planning of the University of the Philippines; Component 2, support of civil society to informal settlers carried out by The Asia Foundation; and Component 3, capacity building on Community-Based Disaster Risk Reduction and Management (CBDRRM), which is assigned to the Disaster Risk Reduction Network Philippines, a coalition of civil society organisations working on disaster risk reduction and climate change adaptation in the country (DFAT 2013).

The Component 1 developed a socioeconomic baseline and supports the DILG in community organising and empowerment activities. It produced a socioeconomic profile of the informal settler families and conducted a study on affordability and willingness to pay for social housing. However, the implementation was delayed for some time as other government agencies such as the National Anti-Poverty Commission, National Housing Authority and the Department of Social Welfare and Development performed the same profiling of informal settlers. In fact, the content and questions in the socioeconomic profiling tool of Component 1 were the same family with other government agencies. Thus, the proponents of the Component 1 had to closely coordinate with relevant agencies to standardise the profiling tool.

Simultaneously, Component 2 strengthened the people's planning process to align the community plans with the requirements of funding sources. Component 3 would be discussed in the succeeding section of the chapter.

Clearly, DRR is the key priority of DFAT in development assistance to the Asia Pacific Region including the Philippines. Generally, DFAT's initiatives on risk reduction are coursed through the Government of the Philippines and its related agencies.

9.3.3 Response: Civil Society Organisations

Although informal settler families have survived various floods in the recent past and have coped with such disasters through their social capital or indigenous coping mechanisms, many high-risk communities still do not know the risks they face as well as the course of actions to take to prevent, mitigate, prepare, respond and recover from catastrophic events. The author's conversations with these families highlight that there is an agreement among urban dwellers of the need for capacity building on risk reduction and climate change adaptation, so that they are equipped with at least the basic knowledge and skills to reduce possible risks. Further, informal settlers in *Barangay 894* emphasised the need for disaster preparedness and early warning system seminars to better equip them for future disasters. This is because even though there are local level government mechanisms to respond to the needs of the informal settler families and the larger community, most of the initiatives are focused on emergency response rather than pre-disaster events (i.e. prevention, mitigation and preparedness).

It is in this context that DRRNet Philippines, a coalition of 28 civil society organisations, implements capacity building on community-based disaster risk reduction and management (CBDRRM) to informal settler families in Metro Manila. DRRNet Philippines comprises of non-government organisations, people's organisations, academic and research institutions and faith-based organisations working on risk reduction and climate change adaptation. The organisation works not only with vulnerable communities but with local and national government as well. Through technical assistance of DFAT to DILG, DRRNet Philippines provided capacity building of informal settler families on CBDRRM covering 115 communities in 10 cities in Metro Manila and the 8 identified priority waterways in the integrated master plan for flood management in Metro Manila (Table 9.1). The two-year capacity building

Table 9.2 Thirty priority communities for CBDRRM training for ISFs

Cities	San Juan	Mandaluyong	Quezon City	Manila	Valenzuela
Communities	Balong Bato	Daang Bakal	Damayang Lagi	185	General T.
	Batis	Poblacion	Dona Imelda	195	Malinta
	Kabayanan		Roxas	602	Marulas
	Progreso		Sta. Cruz	607	Ugong
	Rivera		Sto. Domingo	894	
	Salapan		Talayan	895	
			Tatalon		

project has four primary objectives: (1) to enhance disaster preparedness and the adaptive capacity of residents in selected communities by conducting a series of capacity building on CBDRRM; (2) to install a community-based early warning system in informal settler families' communities and capacitate local communities in monitoring and maintaining them; (3) to support the communities in activating the Barangay Disaster Risk Reduction and Management Committee that would facilitate DRR work at the community level; and (4) to produce the Barangay Disaster Risk Reduction and Management Plan (BDRRMP) formulated by the communities and ratified by the Barangay Council with budget allocation for key activities and investments (DRRNet Philippines 2013).

Thirty priority communities have been selected to first undergo capacity building (Table 9.2). The selection was based mainly on the vulnerability and exposure of the communities, available socioeconomic and risk assessment data, as well as the recommendations of the DILG and the respective community leaders. Capacity building includes three components: community risk assessment workshops (2–3 days), early warning workshops and BDRRM planning (4 days). Towards the end of the activities in each community, a simulation drill to test the feasibility and effectiveness of the BDRRM and contingency plans was carried out during the National Disaster Consciousness Month.

The network proactively engaged the government (community-, city- and national-level agencies) in the undertaking. In fact, it deems that the development initiative to increase the adaptive capacity of the informal settlers will not progress without the active involvement of all of the aforementioned levels. For instance, before the network was able get the data on the informal settler families that they trained, they needed the figures on informal settlers from the national government or the city government involved. Similarly, to engage the families, it had to make several courtesy calls to the city mayors of the 10 cities and work with relevant agencies within the cities to ensure that there is no duplication of work and that the initiative complements existing programmes of the local government.

A total of 30–60 participants from each community participated in the series of capacity building that includes: residents from high-, medium- and low-risk areas of the community; officials; home owner's association presidents; sectoral representatives for women, youth, senior citizens and persons with disabilities;

public and private schools' representatives; and active civil society organisations' representatives in the community.

The process employed by DRRNet Philippines in the capacity building activities was participatory and empowering. In fact, the communities were encouraged to actively share their experiences on previous disaster events that occurred in the community as well as their vision towards a safer future. In addition, to ensure that the informal settler families have a comprehensive understanding on the importance of CBDRRM, the DRRNet Philippines had prepared a module on urban resiliency, which covers their needs and vulnerabilities. This includes an introduction to CBDRRM, basic concepts of disasters, the legal basis of disaster risk reduction in the Philippines and the tools for community risk assessments among others.

9.4 Challenges in the Initiatives

As mentioned before, based on the Philippine Government's blueprint, 104,000 informal settlers are expected to be relocated to safer, decent and affordable housing prior to the end of President Aquino's term in 2016. However, according to DILG NCR Regional Disaster Renato Brion, due to bureaucratic and inter-agency hurdles, the targeted numbers of informal settler families have not been relocated. He further notes that the relocation areas lack basic necessities such as potable water and electricity and are far from markets and work places. Nonetheless, many informal settlers insist on returning to dangerous areas that they used to live in.

The national and local level elections in May 2013 also resulted in new officials being appointed in some localities. For instance, city mayors were newly elected in four of the areas where the DRRNet Philippines' project on informal settler families was implemented. Hence, in addition to the need of making courtesy calls, they also had to wait until the newly elected officials assume office before they could proceed. Likewise, after the community-level elections on 28 October 2013, new sets of coordination had to be carried out as a number of community officials were replaced. As a result, inception meeting and orientations also had to be repeated. In addition, the initial community assemblies carried out by the network to provide basic disaster preparedness tips in time for the rainy seasons were not free from politicking. In fact, in several communities, community officials campaigned to their residents during network assemblies, which became venue for officials to boast about their accomplishments not just in disaster preparedness but all throughout their term.

Apart from these political and bureaucratic impediments, it is challenging to know whether the informal settler families are proactively involved and are benefiting from the initiative. These families are among the most disadvantaged and marginalised groups in the country. Their social exclusion is exacerbated by their vulnerability to disasters as evident in the recent catastrophes in Metro Manila. It is imperative that those who have access and control to resources carry out doable interventions to improve their conditions. If not, the informal settler families would continue to be victimised not only by "natural" disasters but also by political and social disasters such as disempowerment, poverty and marginalisation among others.

The whole undertaking to ensure the safety of informal settler families in Metro Manila provides three prospects. First is the promise of secure, decent and affordable housing. The author is also an informal settler and up to now there has been no assurance that his family will be relocated to better housing. In addition, not all informal settlers have the means, resources and social capital to transfer to more stable housing. Therefore, genuine consultation involving these families before relocation and the promise of a better life that comes with the housing development are crucial. Consultation should result in a People's Plan that truly reflects their needs. While that plan is being concretised, bureaucratic hurdles need to be fixed. For instance, there is a need to ensure that the housing agencies take accountability and act accordingly to deliver the targets. The funds for housing are available although it's delayed/obstructed due to politics and practices of the bureaucracy. If deemed important, there is a need to gather support system from parallel structures that would bring the problem to the President. However, because of political issues and subdivision among the supporters of the President, the initiative is moving but slowly and way behind the target.

Alternatively, given that the supply of housing does not meet the demand, the informal settler families need to explore alternative housing development schemes, such as seeking support of local government units and private sectors. Although there are initiatives to encourage stakeholders to begin relocation projects, there is still a need for informal settlers to look for representatives to take their issues to the relevant authority and agency.

Secondly, the capacity building towards enhancing their adaptive capacity and disaster resilience should be maximised by the informal settler families while they are still in their areas. The DRRNet Philippines prefers to avoid the issue of relocation and for them what is important is that the families are being capacitated on risk reduction. The informal settlers should actively participate in the capacity building since this concerns their safety and security.

Thirdly, the informal settler families are being given an opportunity to make their local authorities accountable to them as well as to be accountable to themselves and their own safety. This is an opportunity for them to demand what is due from their authorities. At the same time, based on their experience of catastrophes, they also need overcome their risk denial culture and relocate for their own safety.

9.5 Conclusions

Informal settling in the Philippines is a complicated matter and addressing it entails collaborative governance and multiple stakeholders' action at all levels. The chapter elaborates on different ongoing community-level and participatory initiatives addressing disaster vulnerability of informal settlers through capacity development.

The government and the DRRNet Philippines seem to collaboratively address informal settlement issues because the very nature of the vulnerability to disasters is strongly linked to informality. In this context, informal settlement and disaster vulnerability co-exist. This means that addressing one would subsequently lead to the other.

On the other hand, certain challenges are also prevalent. There are evident bureaucratic and systemic hurdles that surround the collaboration. For instance, there is a need for government to have a full ownership of the initiative. There is a tendency for the agency to highly depend on DRRNet Philippines in the project direction and implementation. Although a non-government organisation is a primary project partner, the government agency should exhibit strong leadership in the project. Another layer of problem exists at the community-level government. As observed in the community capacity building activities, most of the participants were associated to the incumbent community officials. However, other members of the community who should be benefiting from the projects were not able to participate. Hence, the community officials should set aside partisan politics and ensure the involvement of high-risk families.

In conclusion, the government and civil society partnership in relation to informal settling in the Metro Manila is a remarkable undertaking towards urban disaster governance. The collaboration endeavours to harmonise and reconcile the efforts of the national government, local government, communities, civil society organisations and the donor partner towards disaster risk reduction. Further, the initiative ventures to improve the coordination among government agencies as well as the interdependence of both state and not-state actors. Different stakeholders investing their time, resources, energy and capacities in disaster governance is a major achievement towards urban resiliency. On the other hand, to make sustainable outcomes from collaborations, the government needs to follow-up and replicate the initiative even without external support of the civil society.

References

Asian Development Bank (ADB). (2006). *Urbanization and sustainability in Asia: Good practice approaches in urban region development.* Mandaluyong City: Asian Development Bank.

Asian Development Bank (ADB). (2010). *Fact sheet: Philippines.* Mandaluyong City: Asian Development Bank.

Asian Development Bank (ADB) (2012). *Green urbanization in Asia: Key indicators for Asia and the Pacific 2012.* Mandaluyong City: Asian Development Bank.

Badilla, R. (2014, January 15). *Flood risk modelling in Pasig-Marikina River Basin.* Paper presented at the Training of Trainers on CBDRRM, St. Agatha, Philippines.

Balgos, B. (2013). From Typhoon Ondoy to the unnamed monsoon: Policy reforms and challenges in the Philippines' disaster management system. In K. Yau (Ed.), *Natural disasters and reconstruction in Asian economies: A global synthesis of shared experience* (pp. 277–301). London: Palgrave Macmillan.

Birkland, T. (2010). *Lessons of disasters: Policy change after catastrophic events.* Washington, DC: Georgetown University Press.

Davis, M. (2006). *Planet of slums*. New York: Verso.
Demographia World Urban Areas. (2013, March). Largest urban a*gglomerations* in the world (9th annual Edn.). www.demographia.com/db-worldua.pdf. Accessed 13 Jan 2014.
Department of Interior and Local Government (DILG) (2013). *Technical management group report.* Quezon City: Department of Interior and Local Government.
Department of Foreign Affairs and Trade (DFAT). (2013). *Technical assistance on securing the safety of Informal Settler Families (ISFs) in Metro Manila.* Paper presented at Project Organization Meeting, Australia Aid, Manila, Philippines.
Disaster Risk Reduction Network Philippines (DRRNet Philippines). (2011). *Primer on the disaster risk reduction and management (DRRM) Act of 2010.* Quezon City: DRRNet Philippines.
Disaster Risk Reduction Network Philippines (DRRNet Philippines) (2013). Component 3 contract: Technical assistance on securing the safety of Informal Settler Families (ISFs) in Metro Manila. Manila: Australian Aid.
Hilario, F., & Abadilla, R. (2014, January 15). *Tropical cyclone severe wind risk modelling.* Paper presented at the Training of Trainers on CBDRRM, St. Agatha, Philippines.
Institute of Philippine Culture (IPC). (2010). *The social impacts of Tropical Storm Ondoy and Typhoon Pepeng: The recovery of communities in Metro Manila and Luzon.* Quezon City: Ateneo de Manila University.
Jha, A., & Stanton-Geddes, Z. (2013). *Strong, safe, and resilient: Strategic policy guide for disaster risk management in East Asia and the Pacific.* Washington, DC: World Bank.
Michael, C. (2014). *World's 10 riskiest cities in relation to earthquakes, hurricanes, cyclones, and tsunamis.* http://www.theguardian.com/cities/gallery/2014/mar/25/earthquakes-hurricanes-cyclones-and-tsunamis-10-riskiest-cities-world. Accessed 13 Apr 2014.
National Disaster Coordinating Council (NDCC). (2009). *Typhoon Ondoy and Pepeng final report.* http://www.ndrrmc.gov.ph/attachments/article/92/Narrative%20Report%20re%20TS%20Ondoy%20and%20TY%20Pepeng.pdf. Accessed 12 Apr 2014.
National Economic and Development Authority (NEDA). (2011). *The Philippine development plan 2011–2016.* Pasig City: NEDA.
National Statistical and Census Board (NSCB). (2012). *Population of the Philippines.* Makati City: NSCB. http://www.nscb.gov.ph/. Accessed 20 Mar 2013.
Philippine Atmospheric Geophysical and Astronomical Services Administration (PAGASA). (2011). *Climate change in the Philippines.* Quezon City: DOST-PAGASA.
Shaw, R., Srinivas, H., & Sharma, A. (2009). *Urban risk reduction: An Asian perspective.* Bingley: Emerald Publication.
Singson, R. (2013). *Metro Manila integrated flood risk management master plan.* http://www.gfdrr.org/sites/gfdrr.org/files/(SESSIO_%203)_Sec._Singson_(Philippines)_052413_DPWH_FLOOD_RISK_MANAGEMENT_(SECRETARY).pdf. Accessed 20 May 2014.
Solidum, R. (2014, January 15). *Earthquake hazards and risk in Metro Manila and vicinity.* Paper presented at the Training of Trainers on CBDRRM, St. Agatha, Philippines.
United Nations (UN). (2011). *United Nations global assessment report on disaster risk reduction.* New York: UN.
United Nations International Strategy for Disaster Reduction (UNISDR). (2013). *Global assessment report 2013.* New York: UN International Strategy for Disaster Reduction.
Warren, J. (2013, October 28). A tale of two decades: Typhoons and floods, Manila and the Provinces, and the Marcos years. *The Asia-Pacific Journal, 11*(43): No. 3.

Chapter 10
Bangkok and the Floods of 2011: Urban Governance and the Struggle for Democratisation

Alin Chintraruck and John Walsh

Abstract Flooding is an endemic problem in Thailand, as it is in most of mainland Southeast Asia. The problem has been intensified by deforestation, urbanisation and global climate change. The floods in 2011 were particularly severe and caused the loss of more than 700 lives and a great deal of economic damage. It was also the cause of political crisis as different factions within the Thai society contested about how the flood issue should be treated and, in particular, who should be given priority in being protected. It is necessary to unpack the divisions within the society to understand why the governance system that was adopted did go into practice and why there has been so much deliberation over whether the post-flood infrastructure system should be implemented and if so, how. The limitations of what a democratically elected government in Thailand can achieve when obstructed by supposedly neutral public-sector organisations, the military and their allies are made clear. Trying to overcome these obstructions continues to prove beyond the ability of what elected governments can achieve in Thailand, and this seems unlikely to change in the foreseeable future. As a result, disaster governance will continue to be used as a pawn on the political chessboard and not as a means of saving lives and livelihoods.

Keywords Democratisation • Floods • Bangkok • Thailand • Urban governance • Climate change

10.1 Introduction

The 2011 floods in Thailand were accounted to be the worst since 1942 or even much earlier. More than 700 lives were lost in Thailand and many more in the neighbouring countries, which were also flooded. The World Bank described it as the third worst economic disaster of the year. It has also been described as the most

---

A. Chintraruck (✉) • J. Walsh
School of Management, Shinawatra University, Bangkok, Thailand
e-mail: alin_nuk@hotmail.com; jcwalsh@siu.ac.th

M.A. Miller, M. Douglass (eds.), *Disaster Governance in Urbanising Asia*,
DOI 10.1007/978-981-287-649-2_10

expensive flood ever recorded, owing to the disruption of global supply chains. From an economic perspective, the problem was that the industrial estates to the north of Bangkok around the provinces of Ayutthaya and Pathum Thani were inundated and all those factories supplying the global Japanese automotive complex, among other industries, were stopped. The effects were noticeable around the world, and the Thai economy—which remains heavily dependent on exports—shrank by 2.8 % for the year as a whole. The overwhelming bulk of the economic losses fell upon the private sector, which was estimated to have lost 1,284 billion baht out of an overall loss of 1,425 billion baht (approximately US$45.7 billion) (World Bank 2011). Further, as observed by Courbage et al. (2012), the floodwater can act as a fertiliser for paddy rice fields and lead to the postponing of harvests, while the flooding of industrial estates can be accompanied by chemical release, waste water installations being damaged or destroyed and other environmental problems which could have future implications.

Loss of confidence among investors has led the government to make extensive and expensive claims about how flood governance would be improved for the future. A total budget of 350 billion baht (approximately US$11.7 billion) has been provided for a comprehensive set of flood management infrastructure facilities development. It is supposed to be centred on two floodways covering an area of 375,000 raid (600 km^2) and linking Nakhon Sawan province to the Gulf of Thailand that can tolerate flooding to the height of 1.5 m (Theparat 2012). It is deemed as crucial to complete these preparations rapidly because of the elevated risks of flood repeats under conditions of global climate change. An analysis of the rainfall records indicates that a repeat event of comparable magnitude is imminent in the next two to three decades (Gale and Saunders 2013). This is consistent with the record of previous large-scale floods in 1983, 1995 and 2006. Failure to prepare adequately for inevitable future events has been a characteristic of the general failure to meet successful disaster governance standards (Ziegler et al. 2012).

Flood management is an integral part of the Pheu Thai administration's 2 trillion baht (US$66.7 billion) infrastructure development plan to lift the country out of the middle-income trap by linking the country more deeply with neighbours and promoting added value in production rather than low-wage cost competitiveness.[1] There are, of course, significant risks attached to an undertaking of this magnitude. For instance, the whole process could be bogged down in court actions as a result of complaints by various civil and political groups. The ability of the democratically elected government to realise its policies is the subtext to this process. Since the

[1] This policy is one of several that has been stopped by court order as part of what is widely viewed by both inside and outside Thailand as another judicial coup.

2006 military coup, and arguably since 2001 when Thaksin Shinawatra's Thai Rak Thai party was first elected,[2] a struggle behind the scenes has been taking place for control of those institutions, which are most influential in determining the distribution of resources and power within the Thai Kingdom. Part of this struggle takes place in plain sight. Nonetheless, more aspects take place not just behind the scenes, but beyond the ability of people to acknowledge and discuss due to powerful social taboos, the culture of impunity and the inadequacies of a media hamstrung by powerful and threatening censorship laws. Disaster governance is one part of this larger fight, and the progress of the struggle is indicative of the larger struggle for control of Thailand. Once the 2014 military coup took place, all court and administrative problems facing the flood management policy (and other policies) disappeared overnight.

This chapter seeks to explore the 2011 flood in the context of the emergent concept of disaster governance that recognises the need of many different stakeholders in the concept of an expanding urban setting. These stakeholders involve the residents of Bangkok and those outside the city, as well as investors and business managers in the industrial estates to the north of Bangkok and directly in the path of the flood waters. They may be considered as legitimate stakeholders. It is argued, however, there are also illegitimate stakeholders seeking to take advantage of a disaster for political gain. Klein (2007) used the term "disaster capitalism" to describe those forces and agencies that descend on a disaster-struck area to reshape it according to a neoliberal ideology that remakes it as a place locked into a subservient position in an advanced capitalist system. In Thailand, that situation is effectively reversed as disaster conservatives seek to use floods and other disasters as a means of shoring up their own power and influence and to reinforce existing social and class relations. There has been a struggle taking place within Thai politics between these conservative forces and their numerous allies lurking in courts, civil service, nongovernmental organisations (NGOs) and the popular media and the majority of people who have repeatedly voted for change. Bangkok has become the absolute focus of the intersection between these forces. Embracing an inclusive disaster governance approach would, as this study indicates, represent both victory and defeat in a battle as part of the campaign taking place to control Thai politics at a deeper level. There is a strong element of class struggle within this conflict, but this is disguised by the presence of diverse coalitions of actors on either side, some of which have switched allegiance, while some appear to be on the wrong side of the divide or just as divided as is the society as a whole. Alliance building by Thai Rak Thai and its successors may be seen as a Gramscian approach to political strategy.

Research for this chapter has involved an extensive series of personal interviews with experienced members of the water industry, service providers, academics, practitioners and government officials. Interviews took place in 2011–2012 as part of a doctoral programme that studied the nature and quality of water governance of Thailand as a whole and the southern island province of Phuket in particular. This

[2] Thaksin Shinawatra is the founder of Shinawatra University in Thailand, to which both authors are affiliated.

interviewing has been supplemented by extensive ethnographic observation of groundwater management and related issues across Thailand and by integration with the secondary data research, which has included academic chapters and reports with the popular press and social media.

This chapter examines the management of the 2011 floods in Thailand and the unfolding prospect of 2013 floods (34 people are reported to have been killed at the time of writing) not just in technocratic terms, but also as a means of examining the political struggles continuing in the country. It continues with a discussion of the nature of disaster governance in an urban setting, followed by a section on disaster governance in the Thai context. It then moves on to discuss more specifically the management of the 2011 floods and its impact in future flood control in Thailand. The final section of the chapter, then, underlines the theoretical issues that help in understanding the nature of contemporary Thai politics—Hegel's owl of wisdom famously took flight at dusk, suggesting that enlightenment only arrives when it is too late to act. In this case, it is hoped that there is still some scope for shining light on these issues and encouraging further transparency and accountability within the system.

10.2 Disaster Governance

Natural and manmade disasters and the availability of technology to counter them are always changing. The most obvious example of this is the case of global climate change, which has contributed to the prevalence of deadly flooding events in the Mekong region. Negative effects from climate change are disproportionately distributed, as well as disproportionately caused. Although most economic losses from climate change have accrued in more developed countries, most deaths have occurred in less developed countries (O'Brien et al. 2006). Vulnerability is higher in poorer countries and communities within the affected areas. In the case of floods in Thailand, poor and slum areas in Bangkok were more badly affected than those which had more resources to deal with the problems caused (Stanton-Geddes 2013).

Although technological innovation is a key in mitigating and withstanding disasters, the soft skills and resources are also considered as important. Pearce (2003), for example, argues that since it is necessary to move from a focus on response and recovery to one of sustainable hazard mitigation, it is essential to integrate disaster management and community planning. The ability of communities to adapt to disastrous situations can also be enhanced by taking actions to promote the resilience of those communities through a mixture of soft and hard resources: economic development, social capital, information communication and community competence (Norris et al. 2008).

New forms of promoting resilience within communities through public education campaigns drawing on health promotion have also been proposed (Paton and Johnston 2001). It has been argued that in the context of Bangkok, resilience promotion has been a vital component of the post-flood rehabilitation of Baan Mankong community, which has featured a heavy reliance on institutions and networks. The project has also made use of citywide upgrading as a strategy for tackling urban poverty and managing disaster risks, as well as a recognition that positive impact on the community and its individual members goes beyond asset reconstruction and towards the creation of a more resilient community to future disasters (Stanton-Geddes 2013). The dynamic processes involved are particularly emphasised in the literature, where building networks of organisations committed to adaptive learning and continual inquiry is preferable to the more static attempt to control possible risks through structural or administrative means (Comfort 2005). This must be understood in the context of the rapid urbanisation of the Mekong region and of Bangkok in particular. This urbanisation has intensified the risk of disasters such as flooding because of the spatial polarisation of increased population levels in high-risk coastal and riparian areas (Douglass 2013). This increased concentration of people is considered to be the single most important factor in future disaster governance challenges (Jha and Stanton-Geddes 2013, p. 17, in Douglass 2013).

10.3 Flood Governance in Thailand

The governance of water has always been a crucial issue for the state of Siam, now Thailand. Historically, the ability of leaders to open up new land for wet paddy rice growing was important in determining the legitimacy of the ruling elite and its ability to expand the amount of territory controlled. Baker and Phongpaichit (2009) use the term "rice frontier" to denote the limit of territory controlled by the state and argue that this extended as far as organised agriculture could be managed. The sparse population in the Mekong region as a whole has represented a constraint on social and economic development that opening new spaces has helped to alleviate.

Without such expansion, Siam would have been obliged to rely on extractivism and international trade (Molle 2005). In the age of modern capitalism, relying on such international trade would have had limited control over Siamese interests. The degree to which power over water and nature demonstrated the vigour of the state, although the concept of the hydraulic empire (Wittfogel 1957) which required a despotic administration and a substantial bureaucracy to organise the kind of large-scale slave or corvée labour that would be required to undertake the infrastructure construction necessary, seems to be overstated. Archaeological evidence tends to indicate that, at least under certain conditions, individual communities could organise their own water management systems and negotiate or coordinate with neighbouring communities to manage slightly larger-scale water governance systems, without the need for top-down management control (Mithen 2013).

In the twentieth century, foreign experts were imported to manage Siam's natural resources according to "modern" and "Western" methods (Loos 2002). This involved a significant expansion of the canal and irrigation system both as a means of increasing state power at a time of continued imperial colonisation and, also, to enhance agricultural productivity and exports (Poupon 2013). King Rama V,

Chulalongkorn, instituted new bureaucratic methods to control the state, including the Canal Department (founded in 1902) that subsequently became the Royal Irrigation Department (RID). The RID plays a crucial role in water governance in modern Thailand. When the absolute monarchy was replaced by a constitutional system after the 1932 Revolution, the power of the monarchy was substantially reduced and replaced by an aristocratic-military elite that proceeded to exert control over all sectors of society and the economy. Patron-client networks were strengthened and used to exert control over all aspects of governmental activity. Nevertheless, every attempt to manage floods is handicapped by the geography of the Chao Phraya and other principal tributaries such as very gentle slopes and lack of downstream discharge ability. These have resulted in upstream floods leading to river banks bursting downstream and, hence, extensive flooding of farmlands and other land on either side (Komori et al. 2012).

Owing to their long experience with regular flooding, Thai people have learned how to adapt themselves to this situation by building stilted houses and developing agricultural systems that are resistant to flooding (Mongkonkerd et al. 2013). However, that adaptive capability has become much less important as large-scale change in land use and environmental degradation have brought about an intensification of the flooding problem to a level that was previously unimaginable.

Disaster governance in Thailand has taken an increasingly prominent role in recent years owing to high-profile events such as floods combined with landslides associated with deforestation (Usher 2009), the 2005 tsunami and the 2011 floods. In addition, the turbulent political history of the twenty-first century has seen incidences of disorder and violence included in the tally of disasters. Currently, Thailand ranks as the country which is seventh most prone to flooding. However, it is predicted that economic development and global climate changes will make it the fourth most affected country in the world by 2030 (UNDP 2012).

Large-scale deforestation has contributed to the threat of flooding that has been endemic in the region and has been exacerbated by the fact that most large urban areas have been built on flood plains. A number of dams have been built to help regulate the flow of water and mitigate the twin threats of drought and floods. These are managed by the RID, which consists of 22 different offices.

As a result of the importance of disasters, the Disaster Prevention and Mitigation Act 2007 was introduced by the military junta that seized control of the county with the 2006 coup. It replaced the Civil Defence Act of 1979 and the Fire Defence Act of 1999 and included within the definition of disasters terrorist and sabotage attacks and airborne military attacks by other nations. The role of the military forces in disaster governance was thereby strengthened according to the law. The prime minister or an assigned deputy manager is designated as the National Commander, and a rigid bureaucratic structure is imposed. The Department of Disaster Prevention and Mitigation (DDPM) was created in 2002 under the umbrella of the Ministry of the Interior (MoI) as part of restructuring the bureaucracy organised by the Thai Rak Thai administration (2001–2006).

10.4 The 2011 Flood and Its Impact

The floods began as a result of the monsoon rains combined with the remnants of several tropical storms. Concerned with the perennial threat of drought, in a country that relies on wet paddy rice agriculture but has comparatively little irrigation (Poupon 2013), the RID maintained water levels in several dams at a very high level. As the water levels increased and much of the north and northeast (Isan) regions became inundated, there was insufficient capacity in holding areas, and the mass of water began to move slowly southwards towards Bangkok. By the middle of October, water was threatening Bangkok and the industrial estates to the north. Walls protecting industrial estates started to collapse, and many factories had to be abandoned. University campuses on the periphery of Bangkok, including Thammasat Rangsit campus and the adjoining AIT, Mahidol University at Salaya and Shinawatra University at Pathum Thani, were all closed for extended periods. Conflicts began to break out when it became evident that some areas were being protected, while a few metres-away communities were flooded by up to 2 m of water for what turned out to be months at a time. There were some scuffles around sandbag walls. Further, when the Khlong Prapa water supply canal protection was breached, water flooded into the city temporarily. Although it was soon brought under control, it was not before panic was sparked and people began stockpiling basic foodstuffs and other supplies and parking their cars illegally on highway flyovers and other elevated roads.

The central government had only recently been elected and, hence, lacked experience in dealing with a disaster of this magnitude. The incoming Prime Minister, Yingluck Shinawatra, had not previously been active in politics. She was chosen to stand for Pheu Thai almost entirely because she was a sister of Thaksin Shinawatra, Prime Minister of the Thai Rak Thai administration of 2001–2006 (and founder of Shinawatra University), whose government was ousted from power by a military coup. There had also been pressure to reward the grassroots redshirt activists who had demonstrated in Bangkok until suffering massacres under the Abhisit regime, by giving cabinet seats to their leaders. The political divisions within the country signified by the red versus yellow conflicts were replicated within the civil service and the important institutions of the state.

The government's central agency, the Flood Relief Operations Centre (FROC), found itself under fire, particularly from the Bangkok-based media, which generally supports the right of centre Democrat Party for which Bangkok-registered citizens regularly vote (while migrant workers are required to return to their registered home communities to vote). Contradictory advice was propagated by the Governor of the Bangkok Metropolitan Administration (BMA) M.R.[3] Sukhumbhand Paribatra, while the richly resourced military forces were regularly shown in high-profile rescue activities.

[3] M.R. is an honorary aristocratic title.

The problems centred on a variety of issues. Firstly, there was the issue of inadequacy of experience, resources and suspicion over motives. The level of corruption in Thailand remains comparatively high, and this has bred a sense of cynicism about any problem in providing material resources. Governor Sukhumbhand tapped into this cynicism when rejecting 800,000 sandbags provided by the central government. The fact that the FROC had to be relocated after its operational base in Don Mueang International Airport fell victim to the encroaching waters did not help the situation. Secondly, there was the provision of confusing and misleading information. Draconian censorship laws and powerful political and social taboos have bred a society in which unsourced rumours and gossip are often considered more reliable than official pronouncements or media communications. With various people saying different and sometimes contradictory things, it is not surprising that many poor decisions were made. Thirdly, there was the issue of effective governance. Government opponents called for the declaration of a state of emergency that would have effectively brought about a situation of martial law. Prime Minister Yingluck, instead, invoked a different law that allowed her to remain in overall control and, ostensibly, to command the military. Intensive politicking and opportunistic gift giving to those affected lent an occasionally farcical element to the crisis.

Above all, the main issue was perhaps the fact that Bangkok was so resolutely protected, while most of the other parts of the country were apparently left to fend for itself. Bangkok is a primate city (McGee 1967). In other words, it houses all of the principal religious, political, cultural, economic and social institutions of the Kingdom. It would have been an act of political suicide for the Pheu Thai government to have permitted any of the central monarchy-linked institutions or facilities to become inundated without a very visible and frantic struggle. However, to have saved Bangkok in a way that made it seem the rest of the country had been sacrificed upset two other major stakeholders: the rural poor whose lives and livelihoods were lost while the flooding continued until mid-January 2012 and foreign investors whose factories had done so much to propel Thailand from low- to middle-income status. Bangkok was, ultimately, saved, since the central areas were protected, although a number of districts were flooded and houses and businesses had to be abandoned for some weeks. However, confidence in the government and the country as a whole had led to a new political crisis.

To deal with upsetting these important stakeholders, the government has outlined, in a series of stages, a significant plan for the future, which includes the following components. Firstly, the country will no longer rely on low labour cost competitiveness. The minimum wage rate would be increased by about 40 % [to 300 baht per day (US$10)], and assistance in adding value to production would be provided in every province and sector. Secondly, an enormous infrastructure development plan would be launched (Table 10.1) connecting the country in a number of different ways with neighbouring countries and the Asian Highway Network as a whole. As part of this concept, those company owners who rather predictably complained that increases in labour costs would destroy their businesses will be provided with roads to overseas industrial estates, where they can recreate their

Table 10.1 Measures taken to promote the economy, 2013 (Post Reporters and Agencies 2013)

Private consumption	Juristic persons are allowed to claim up to two times actual expenses incurred from organizing training-related seminars in their corporate tax filings in the 2013 and 2014 accounting years. Sales campaigns for low-cost energy-saving products will promote energy-saving electrical appliances with the government's financial support.
Private investment	Tourism operators investing in buying or repairing assets, furnishings and furniture, except vehicles, will be allowed faster depreciation, with up to 60% of the value of the improvements taken into account in the first year and the remainder averaged over five years. The measure will apply only for the 2013 and 2014 accounting years. The launch of the second phase of the eco-car project with greater investment privileges than in the first phase. Easing rules and regulations for investors who set up food factories, sugar mills and ethanol factories in compliance with the agricultural zoning policy. To provide more credit and credit guarantees to small and medium-sized enterprises with cooperation from commercial banks and specialized financial institutions, plus microfinance for OTOP[a] and community enterprises.
Public spending	To accelerate disbursement of the fiscal 2013 budget allocated to provinces, local administrations and off-budget funds.
Exports	To step up promoting exports to potential destinations, particularly other ASEAN countries and certain provinces of China. To promote the use of the baht in cross-border trading. To promote trade financing through the Export-Import Bank of Thailand in a move aimed at curbing financial risks and support financial liquidity for business operators. To provide multiple-entry visas that are effective over the course of several years to international tourists as well as promote Thailand as a shopping paradise.

exploitative business practices with the blessing of the government. Finally, important institutions would be brought under the rule of the democratically elected government on a case-by-case and a step-by-step basis. The latter will be considered in the remainder of this chapter, with a specific focus on the implications for flood management and disaster governance.

10.5 The Struggle for Institutions

On 30 January 2013, at a keynote address at the first Shinawatra University International Conference at Pathum Thani, Deputy Prime Minister and Minister of Education Phongthep Thepkanjana argued that it was unfair that he, as a politician, was forced to make full and open submission of all assets owned with the threat of disqualification or even criminal action in the event of any nondisclosure, whether deliberate or accidental. By contrast, members of unelected and unaccountable institutions were free from this onerous and potentially embarrassing scrutiny.

Indeed, Mr. Phongthep might (although he did not) go on to observe that many important state activities are shielded from public scrutiny of any sort. Further, the ability of institutions and the individuals within them to associate themselves with the monarchy—or the network monarchy, as McCargo (2005) has described it—means that they can become virtually immune to any form of criticism in contemporary Thailand. For instance, naming a dam as the Bhumibol Dam or by using the Royal in RID means management and decisions made with respect to those institutions become effectively beyond criticism. Finding ways to make these institutions accountable is one of the central struggles for control of the Thai polity currently taking place.

After the 1997 Asian financial crisis—which started in Thailand and means that many Thai people still like to talk of the Tom Yum Goong crisis as if nothing important happened elsewhere[4]—the ruling Democrat Party largely embraced the discourse of neoliberalism peddled by the International Monetary Fund (IMF) as the price for receiving the money necessary to solve the short-term exchange rate liquidity problems facing the country (Hewison 2005). Democrats reinvented themselves as technocrats, taking the difficult and often unpopular decisions necessary to keep the country on the right track. A moral tone was attached to this stance, just as it has been by various European governments currently perpetrating the crisis of austerity. Once the Democrats were swept from power at the crushing 2001 victory for the Thai Rak Thai party, it became very convenient for political opponents to label the new politics as "populism" or even "business populism" (Phongpaichit and Baker 2005). Although it does not necessarily have a moral component or belong to either the right or the left (Laclau 2007) and is, indeed, characterised by a great deal of smoke and mirrors about its actual meaning (D'Eramo 2013), populism is now almost universally used in political discourse in Thailand. It means a brand of politics which is, at best, irresponsible and, more likely, a deliberate attempt to buy power through handing out the resources of the state as a form of vote buying. As a result, opponents of Thai Rak Thai and its successors have customarily taken a lofty moral tone that combines the accusation of corruption with the charges of irresponsibility, wastefulness and ochlocracy—that is, the fear of the mob given control of the reins of government (Dunn 2006). These accusations were used in the organised yellow-shirt demonstrations that ultimately resulted in the 2006 military coup and have subsequently been redeployed to dismiss the ideology and intellectual underpinning of the Yingluck Shinawatra administration. It has been necessary for the government to try to take control of those important institutions which, during the 2011 flood, failed to cooperate with the central government using some stealth because of the populism-laced diatribes that would accompany any open movement. For example, NGOs, which have in many cases been captured by anti-Pheu Thai interests, launched various attempts to have the bidding process for the water management contracts blocked by court decisions, on the basis that it either contradicted the constitution introduced by the military junta in 2007 or led to corruption (Bangprupa and Jikkham 2013, p. 2).

---

[4] Similarly, the global economic crisis of 2008 is known as the Hamburger Crisis.

Meanwhile, when Prime Minister Yingluck made a speech in Mongolia at the 7th Ministerial Conference of the Community of Democracies arguing that there are still antidemocratic elements in Thailand and that some independent organisations set up under the 2007 constitution were working against the interests of the people, there was an uproar and 58 appointed senators threatened to make her testify in the Senate in front of four separate committees: foreign affairs; religious affairs, moral and ethics; human rights, freedom and consumer protection; and, finally, corruption and good governance (Post Reporters 2013, p. 3). Similarly, friction between the Ministry of Finance, Kittiratt Na Ranong (a former president of Shinawatra University) and the Bank of Thailand over the appropriate interest rate to use at a time of international economic crisis and with the baht surging against the dollar has been portrayed in the media as a clash of personalities. On the one side is the impersonal civil servant doing his best for the country as a whole and possessing wisdom and knowledge beyond the public's awareness and, on the other side, the venal politician whose motivations are suspicious. A tentative rapprochement between the two sides has apparently been reached (Yuthamanop 2013, p. 1). Personal negotiation between two offices is one matter, but in other cases it has been necessary to outflank entrenched interests by altering the governance structure of the institution or by involving the private sector. In the first case, one approach has been to appeal to a higher level of institution—transnational government. To counter what some people perceive as problematic decisions in Thai courts which cannot be challenged or even questioned, many deputy prime ministers as well as the current Minister of Labour Chalerm Yubamrung suggested reporting decisions made by the Supreme Court's Division for Political Office Holders to a panel on human rights organised by the United Nations (Ruangdit 2013b, p. 1). The Water and Flood Management Committee has been established to oversee all such preparations in the country, and it managed to place a member at the RID, who now must give accounts of all decisions that it makes. Table 10.2 indicates the main provisions for flood and water governance proposed.

Table 10.2 Principal components of the Pheu Thai government's flood management infrastructure development proposal (Adapted from Ruangdit 2013a, p. 1)

Package A (Chao Phraya River Basin)		Budget (billion baht)
A1	Reservoir construction	50
A2	Land utilization and town planning	50
A3	Flood retention areas	60
A4	Waterway improvements	7
A5	Floodway construction	120
A6	Data management and warning systems	3
Package B (Other River Basins)		
B1	Reservoir construction	12
B2	Land utilization and town planning	10
B3	Waterway improvements	10
B4	Data management and warning systems	2

In terms of the private sector, the government has contracted out important services to private-sector bodies as a means of counteracting obstructive public-sector agencies. Both reforestation and water provision have been used to improve the quality of services provided, while the auctioning of the water management projects indicated the commitment of the government to use the private sector to ensure that its most vital policies are conducted by organisations over which it can enforce power effectively. The policy initially emerged under the previous regime, somewhat paradoxically. It occurred as a result of the 1997 crisis, when Thailand required the intervention of the IMF at a time when they were imposing conditionalities, on the recipients of its funds, according to its neoliberalist agenda that included the privatisation of a number of government services. This was a controversial issue in Thailand since unemployment rose sharply as a result of the forcible closure of a number of factories and other companies deemed to be unsustainable and, also, because of the perceived assault on national pride and self-esteem (cf. Haggard 2000). Consequently, when the first instance of water privatisation followed in the province of Pathum Thani (Zaki and Nurul 2009), there was suspicion expressed by various parts of the society. However, the process was completed smoothly and a number of other incremental privatisations have followed, linked with the role of East Water, which is the principal organisation tasked with administering and overseeing corporatised and privatised water privatisation projects in Thailand (Chintraruck and Walsh 2014).

10.6 Conclusion

Flood governance in Thailand takes place on a number of different levels. There is the hard-science level which deals with run-off areas, monkey cheeks and dam measurements. There is the soft-science level which deals with the needs to help communities adapt to climate change and its impacts, as well as to promote resilience and competence. However, at the political level, a struggle for control over resource management is happening. There are, clearly, difficulties in trying to reconcile the actions of these different levels—and as a result, there is waste and inefficiency. It is quite well established that intra-agency competitions have led to lower levels of efficiency and those on the margins of society (e.g. those affected by floods in remote areas) can be left to fend for themselves (Manuta et al. 2006). This is a general feature of water governance in Thailand, which includes many different agencies working for cross-purposes with each other. The current processes may be playing a role in helping to simplify and improve this situation, but actual results remain to be seen.

This chapter has intended to show that Bangkok is a politically divided city and that various strategies are being employed to overcome these rifts with a view to putting into practice integrated and inclusive strategies—in this case for disaster governance. The strategies include the use of the private sector seeking the use of international norms and systems to supplement or replace local methods and the painstaking attempts to make the unaccountable accountable by requiring those institutions to make their controversial or unorthodox decision-making/justifications public and transparent. Recent events indicate how much distance remains to be travelled in this regard, as the highest courts in the land continue to issue verdicts that appear to ignore the law while relying on the supposed moral authority of judges. It remains to be seen whether the significant progress required can be made without violence.

Thailand's economic strategy continues to rely on providing confidence to international investors and more securely connecting the places of production of goods and services and the places of their consumption locally and globally. In the past, it might have meant riding roughshod over ordinary people who were required to make way for those connections to be made. Democratisation and the spread of low-cost mobile telecommunications technology and Internet access have made that a less sustainable course of action, albeit not one that is beyond the bounds of possibility. The application of transnational legal and regulatory regimes, such as the ASEAN Economic Community due to start in 2015, might help in restricting the space in which transgression can occur in secrecy. Disaster governance is, therefore, both a means to an end in broadening the ability of the accountable state to introduce inclusive policies and systems and an end in its own right.

References

Baker, C., & Phongpaichit, P. (2009). *A history of Thailand*. Cambridge: Cambridge University Press.

Bangprupa, M., & Jikkham, P. (2013, May 3). Water project bids to proceed after court rejects challenge. *Bangkok Post*.

Chintraruck, A., & Walsh, J. (2014). East water. In G.D. Sardana (Ed.), *Managing world class operations* (pp. 248–256). New Delhi: Bloomsbury Publishing.

Comfort, L.K. (2005). Risk, security, and disaster management. *Annual Review of Political Science, 8*, 335–356.

Courbage, C., Orie, M., & Stahel, W.R. (2012). 2011 Thai floods and insurance. In C. Courbage & W.R. Stahel (Eds.), *Extreme events and insurance: 2011 annus horribilis*. The Geneva reports: Risk and insurance research no.5. Geneva: The Geneva Association.

D'Eramo, M. (2013). Populism without the people. *New Left Review*, 82, 5–28.

Douglass, M. (2013). The urban transition of environmental disaster governance in Asia. *Asia research institute working paper series, 210*. Singapore: National University of Singapore.

Dunn, J. (2006). *Setting the people free: The story of democracy*. London: Atlantic Books.

Gale, E., & Saunders, M. (2013). The 2011 Thailand flood: Climate causes and return periods. *Geophysical Research Abstracts*, 15. http://www.geophysical-research-abstracts.net/gra_volume_15.pdf. Accessed 30 June 2014.

Haggard, S. (2000). *The political economy of the Asian financial crisis*. Washington, DC: Institute for International Economics.

Hewison, K. (2005). Neo-liberalism and domestic capital: The political outcomes of the economic crisis in Thailand. *The Journal of Development Studies*, 41(2), 310–330.

Jha, A., & Stanton-Geddes, Z. (2013). *Strong, safe, and resilient: Strategic policy guide for disaster risk management in East Asia and the Pacific*. Washington, DC: The World Bank.

Klein, N. (2007). *The shock doctrine: The rise of disaster capitalism*. New York: Picador.

Komori, D., Nakamura, S., Kiguchi, M., Nishijima, A., Yamazaki, D., Suzuki, S., et al. (2012). Characteristics of the 2011 Chao Phraya River flood in Central Thailand. *Hydrological Research Letters*, *6*, 41–46.

Laclau, E. (2007). *On populist reason*. London: Verso.

Loos, T. (2002). *Subject Siam: Family, law and colonial modernity in Thailand*. Chiang Mai: Silkworm Books.

Manuta, J., Khrutmuang, S., Huaisai, D., & Lebel, L. (2006). Institutionalized incapacities and practice in flood disaster management in Thailand. *Science and Culture* (January–February), 10–22.

McCargo, D. (2005). Network monarchy and legitimacy crises in Thailand. *The Pacific Review*, 18(4), 499–519.

McGee, T.G. (1967). *The Southeast Asian city: A social geography of the primitive cities of Southeast Asia*. London: G. Bell and Sons Ltd.

Mithen, S. (2013). *Thirst: Water and power in the ancient world*. Cambridge, MA: Harvard University Press.

Molle, F. (2005, December 1–4). *Elements for a political ecology of river basins development: The case of the Chao Phraya River Basin, Thailand*. Paper presented at the 4th international water history association conference: Water and civilization, UNESCO Headquarters, Paris, France.

Mongkonkerd, S., Hirunsalee, S., Kanegae, H., & Denpaiboon, C. (2013). Comparison of direct monetary flood damages in 2011 to pillar house and non-pillar house in Ayutthaya, Thailand. In *Proceedings of the 3rd international conference on sustainable future for human security (SUSTAIN)*, 17 (pp. 327–333). Amsterdam: Elsevier B.V.

Norris, F.H., Stevens, S.P., Pfefferbaum, B., Wyche, K.F., & Pfefferbaum, R.L. (2008). Community resilience as a metaphor, theory, set of capacities, and strategy for disaster readiness. *American Journal of Community Psychology*, 41(1–2), 127–150.

O'Brien, G., O'Keefe, P., Rose, J., & Wisner, B. (2006). Climate change and disaster management. *Disasters*, 30(1), 64–80.

Paton, D., & Johnston, D. (2001). Disasters and communities: Vulnerability, resilience and preparedness. *Disaster Prevention and Management*, 10(4), 270–277.

Pearce, L. (2003). Disaster management and community planning, and public participation: How to achieve sustainable hazard mitigation. *Natural Hazards*, 28(2–3), 211–228.

Phongpaichit, P., & Baker, C. (2005). "Business populism" in Thailand. *Journal of Democracy*, 16(2), 58–72.

Post Reporters (2013, May 2). Senators lash PM over speech. *Bangkok Post*.

Post Reporters and Agencies (2013, August 7). Stimulus packages backed. *Bangkok Post, Business B1*.

Poupon, R. (2013). *The Thai food complex: From the rice fields to industrial and organic foods*. Bangkok: White Lotus Press.

Ruangdit, P. (2013a, February 6). S. Korean firm leads the pack. *Bangkok Post*.

Ruangdit, P. (2013b, May 22). Chalerm wants UN check on Thai court. *Bangkok Post*.

Stanton-Geddes, Z. (2013, July 23). Bangkok post-2011 floods: How about the poor? *World Bank Blogs*. http://blogs.worldbank.org/sustainablecities/bangkok-post-2011-floods-how-about-poor. Accessed 30 June 2014.

Theparat, C. (2012, May 14). Floodways projects envisioned as permanent solution. *Bangkok Post, Business B1*.

United Nations Development Project (UNDP) (2012). *Thailand and UNDP launch initiative to boost disaster management capacity*, UNDP Press Release. http://www.undp.org/content/thailand/en/home/presscenter/pressreleases/2012/09/21/thailand-and-undp-launch-initiative-to-boost-disaster-management-capacity/. Accessed 30 June 2014.

Usher, A.D. (2009). *Thai forestry: A critical history*. Chiang Mai: Silkworm Books.

Wittfogel, K. (1957). *Oriental despotism: A comparative study of total power*. New York: Random House.

World Bank (2011). *The World Bank supports Thailand's post-floods recovery efforts*. http://www.worldbank.org/en/news/feature/2011/12/13/world-bank-supports-thailands-post-floods-recovery-effort. Accessed 30 June 2014.

Yuthamanop, P. (2013, May 6). Ministry, BOT ink accord. *Bangkok Post*.

Zaki, S., & Nurul A.A.T.M. (2009). Does basic services privatisation benefit the urban poor? Some evidence from water supply privatisation in Thailand. *Urban Studies*, 46(11), 2301–2327.

Ziegler, A.D., Lim H.S., Tantasarin, C., Jachowski, N.R., & Wasson, R. (2012). Floods, false hope, and the future. *Hydrological Processes*, 26(1), 1748–1750.

Chapter 11
Is It Possible to Integrate Disaster Governance into Urbanization? Evidence from Chinese Townships Hit by 2008 Wenchuan Earthquake and 2013 Lushan Earthquake

Qiang Zhang, Yameng Hu, and Qibin Lu

Abstract This research contributes to fill the void in the knowledge of the evolution of disaster governance of Asia during urban transition, by studying the changes of disaster risk management (DRM) capacities of townships in southwestern China. These townships have particular significance in China's urban spectrum. While a number of townships were hard hit twice by 2008 Wenchuan Earthquake and 2013 Lushan Earthquake, it was a precious opportunity to examine the effect of integrating disaster governance into sustainable urbanization. The Wenchuan Reconstruction Plan had claimed the goal of building DRM capacity along the urbanization process of the post-disaster areas during the 5 years. This research finds that although with tremendous efforts, the fundamental issues of DRM remained unsettled in 2013 due to a failure of integrating the two policy systems of urbanization and disaster governance. The major contributions of this research are to link disaster governance and urbanization through DRM Capacity Framework and to put forward a new framework for forming a coalition between urbanization and disaster governance guided by governance approaches in strategic, policy, and program levels.

Keywords Disaster governance • Urbanization • Wenchuan Earthquake • Lushan Earthquake

Q. Zhang (✉) • Y. Hu • Q. Lu
School of Social Development and Public Policy, Beijing Normal University, Beijing, China
e-mail: qz@bnu.edu.cn; huyameng.bnu@hotmail.com; luqibin@bnu.edu.cn

M.A. Miller, M. Douglass (eds.), *Disaster Governance in Urbanising Asia*,
DOI 10.1007/978-981-287-649-2_11

11.1 Introduction

Over the last three decades, China has enjoyed rapid urban growth. China's urbanization rate[1] has increased from 17.92 % in 1978[2] to 53.73 % in 2013.[3] As an engine to promote economic development and industrialization of the country, urbanization also causes a variety of problems. The leading issue of importance is of one spatial inequality; this includes the coastal-inland development gap and urban-rural divide. All over the country, the rural poor account for 91 % of the total poor. The western provinces account for over 50 % of China's poor, while their population only represents around 20 % of the national population (Lall and Wang 2012). Secondly, migrating or floating population has become one of the toughest challenges facing China's urbanization. Attracted by the urban-rural income difference and made redundant by surplus farming labor, China's rural residents have turned to work as migrant workers in cities. However, restricted by the system—China's permanent residential registration system, migrant workers do not have access to public services, including affordable housing, healthcare, children's education, etc. With this group mounting to 37.8 million in 2012,[4] their massive unfulfilled needs will become an ever-pressing issue, which will threaten not only social security of the country but also its sustainable development. Moreover, China's sustainable urbanization process is also stalked by resources constraints, municipal debts, transportation, environment pollution, etc.

Consistent with the country's economic growth, China's urbanization also shows an observable imbalance between eastern and western regions[5] (Pan and Wei 2010). 28 of the 38 cities with more than 2.5 million residents are located in the eastern part, while the western region is dominated by smaller cities with population less than one million. In 2011, the average urbanization share of the eastern, central, and western areas are 56.4 %, 53.4 %, and 44.9 % respectively (Yao and Xu 2013).

While the western provinces are striving to increase urbanization rate by constructing infrastructures and creating employment, they also face another vital chal-

[1] We adopt the definition of Urbanization Rate used by National Bureau of Statistics of China as the percentage of resident population in urban areas compared to total population.
http://news.xinhuanet.com/fortune/2013-04/15/c_124582332.htm (accessed on 7 October 2013).

[2] Xinhua News Website:
http://news.xinhuanet.com/fortune/2011-02/14/c_121072377.htm (accessed on 7 October, 2013).

[3] People.cn Website:
http://politics.people.com.cn/n/2014/0120/c70731-24170142.html (accessed on 10 February 2014).

[4] Data announced by National Statistics Bureau. The Central Government Website:
http://www.gov.cn/gzdt/2013-05/27/content_2411923.htm (accessed on 7 January 2013).

[5] According to China's Western Development Program, China's western regions include Neimenggu, Shaanxi, Chongqing, Guizhou, Guangxi, Ningxia, Gansu, Sichuan, Yunnan, Qinghai, Xizang, and Xinjiang. Western Development Program Website: http://www.chinawest.gov.cn/web/Column1.asp?ColumnId=6 (accessed on 7 October 2013).

lenge. That is, throughout history, the majority of China's epicenters have agglomerated in western areas (Shi et al. 2011), making these areas more prone to disasters (including earthquakes and the cascading events such as debris flow, landslides, etc.) and calling for more capacities in disaster risk management (DRM).

Though a disaster can cause enormous losses to human beings, there is also a growing recognition that it also brings an opportunity to facilitate long-term economic, social, and physical development (Berke et al. 1993). On May 12, 2008, an earthquake measuring 8.0 on the Richter scale hit Wenchuan County, killing 69,226, wounding 374,643 and affecting more than ten million people (The State Council 2008b). While the Wenchuan Earthquake resulted in a direct economic loss of 121.6 billion USD (845.1 billion CNY),[6] it received as much as 244.6 billion USD (1.7 trillion CNY) in reconstruction investment coupled with policy support in finance, taxation, land, and industry (The State Council 2008a). The Reconstruction Plan of the Wenchuan Earthquake issued by the State Council in September 2008 stated the goals to improve the livelihood and socio-economic development of the affected areas in combination with promoting urbanization progress. In 2011, the 12th Five-Year Urbanization Plan of Sichuan Province re-emphasized the integration of DRM into infrastructures including schools, hospitals, and dams (Sichuan Provincial Government 2011). With substantial efforts and resources, the urbanization rate of Sichuan Province improved from 37.4 % in 2008[7] to 41.83 % in 2011.[8] On April 20, 2013, the Lushan Earthquake, measuring 7.0 on the Richter scale, hit Sichuan Province again and some townships hard-hit by the Wenchuan Earthquake met heavy suffering once more. The 2013 earthquake became a precious opportunity to examine if disaster governance had been integrated successfully into the urbanization process in these areas, measured by the possible changes of local DRM capacities.

The purpose of this article is to explore the integration of urbanization and disaster governance in post-disaster areas of Wenchuan Earthquake and Lushan Earthquake and put forward a framework to promote the coalition of these two systems. The next section reviews disaster risk management and key effective measures taken by various countries and international organizations. It also explores core elements of *disaster governance* as an emerging concept. The third section introduces China's DRM system and the role of townships in the country's urban spectrum and DRM system. This section follows with a review of the socio-economic conditions of research areas and research methods. The researchers conducted an in-depth structured interview as the main method to assess possible changes of local DRM capacities after 5 years of recovery. Interviews were carried

[6]The average USD to CNY exchange rate is 6.95 in 2008, according to China Statistic Yearbook 2012. This paper uses this rate to convert currency in CNY to USD.

[7]Sichuan Provincial Government Website: http://www.sc.gov.cn/zwgk/zwdt/bmdt/200904/t20090407_665982.shtml (accessed on 7 October 2013).

[8]China News Website: http://www.chinanews.com/cj/2012/06-04/3936370.shtml (accessed on 7 October 2013).

out in 2008 and 2013, both at the stage of responding to the earthquake. 16 samples were selected from Beichuan County and 16 samples from Lushan County and Baoxing County. Beichuan was hardest hit in 2008, while Lushan and Baoxing were hardest hit in 2013. In this chapter, we employ the Disaster Risk Management Framework, including legislation, contingency plan and institutions, risk analyses and early warning systems, emergency response, logistics management, social participation, education and training, and recovery planning as the assessment tool. We then present our findings and analyze the causes of challenges presented to townships' DRM capacities. The last section summarizes the findings and put forward an integration framework of urbanization and disaster governance at strategic, policies, and program levels.

11.2 From Disaster Risk Management to Disaster Governance

With growing economic and social losses brought about by disasters, disaster risk management (DRM) has gained increased attention globally as an approach to not only build resilience and reduce effects of adverse events but also allow residents to take advantage of opportunities for improvement (World Bank 2013b; Manyena 2006). The most commonly used definitions of DRM are from the World Bank and the United Nations. The World Bank adopts the DRM definition of the Intergovernmental Panel on Climate Change (IPCC) as:

> Processes for designing, implementing, and evaluating strategies, policies, and measures to improve the understanding of disaster risk, foster disaster risk reduction and transfer, and promote continuous improvement in disaster preparedness, response, and recovery practices, with the explicit purpose of increasing human security, wellbeing, quality of life, and sustainable development (IPCC 2012).

The United Nations defines disaster risk management as:

> The systematic process of using administrative decisions, organization, operational skills and capacities to implement policies, strategies and coping capacities of the society and communities to lessen the impacts of natural hazards and related environmental and technological disasters. This comprises all forms of activities, including structural and non-structural measures to avoid (prevention) or to limit (mitigation and preparedness) adverse effects of hazards (UNISDR 2004).

Both definitions note that improving DRM capacities will require various types of efforts such as policies, regulations, and education to achieve a comprehensive set of goals. To accomplish these goals, a large number of international organizations, think tanks, and scholars created frameworks to guide the actions of the public, private, and nonprofit sector as well as residents (Overseas Development Institute 2012). The most important strategy and challenge is to design an institutional framework and mechanism to build long-term DRM; and governments play a critical role in managing systemic risks (World Bank 2013c). Leading institutions for DRM must have the necessary authority to coordinate powerful sectoral ministries, and there should be cross-departmental collaboration at the highest possible level of government. It is also important to provide proper incentives for people in these institutions to work out disaster reduction plans (World Bank 2013a). Moreover, legislative revamping, perhaps even a drastic one, is essential to enhance the effectiveness of disaster management institutions (Gopalakrishnan and Odaka 2007). Many scholars (Brennan 2003; Benouar 2002; Mili 2003; Wamsler 2006) proposed that DRM should be *mainstreamed* into development planning, such as urban settlements and housing, agriculture and aquaculture, road construction, school design, etc.

Second, financial resources are critical if policies and planning are to be implemented effectively. To guarantee against potential budget volatility in the aftermath of disasters, contingency funds should be established, at both national and local levels (World Bank 2013a). Furthermore, it is vital to develop a long-term financing framework and encourage vulnerable populations to create self-financing methods (World Bank 2013b; Vatsa 2004). Additionally, risk insurance through public-private partnership might also have a huge potential to ensure that financial needs of recovery are met (Fisher and Surminski 2012).

Third, quality, transparency, and availability of information affect every phase of DRM. For instance, risk information and modeling systems are essential during risk identification period (World Bank 2013a, c). Sharing risk information with all relevant government agencies reaching all levels and sectors, as well as communities and the private sector, is conducive to gathering efforts from all stakeholders. Moreover, data on hazards, exposures, vulnerabilities, and losses are crucial to risk assessment, as is the development of risk financing tools and strategies (OECD 2012).

Lastly, DRM calls for actions and responsibilities from all members of the entire society at all levels. Despite cross-sectoral cooperation of government, companies and nonprofit organizations, efforts ranging from individuals to households to international communities are all valuable (World Bank 2013b; Paton and Johnston 2001). Local participation is especially important in the building of long-term resilient communities (World Bank 2013c).

While DRM contains concrete measures and focuses more on the efforts of the government, there is a growing attention on disaster governance. This suggests to implement disaster risk reduction by using governance practices (Tompkins et al. 2008; Ahrens and Rudolph 2006; Blackburn 2014). Participation of multiple stakeholders or multi-actors is a key element in disaster governance. The interdependence and intercommunication of residents, volunteers, and nonprofits are stressed not only in decision-making but also critical in emergency times and damage reduction (Craye et al. 2009; Johnson 2005). Despite participation, horizontal and vertical integration are also important to respond to newly emerging disaster risks that have the nature of complexity, uncertainty, and ambiguity (Nagasaka 2006; Ikeda 2006; Burby and Linda 1994). Networks and partnerships of multiple organizations (spanning pre-disaster, trans-disaster, and post-disaster periods) are essential in disaster governance (Assmuth et al. 2009, 2010; Tierney 2012; Djalante et al. 2011). However, disaster governance as an emerging concept has yet to be analyzed in terms of practical feasibility (Jones et al. 2013).

11.3 Disaster Risk Management in the Context of China's Township

Established at the end of 2007, China's Emergency Management System—the national strategy to guide DRM at all levels of government—consists of (1) contingency plan system, (2) legal system, (3) institutions, and 4) mechanisms. The contingency plan system consists of the Master State Plan for Rapid Response to Public Emergencies, special state plans for responses to emergencies, and contingency plans of state departments, local governments, enterprises, and public institutions. The legal system designed for emergency management is comprised of the Law of Response to Public Emergencies of P.R.C.—China's first basic law for emergency management, as well as 35 laws, 37 administrative regulations, 55 departmental rules, and 111 regulatory documents for responding to public emergencies (Gao 2008). 31 provinces have specific institutions to strengthen their role during an emergency management. Lastly, emergency management mechanisms were established to address monitoring, early warning and alert systems, emergency decision-making and coordination, social engagement, the allocation and commandeering of supplies, social security, community management, state-social coordination, international coordination, etc. (Fig. 11.1).

As Fig. 11.2 shows, townships are at the bottom of China's government system and are directly responsible for rural areas within their administrative regions. The four units of townships, namely, the Office for Administrative Affairs, Office for

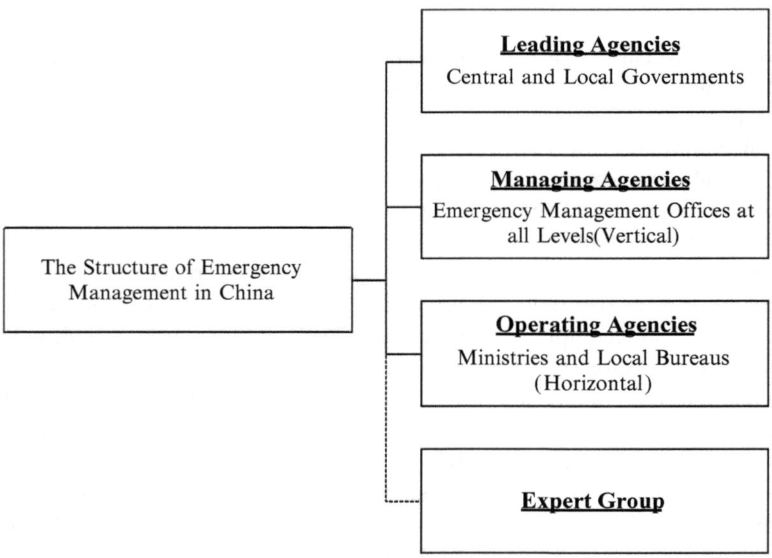

Fig. 11.1 The structure of emergency management in China (*Source*: Adopted from China National Emergency Response Plan, http://www.gov.cn/yjgl/2006-01/08/content_21048.htm (Accessed on 14 November 2013))

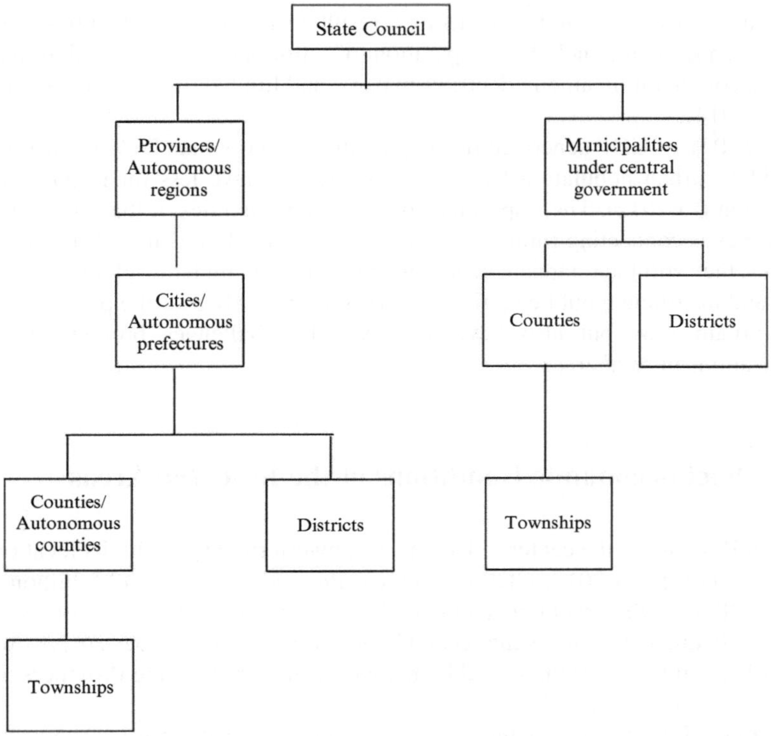

Fig. 11.2 Levels of government under the State Council of China (*Note*: The Figure is depicted by researchers according to Article 30 in *Constitution of the People's Republic of China* (Adopted on December 4, 1982))

Economic Development, Office for Social Affairs, and Branch of Finance and Budget, are responsible for implementing all the policies and regulations from the upper-level county, municipal, provincial, and central governments. In this light, township governments are responsible for DRM within their sovereignty (including urban and rural areas) even they do not have specific institutions in charge of the issue. Townships' DRM responsibilities[9] (The State Council of P.R.C 2007) include establishing contingency plans, defining leading organizations and officials responsible for DRM, organizing public education and drills with the engagement of residents, convening DRM human resources, and enforcing infrastructure construction. However, at all levels of the governments, DRM responsibilities are distributed among various departments and there is a lack of an enabling specific entity to take charge of DRM. Hence policies and regulations are implemented inadequately. In practice, decision-making is subject to individual interests, departmental interests,

[9]The Central Government Website:

http://www.gov.cn/gongbao/content/2007/content_744155.htm (accessed on 4 November 2013).

individual capabilities, and other factors, resulting in various difficulties in emergency management, including separation of information and decision-making, blocked collaboration among decision-makers, and limited decision-making power (Zhang 2014).

Over 20 % of GDP generated from agriculture in more than 56 % townships and over 51 % urban population live in townships that have less than 200 thousand population (Li 2012). The importance of townships also lies in that townships are crucial nexus connecting rural and urban areas. Although they have been classified as cities, they still face stagnant economic development, underdeveloped infrastructures, and insufficient public services. Townships not only bear the pressure to promote urbanization but also have to tackle the challenges brought about by urbanizations in rural areas.

11.4 Socio-economic Conditions of the Research Areas

Sichuan Province is the center of China's southwestern region with its location and economic power. In 2012, GDP of Sichuan Province reached 343.5 billion USD (2387 billion CNY), ranking it 8th of the country and 1st in the southwestern region.[10] Sichuan Province is the second largest *migrant worker exporter,* as 26 % of its whole population left their rural hometowns and worked in the Provinces' urban areas or coastal provinces.[11]

As Fig. 11.3 shows, Beichuan County is located in the northwestern part of Sichuan Province, which is 160 km from Chengdu, the provincial capital. Beichuan County is a mountainous area and in 2012, 36.65 % of its population was made up of ethnic minority groups.[12] Lushan County is located on the western edge of Sichuan Basin. Baoxing County is on the eastern side of Lushan and 200 km away from Chengdu. It has a famous tourist site as the natural habitat for pandas and has mountainous landscape. Ethnic minority group accounted for 18.17 % of Baoxing's population in 2010, of which Tibetan people make up 96.64 %.[13] The major contribution to economic growth of the three townships still stems from agriculture. As these areas lagged behind the national urbanization process, the leadership laid great emphasize on promoting economic development and urbanization. Table 11.1

[10] People.cn Website:
http://finance.people.com.cn/GB/8215/356561/359047/ (accessed on January 20 2014).

[11] Central Government Website:
http://www.gov.cn/jrzg/2012-08/15/content_2204588.htm (accessed on January 20 2014).

[12] Beichuan Government Website:
http://www.beichuan.gov.cn/index.php?m=content&c=index&a=lists&catid=259 (accessed on January 20 2014).

[13] Baoxing Government Website:
http://www.baoxing.gov.cn/2011/1122/377.html (accessed on January 20 2014).

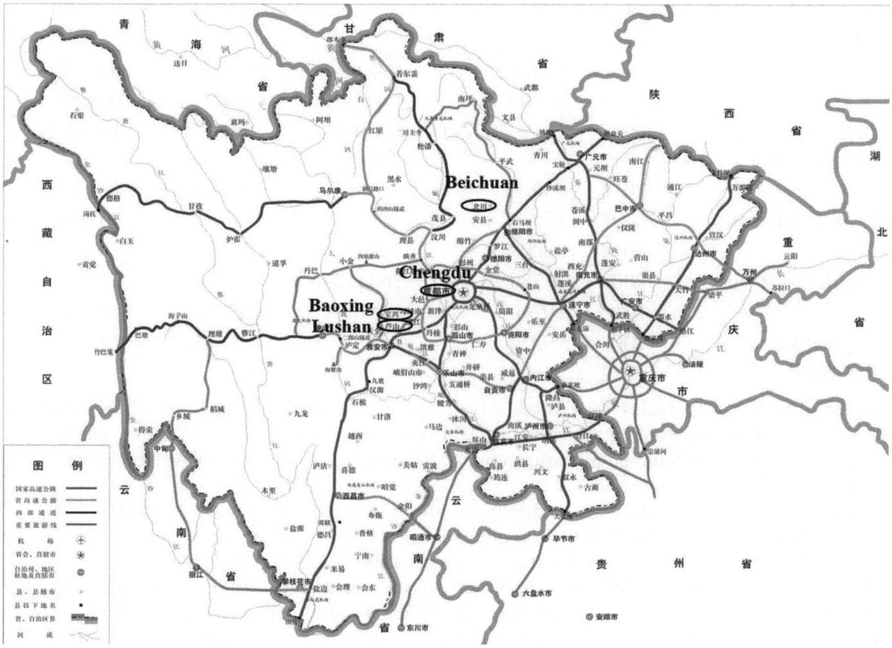

Fig. 11.3 Locations of Beichuan, Lushan, and Baoxing County (*Source*: The Transportation Bureau of Sichuan, 2010. Ministry of Transport Website: http://www.moc.gov.cn/xinxilb/xxlb_fabu/fbpd_sichuan/201001/t20100113_651241.html (Accessed on April 15, 2014))

shows the GDP, population, and urbanization rate of Sichuan Province and the three counties.

11.5 Research Methods

This research was carried out in townships[14] (*xiang zhen*) of the hardest hit areas of both 2008 and 2013 earthquakes, as townships have particular importance both within China's urban spectrum and within its vertical DRM system. This study employed an in-depth structured interview as the main method to assess local DRM capacities. The interviews of township governors were carried out with the aim of obtaining thorough information about their actions related to disaster preparedness, immediate response, and recovery planning. The interview structure was formulated in accordance with Disaster Risk Management Capacity Framework (Federal

[14]According to the State Council (1963), townships are localities with populations of over 2,500. Minister of Civil Affairs Website: http://qhs.mca.gov.cn/article/zcwj/qhgl/200711/20071100004365.shtml (accessed on 7 October 2013).

Table 11.1 GDP, population, and urbanization rate of Sichuan Province and Beichuan, Lushan, and Baoxing County

		Sichuan province	Beichuan county	Lushan county	Baoxing county
GDP billion USD (billion CNY)	2008	181.3 (1260.1)	1.5 (10.6) (2007)	1.9(13.1)	1.6(11.1)
	2012	343.5 (2387.3)	4.6(31.7)	3.7(25.4)	2.9(20.0)
Urbanization rate	2008	35.6 %	/	24.5 %	30.8 %
	2012	43.5 %	32.04 %	28.7 %	35 %
Population (thousand)	2008	81380	160.2(2006)	120.1(2007)	57.8(2007)
	2012	80760	241.3	120.09	58.7
Migrant workers (thousand)	2008	20234	/	25.1(2007)	2.6
	2012	21000	55	28	12.3

Source: The Ya'an Statistics Year Book 2013; Beichuan Statistics Year Book 2007; Beichuan Government Website; Baoxing Government Website; Lushan Government Website; The Central Government Website; National Statistics Bureau Database; China Institute of Water Resources and Hydropower Research Website

Beichuan Government Website: www.beichuan.gov.cn (accessed on January 20 2014)

Baoxing Government Website: www.baoxing.gov.cn (accessed on January 20 2014)

Lushan Government Website: www.yals.gov.cn (accessed on January 20 2014)

The Central Government Website: http://www.gov.cn/gzdt/2009-01/09/content_1200705.htm, http://www.gov.cn/jrzg/2012-08/15/content_2204588.htm (accessed on January 20 2014)

National Statistics Bureau Database: http://data.stats.gov.cn/workspace/index?m=fsnd (accessed on January 20 2014)

China Institute of Water Resources and Hydropower Research Website: http://www.iwhr.com/zgskyww/ztbd/wcdz/zqslgc/webinfo/2008/05/1273896623312758.htm (accessed on January 20 2014)

Note: Due to data unavailability and serious damage caused by Wenchuan Earthquake, we could only use data of previous years for 2008 data of some categories

Emergency Management Agency and National Emergency Management Association 2001; Zhang 2010) as Table 11.2 shows. The interviews in 2008 and 2013 were both conducted at the same stage of responding to earthquakes (Tables 11.3 and 11.4).

11.6 Findings

11.6.1 Contingency Plan and Institutions

According to the National Emergency Management System, township authorities bear the responsibility of establishing contingency plans at the township level (The State Council 2007). The importance of contingency plans (CPs) was reinforced after the 2008 Wenchuan Earthquake. However, there are still problems that require attention to improve the effects of CPs.

First, as townships' CP systems comprise of separate plans for specific disasters, both sets of interviews indicated that while CPs of most townships offer detailed

Table 11.2 Disaster risk management capacity framework

Category	Description
1. Legislation	1.1 Establish and enforce laws on DRM
2. Contingency plan and institutions	2.1 Contingency plans: contents and coverage
	2.2 The implementation and effects of the plans
	2.3 Specific institutions/persons responsible for DRM
3. Risk analyses and early warning systems	3.1 Established mechanisms and resources (human, equipment, funding, expertise, etc.) for risk analyses and alerts
4. Emergency response	4.1 First responders for rescue and relief and their actions
	4.2 The involved groups and their participation and coordination
5. Logistics management	5.1 Resources (human, equipment, funding, expertise, etc.)
	5.2 Infrastructure for DRM
6. Social participation	6.1 Participation of residents in DRM
	6.2 Participation of nonprofits in DRM
	6.3 Participation of the private sector in DRM
7. Education and training	7.1 Mechanisms and forms of education and training for DRM
	7.2 Involvement of government officials and residents
8. Recovery planning	8.1 The planning process and stakeholder participation
	8.2 Resource management for recovery

Note: As townships do not have legislative rights, we did not assess legislation in our interviews

Table 11.3 Impact of Wenchuan and Lushan Earthquakes

Category	Wenchuan earthquake	Lushan earthquake
Time	14:28 am (GMT+8), May 12, 2008	8:02 am (GMT+8), April 20, 2013
Magnitude	8.0 on Richter scale	7.0 on Richter scale
Epicenter	Yingxiu Township, Wenchuan County, Sichuan Province	Longmen Township, Lushan County, Sichuan Province
Source	Lianhuaxingou of Niuquangou	Xichuangou, Shuangshi Township
Depth	10 km	13 km
Intensity	11 °	9 °
Casualties	69,226 killed, 17,923 missing, 374,643 injured	196 killed, 2 missing, 11,485 injured
Affected population	More than ten million	2,184

Source: The State Council of P.R. China 2008a, b, 2013a, b

Table 11.4 Townships interviewed

Interviewed time	County	Townships
June–July, 2008	Beichuan	Guixi, Chenjiaba, Kaiping, Qingpian, Baishi, Guanling, Yuli, Xuanping, Baini, Duba, Dunshang, Xiaoba, Qushan, Tongkou, Leigu, Badi
August, 2013	Lushan	Luyang, Feixianguan, Shuangshi, Taiping, Dachuan, Siyan, Qingren, Longmen
August, 2013	Baoxing	Muping, Lingguan, Fengtongzhai, Qiaoqi, Yongfu, Mingli, Wulong, Daxi

guidance for dealing with common hazards, such as floods, droughts, forest fires, landslides, and mudslides, townships however lack CPs for earthquakes. As the governor of BC13 Township[15] said, "I started to work at this position in September 2006 and only heard about preparing for earthquakes for one time in a conference in 2007. This topic was not mentioned again" (BC13 Governor 2008). This situation remains noticeably unchanged even after the 2008 Wenchuan Earthquake. According to governor of LS03 who stated, "Our contingency plans cover floods and (small) geological disasters. We could not make plans for catastrophes such as earthquakes" (LS03 Governor 2013). The progress made after 2008 is mainly on CPs for cascading disasters, which a few townships established to help avoid casualty in the aftermath of calamities. In these townships, such plans helped achieve zero casualty in the cascading disasters of the Lushan Earthquake.

Second, the implementation of CPs is highly dependent on roads, electricity, and telecommunications, the basic infrastructure components which are extremely fragile in times of disasters. "Roads and telecommunications were absolutely cut off. How could we share information?" (BC11 Governor, 2008). Moreover, the CPs mainly consider township governments and fail to address how the infrastructure should be fixed with the support of related departments under the administration of counties, including the county transportation department, local branches of the national power plant, etc. (LS03 Governor 2013).

Despite the CPs, institutions present a difficult challenge with respect to the implementation of DRM. As there are only four units of township authority, no specific office is in charge of DRM. Therefore, not only do the townships lack first responder ability when a disaster hits, but they also lack a specific unit for disaster preparedness, relief, mitigation, and recovery (BC01 Governor 2008). Before the Wenchuan Earthquake, adaptive measures were adopted by county authorities mandating that the counties' Office for Work Safety act as an informal coordinating unit of disaster management of the townships (BC01 Governor 2008; BC03 Governor 2008). Since 2010, county governments have organized annual emergency management conferences and convened township officials to guide their DRM work (LS05 Governor 2013). However, the fundamental institutional structure has still not been revamped.

In addition to the lack of an administrative body, the townships do not have specific rescue teams either. Every village organized its own emergency rescue team, mainly comprised of militia members and other villagers (BC06 Governor 2008) who were trained to cope with floods and forest fires (BC12 Governor 2008). After the Wenchuan Earthquake, there were growing numbers of militia members on the rescue teams who played positive roles in the rescue of the 2013 Lushan Earthquake.

[15]At the request of our interviewees, we maintain the anonymity of the names of townships and governors. We use BC+Number to represent the townships interviewed in the 2008 Wenchuan Earthquake and LS+Number to represent those in the 2013 Lushan Earthquake.

11.6.2 Risk Analyses and Early Warning

Aimed at geological disasters, the traditional township early warning system operates by employing residents as monitors to watch for high-risk spots near their own houses and use certain instruments to alert others (BC01 Governor 2008; BC06 Governor 2008). However, those townships lack human resources and financial and equipment support to support this system. Firstly, there is a lack of sufficient and professional labor responsible for risk analyses and early warning. Prior to 2008, the monitors, who were residents, either volunteered or worked part-time. After the Wenchuan Earthquake, a four-level monitor system was established, in order for the community within the village, the township, and the county to provide specific people who could cooperatively monitor high-risk areas (LS02 Governor 2013). At the bottom (community) level, there was a geological hazard monitor and a road safety monitor who worked either full-time or part-time (LS06 Governor 2013). However, these monitors performed their responsibilities based on experience rather than on professional knowledge, and they seldom received training. Secondly, the stipend for the monitors was very low, only 0.1 USD (1 CNY) per day and 0.4 USD (3 CNY) to 1.4 USD (10 CNY) per day before and after the Wenchuan Earthquake, respectively (LS07 Governor 2013). Under such circumstances, it is difficult to recruit monitors and to provide proper incentives for them to perform satisfactory outcomes. Thirdly, as monitors are provided with no equipment to conduct risk analyses, they must rely solely on their eyes to define risks. Furthermore, the equipment used to inform villagers before the 2008 Earthquake was extremely crude, that is, in many cases, only a gong[16] (BC01 Governor 2008) which could only alert nearby neighborhoods rather than communicate over long distance. During the 2013 interviews, a few townships had two-way radios for communication, but such equipment is only found in townships that are highly prone to debris flows and landslides (LS02 Governor 2013).

In addition to insufficient labor, expertise, financial support, and equipment, a lack of awareness regarding disasters on the part of local residents adds to the challenges associated with effective DRM. In some highly vulnerable and at-risk locations, even when risk analysis recommends that residents move to other areas, some of these residents accept relocation only after suffering great losses due to a catastrophe or to numerous disasters. "We successfully relocated some villagers after the county was caught by the most serious flood in recent 50 years" (BC03 Governor 2008).

11.6.3 Emergency Response

Township authorities and officials are the main groups responsible for directing DRM. Normally, a township has approximately 30 officials who, on the event of an earthquake, would be grouped into several task forces. During the rescue period,

[16] A gong is a traditional Chinese instrument that creates loud noise through striking and can be regarded as a portable drum.

there are three task groups: a rescue group, an information collection and communication group, and a supply management group. Later, more groups emerged, such as a sanitation group, a relocation group, and a security group (LS01 Governor 2013). While facing the challenge of a lack of labor, officials would develop *temporary party*[17] *member mechanism* to absorb more manpower to perform administrative work.

Local residents were the most effective rescue force in both the 2008 and 2013 earthquakes. In the Wenchuan Earthquake, 70,000 of the 87,000 evacuees were local residents who saved themselves and also helped others. Aside from the residents and local emergency rescue teams, rescue labor included external military forces, armed police, and firemen. However, the contribution of these three forces was limited due to their lack of related training, experience, and equipment.

Within the government system, there were frequent multilayered cross-regional and cross-departmental coordination and cooperation. Apart from township authorities, government officials at the county and municipal levels also worked to manage disaster rescue and relief. Another form of multilayered coordination in 2008 existed in the paired assistance reconstruction model,[18] which required provincial and municipal officials of the assisting provinces to work with county and township governors of post-disaster areas, respectively. Such cooperation was strengthened in 2013 as these officials responded to the disaster more quickly and worked in post-disaster areas for longer periods of time. Cross-regional cooperation was a frequent occurrence during the rescue period, as nearby townships shared emergency supplies and information. Township officials, military, police, and firemen also engaged in cross-departmental cooperation by working together and convening conferences.

11.6.4 Logistics Management

The logistics management system includes emergency supplies, such as food, water, funding, tents, medicine, and rescue equipment as well as infrastructures such as roads, emergency shelters, electricity, and telecommunications.

While townships face severe challenges dealing with emergency rescue resources, the lack of financial support for supply acquisition is the most critical problem. Townships do not have fiscal power and there is no budget for emergency supply acquisition (LS06 Governor 2013). With respect to a contingency fund, some townships have no funds (LS11 Governor 2013), while others have only minimal funds (LS12 Governor 2013).

Neither the township authorities nor the villagers reserved sufficient and effective food, water, medicine, tents, and rescue equipment. "We have not reserved any

[17] Communist Party of China.

[18] In paired assistance reconstruction, provinces of the eastern region provide financial, human, and technological resources to help the post-disaster areas.

food and water. The whole township of over 3,000 residents only collected 7 boxes of instant noodles, cookies and bottled water" (BC03 Governor 2008). "I hope in the future, our township will have two-way radios and vehicles for rescue" (LS15 Governor 2013). Because of the severe impoverishment, there were areas that had few stores to provide food and water. Furthermore, the delay in receiving the supplies made the situation worse, as most townships did not receive supplies until 3–5 days following the Wenchuan Earthquake.

Although a few townships began to stockpile tents and tarpaulins after the 2008 Earthquake, such acquisitions were inadequate "official response" as the amount was too little to carry out an effective role, when the Lushan Earthquake hit. For instance, LS01 Township actually had only ten tents in reserve, while LS06 Township had only one. The lack of food reserves was also due to the absence of a flexible reserve mechanism in some townships. As one governor said, "*We could not reserve enough food because it easily goes bad*" (LS04 Governor 2013).

A degree of progress is evident in the mechanics of acquiring supplies through the establishment of public-private partnership (PPP). Some townships signed PPP contracts with local stores, which thereby allowed the authorities to use commodities under emergency circumstances and make payments later (LS02 Governor, 2013; LS14Governor, 2013; LS15 Governor, 2013).

As a couple of townships were located in mountainous areas, it was difficult to build large-scale emergency shelters. The governor of BC11 (2008) stated that they could only build temporary emergency shelters with tents and tarpaulins in flat places after the earthquake. The importance of emergency shelters was stressed after the 2008 earthquake, as our interviews found that the majority of townships visited constructed gyms and "town squares" to function as emergency shelters.

Transportation, electricity, and telecommunication remain difficult issues with respect to DRM. In both earthquakes, these areas of infrastructure crashed within a very short time, and it took a significant period of time for them to recover. The majority of the townships interviewed in 2008 and 2013 were *islanded* without emergency supplies and information. While recovery following the Wenchuan Earthquake claimed to enhance disaster response capacity of the infrastructure systems, it seems that no fundamental improvements have been achieved. Though roads were reconstructed and strengthened to better withstand the ferocity of disasters, the standard width of 3.5 m was not changed, impeding large rescue vehicles from passing and reaching affected areas (LS10 Governor 2013). While a few townships were able to recover electricity and telecommunications within 3–5 days after the Lushan Earthquake, many others had to wait passively for recovery.

11.6.5 Social Participation

Social participation consists of the engagement of local residents, nonprofit organizations, and the private sector. The participation of local residents in DRM was mainly through serving as risk monitors and joining the militia, as well as conducting self-evacuations. However, residents rarely joined in other phases of DRM such as disaster preparedness, mitigation, and recovery. The private sector contributed by participating in PPP, donations, and their employees coming forward as professional volunteers.

The year 2008 was referred as the Year of the Civil Society (Shieh and Deng 2011) as nonprofits' participation in providing public services began to soar after the Wenchuan Earthquake. However, there were few well-established, large-scale nonprofits involved in the 2008 Earthquake, and most volunteers lacked the experience to carry out a rescue operation efficiently. On the other hand, in 2013, many formal and renowned nonprofits played very positive roles in various aspects of disaster response including rescue, relocation, and service delivery, such as building kindergartens and libraries and broadcasting movies. Most of these nonprofits, however, were from areas outside the impacted regions, and there were no local nonprofits. "We need them (the nonprofits) to stay here for the long run," as the governor of LS09 Township said.

In 2008, very few township governors understood the concept of a nonprofit organization. To cooperate and communicate better with nonprofits in 2013, the Sichuan Provincial Government and the Ya'an Municipal Government established the Nonprofit and Volunteer Service Center for the Ya'an Disaster Relief with two branches; one in Lushan County and one in Baoxing County, to provide information and coordinate the activities and resources of the nonprofits. However, our interviews revealed that township governors still did not have a clear understanding of the definition and roles of nonprofits. They also held overly vigilant attitudes toward nonprofits, attempting to control them rather than to cooperate with them or provide services to them. One governor claimed, "We certainly welcome nonprofits that could help us. But they must obey our (township authority's) rules and management" (LS 13 2013).

11.6.6 Education and Training

The interviews revealed two problems related to the education and training intended to cultivate disaster preparedness in both 2008 and 2013. First, there was a lack of participatory and engaging approach to conduct education and training. Township governments implemented these activities through annual conferences and drills. However, as the drills and conferences organized by township governments through compulsory orders and the forms of the drills were monotonous, they resulted in resistance from the residents. Moreover, it was difficult to evaluate the effect of education and training on the local residents. To some degree, the activities were regarded simply as evidence of the administrations having carried out their responsibilities.

11.6.7 Recovery Planning

As both sets of interviews were conducted before recovery efforts began, we analyzed related issues based on how township governors envisioned recovery plans and how they predicted potential challenges. At the macro level, there was a lack of feasible recovery planning as well as a failure to incorporate a participatory approach in the planning. Townships faced impractical orders from upper-level government that focused on pursuing rapid reconstruction regardless of the difficulties the areas encountered, such as bad weather, insufficient manpower, economic conditions, etc. Our interview with the BC11 governor reflected these problems as the governor said, "We are required to complete permanent relocation before the 2009 Chinese New Year, but it is really a mission impossible! They (the decision makers) did not know our customs, that residents here would not cut trees until October. But then we would not have enough time to build houses" (BC11 Governor, 2008). Moreover, as township authorities were unable to participate in the planning process, they were unable to voice demands of the townships and the residents. Under such circumstances, it is highly unlikely that the local concerns were considered in plans.

At the microlevel, several factors prevented residents from rebuilding their own houses, including insufficient human and financial resources. Despite the huge economic input for recovery, amounting to 244.6 billion USD (1.7 trillion CNY) in 2008,[19] a very small part of that money was allocated to residents as compensation to rebuild their houses. In 2008, every household whose house had been collapsed or destroyed was eligible to receive 1438.8 USD (10,000 CNY) but the average cost to build a house was more than 21,582.7 USD (150,000 CNY), and the soaring prices of construction materials made the situation even worse. Therefore, the majority of the residents decided to complete the construction by working in cities as migrant workers, thereby depleting the already short supply of manpower and technicians (Fang et al. 2009). Some families had not even repaid the loans incurred during Wenchuan reconstruction when Lushan reconstruction began (Table 11.5).

11.7 Analyses

The findings revealed that townships' DRM capacities in the hard-hit post-disaster areas of Sichuan Province have not been significantly enhanced and that there are serious problems in every dimension of DRM. Our analyses focused on the role of governance in promoting urbanization and DRM capacities as well as the influence of urbanization on DRM of the vulnerable townships.

While China is achieving outstanding urbanization progress in the last decade, there are severe challenges in various fields. These problems, including the coastal-inland development gap and urban-rural divide, unequal public services between

[19] The entire investment in Lushan Recovery was not announced.

Table 11.5 Summary of key findings

Category	Improvement	Remaining challenges
1. Contingency plan and institutions	A few townships developed CPs that helped avoid casualties in cascading disasters	Townships lacked specific CPs for earthquakes
		CPs were highly reliant on infrastructures
		No specific government unit responsible for DRM at township level
		No professional emergency rescue force
2. Risk analyses and early warning	No observable improvement	Insufficient labor and expertise
		No professional equipment
		Lack of financial support
3. Emergency response	Frequent and effective multilayered, cross-regional, and cross-departmental coordination and cooperation within the government sector	Rescue forces lacked training, experiences, and equipment
		Self-rescue of local residents played important roles
4. Logistics management	A few townships signed PPP contracts to ensure emergency supplies	Lack of financial support for purchasing and reserving emergency supplies
		Slow supply after disasters
		Insufficient emergency shelter, fragile roads, weak telecommunication, and electricity plants
5. Social participation	Increased nonprofit participation in post-disaster social service delivery	Mere participation of residents in the DRM except rescue
		Government held cautious attitude toward nonprofits
6. Education and training	No observable improvement	Lack of participatory and effective approaches to engage residents
7. Recovery planning	No observable improvement	Impractical planning for townships to realize goals stated
		No participation of township officials and residents in decision and planning processes
		Lack of manpower at the household level to rebuild houses
		Insufficient funding for households to cover reconstruction expenses

urban and rural (and migrant) population, municipal debts, transportation, pollutions, and so on, are presenting increasing challenges to China's sustainable development. While urbanization is a national strategy to drive economic growth and domestic consumption, urban planning is mainly conducted by the top-down political system with mere participation of other sectors. Even the latest National

New-type Urbanization Plan (2014–2020) remains a top-down approach to promote urbanization (National Development and Reform Commission 2014). The problems in urbanization together with the absence of governance approach in the urbanization process partially account for low DRM capacities in the research areas.

The role of human resources in DRM is extremely important since first responders and decision-makers are key components in crisis management. First responders mainly consist of local police, firefighters, emergency medical professionals, and civilians. Being caught up in or near the location of a disaster, their assessment and actions of the disaster are critical to affect DRM progress. First responders need to acquire vital information, to manipulate inter-departmental cooperation and cooperation between nonprofits and local government, and to be well trained to deal with natural disasters (Kim et al 2007; Janssen 2010). Christensen (2009) coined the term Crisis Leadership to describe leaders' capacities to mobilize resources to perform optimal outcomes in face of hazards and to affect others in a short term. Crisis leadership has its special significance in China's western townships with limited resources and underdeveloped infrastructure. These townships rely on leadership to mobilize community resources and knowledge, to reduce losses, to engage residents, and to build disaster preparedness and community resilience (Zhang 2010). However, urbanization deprives the western townships of the most needed labor and adds extra difficulties for these vulnerable areas. As the townships are less urbanized and suffer from urban-rural income divide, nearly half of the laborers are attracted to work in more developed areas, such as municipalities, provincial capitals, and coastal provinces. For instance, the numbers of migrant workers from Lushan and Baoxing are 28 thousand and 12.3 thousand, accounting for 23.3 and 21 % of the townships' whole population. Our study even found that most of the township officials of Lushan and Baoxing lived in the Ya'an municipality and drove to work during the weekdays. This is made worse, as the Hukou system prevents the migrant population from enjoying the same public service in the areas they work, they have to leave the other family members—the elderly and children—in the rural hometowns. Under such circumstance, the townships are short of leadership and first responders and other human resources for disaster preparedness, response, relief, and recovery. Findings revealed that there was a lack of capable and adaptive individuals to carry out risk analysis, early warning and create participatory educational approaches to engage residents. When a disaster hit, the leftover elderly and children, who accounted for the majority of population, added to the pressure of emergency rescue. Human resource shortage impeded the reconstruction process as most migrant workers returned to cities when the relief period wrapped up and the townships lacked manpower to build collapsed houses.

Apart from human resources, urbanization also deprives the townships of financial resources in combination with China's fiscal system. According to the township's fiscal system (*xiang cai xian guan*), which was newly established in 2008, the right to collect revenues and use township budgeted money belongs to the county government. Therefore, township finances are subject to strict monitoring by the county government. However, the townships still have to pay the costs incurred to deliver public services (Hou and Yang 2008; Yao 2009). These situations together with low revenues caused by urban-rural inequality, township governments often have a limited budget for disaster risk management as money must go to priorities such as economic development and other issues. There was also a lack of financial support to employ and train full-time local workers and experts for risk assessment, monitoring, and post-disaster recovery. The limited money available could only be allocated to establish contingency funds and public-private partnership and reserve emergency supplies such as food and water, medicine, two-way radios, and monitoring equipment.

Furthermore, urban planners of the townships do not take full consideration of DRM and fail to grasp the real needs of townships due to a lack of participation of the local residents in the planning process. Both the Wenchuan Reconstruction Plan and the Lushan Reconstruction Plan were organized and developed by the central government and by experts and did not include participation of township authorities and residents. Though the reconstruction plans stressed to improve DRM during recovery, it is still extremely difficult to incorporate enhancing DRM into every aspect of the urbanization process such as the selection of a suitable relocation site and the implementation of construction standard. Laws and regulations on land use, telecommunications, and urban planners do not address the concerns with regard to disaster preparedness and response (Zhang 2014). An example would be the strengthening of road shoulders after the Wenchuan Earthquake; this effort did not include widening of the road to enable rescue vehicles to pass. Implementation of building standards was affected by the lack of government funding and household financial ability, both are consequences of urbanization. On the one hand, due to unequal allocation of financial revenue, township governments were unable to provide sufficient funding to implement regulations of building code. On the other hand, as households were still struggling for their livelihoods, they had neither the awareness nor the economic ability to obey building code. Having very few qualified builders to build appropriately approved houses made the situation worse. The reconstruction plans did not address the endogenous factors for long-term local development that could boost sustained economic growth and residents' living standards without continuing large-scale migration. Though the Wenchuan Reconstruction Plan emphasized the development of local industries, the mainstream implementation strategy still focused on the construction of infrastructure without considering the expertise and needs of the laborers, such as providing competitive wage.

Apart from urbanization, lacking disaster governance also attributes to problems found in local DRM capacities. At the local government level, due to insufficient legislation and regulation and failure to move away from the traditional pattern of management system in setting up a unified emergency management organization, emergency management functions are still distributed among various departments. This has resulted in a lack of implementation capacities to mobilize human and financial resources for the thoroughly integrated cycle of disaster risk management. The disaster management system has not been implemented in various departments and at all levels. Lacking such a system impedes the improvement of DRM capability.

Additionally, a polycentric disaster governance system has not been established. The traditional top-down system—which means local governments follow orders of central government and governments of lower levels follow orders of those of higher levels—prevails and prevents the township governments and local residents to vigorously engage in DRM cycle, especially in the recovery period. The cross-sectoral partnership between local governments and nonprofits played increasing but insufficient roles. While there were increasing cooperation between the governments and the nonprofits, there's still insufficient support from governments to sustain such partnership.

An important aspect of regarding disasters as a window of opportunity is that disasters might facilitate learning and accelerate changes in rules and regulations to adapt to future calamities (Naess et al. 2005; Baker and Refsgaard 2007) as institutional learning emerges out of society's response to previous crises (Adger et al. 2005; Pelling and Dill 2010). Participation, social capital, networks, and multi-sectoral cooperation are essential components that promote social learning process. Participation in planning incorporates diverse views from various stakeholders so that lessons learned during disasters could be appropriated in further disaster prevention and management (Baker and Refsgaard 2007). The social networks and interactions among various stakeholders provide opportunities for not only reproducing knowledge and refining previous practice but also enhancing innovation and capacity to learn (Melé 2003; Pelling and High 2005a, b; Waugh and Streib 2006). Furthermore, when collaborating together, government agencies, nonprofits, corporate, and communities can learn from one another (Berkes 2007). On the contrary, a centralized system could cause failure in utilizing local knowledge, obtaining adequate understanding of local environments, and developing adaptive strategies when a disaster strikes (Adger et al. 2005). Due to insufficient participation and cross-sectoral cooperation, the substantial lessons learned from Wenchuan Earthquake were seldom converted into institutional changes that could enhance local DRM capacities.

11.8 Conclusion

As the nexuses of urban and rural areas, China's townships have particular significance in the country's urban landscape and government system. Southwestern townships, however, face greater challenges from both unequal economic and social development and from disaster risks. In this research, we explored the disaster risk management (DRM) capacities of earthquake-stricken townships in Sichuan Province in 2008 and 2013 and analyzed how rapid urbanization has affected these areas' DRM capacities.

One of the most challenging issue of China's rapid urbanization process is large-scale migration, as vast rural population leave their family members—the elderly and children—and work in the cities without having access to the same public services as urban residents. The migration deprived China's southwestern townships the inevitable human resources to disaster preparedness, rescue, and recovery since there lack sufficient and capable first responders and leaderships to conduct early warning and risk analysis, to organize education and training, and to reconstruct houses with proper building standard. Insufficient financial resources is the next challenge; this is caused by the fiscal structure of townships and urban-rural income gap, which prevented the townships from employing DRM professionals, reserve emergency supplies, and establish contingency fund and public-private partnership. Thirdly, recovery planning to promote urbanization of townships failed to take into account of enhancing DRM. For instance, implementation of the building regulation was impeded by the lack of human and financial resources. Therefore, infrastructures rebuilt after Wenchuan Earthquake were still too fragile to withhold the Lushan Earthquake. The planners also failed to catalyze endogenous economic development of the underdeveloped townships and to create local employment to attract migrant workers back to their rural hometowns. Lastly, the partial urbanization that promotes urban sprawl rather than human development could not create an enabling environment for residents to take active participation into disaster preparedness and recovery planning. Due to insufficient participation and cross-sectoral cooperation, the substantial lessons learned from Wenchuan Earthquake were seldom converted into institutional changes that could enhance local DRM capacities. The findings of our research reveal that although the central government and the Sichuan provincial government have put in enormous efforts into promoting urbanization and enhancing DRM capacities during earthquake recovery, the post-disaster townships have yet to improve the local DRM capacities under a variety of influence by urbanization.

Learning from the Advocacy Coalition Framework by Sabatier (1998), we suggest the integration of urbanization and disaster governance as a coalition of two policy systems. As Fig. 11.4 illustrates, the integration of the two systems requires joint efforts at three levels. At the strategic level, both urbanization and disaster governance should be established as core policy beliefs (Sabatier 1998) which include the valuable practices of governance, especially decentralization and participation, as well as multilayered and multi-sectoral cooperation, to establish institutions, legislation, and mechanisms to mobilize capacities and resources of communities and the civil society to take active part. At the second level, urban planning and DRM policies should be created and enforced. China has a well-established National New-Type Urbanization Plan (2014–2020) that emphasizes to enhance the country's overall DRM capacities through improving risk analysis and early warning system, enforcing building code, and organizing education and training programs. This plan also includes engaging the participation of the nonprofits and private sector. The DRM system, including institutions, legislation, and mechanisms, has been thoroughly established in 2007. But the strategic level and the program level still call for more endeavors. For instance, disaster governance should be set up as a national strategy so that local governments become aware of the importance of this issue and take DRM into consideration when enforcing policies and allocating resources. At the program level, lacking effective and efficient tools to promote sustainable urbanization and enhance DRM capacities remain as challenging issues.

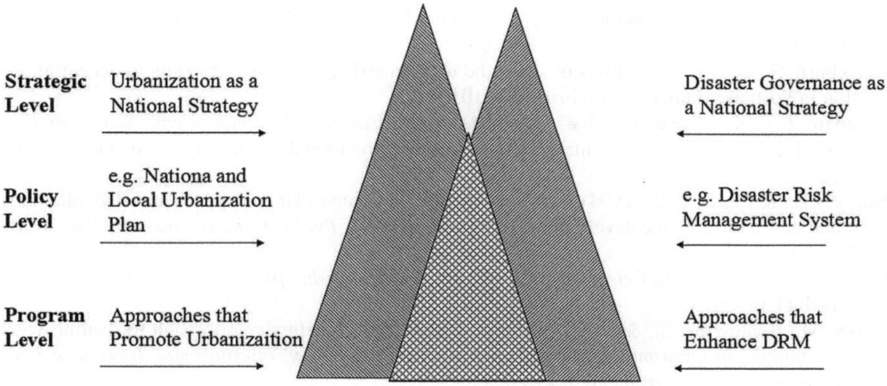

Fig. 11.4 Integration of urbanization and disaster governance

Our empirical and theoretical contributions were addressed in the following ways. First, our study provides the unique and valuable experiences of China's western townships to the international community of disaster research. Second, we use the Disaster Risk Management Capacity Framework to measure the outcomes of employing approaches of disaster governance. The components of the framework serve as important and feasible linkages of disaster governance and urbanization. Third, we propose the framework of Integration of Disaster Governance and Urbanization to combine the complicated policy systems at strategic, policy, and program level, which would provide guidance for policy makers. Future research could explore the roadmap of mainstreaming disaster governance into the holistic governance system.

References

Adger, W. N., Hughes, T. P., Folke, C., Carpenter, S. R., & Rockström, J. (2005). Social-ecological resilience to coastal disasters. *Science*, 309(5737), 1036–1039.

Ahrensn, J., & Rudolph, P. M. (2006). The importance of governance in risk reduction and disaster management. *Journal of Contingencies and Crisis Management,* 14(4), 207–220.

Assmuth, T., Hildén, M., & Benighaus, C. (2010). Integrated risk assessment and risk governance as socio-political phenomena: A synthetic view of the challenges. *Science of the Total Environment,* 408, 3943–3953.

Baker, D., & Refsgaard, K. (2007). Institutional development and scale matching in disaster response management. *Ecological Economics,* 63(2), 331–343.

Benouar, D. (2002, December 9–11). *The need for an integrated disaster management strategy in North Africa towards poverty reduction: A case study of Algiers (Algeria).* Paper presented at the urban development for poverty reduction- towards a research agenda, Washington, DC, USA.

Berke, R. P., Kartez, J., & Wenger, D.R. (1993). Recovery after disaster: Achieving sustainable development, mitigation and equity. *Disasters*, 17(2), 93–109.

Berkes, F. (2007). Understanding uncertainty and reducing vulnerability: Lessons from resilience thinking. *Natural Hazards*, 41(2), 283–295.
Blackburn, S. (2014). The politics of scale and disaster risk governance: Barriers to decentralisation in Portland, Jamaica. *Geoforum*, 52, 101–112.
Brennan, T. (2003, December 2–4). *Mainstreaming disaster risk management: Some possible steps*. Paper presented at the international conference on total disaster risk management, Kobe, Japan.
Burby, J.R., & Linda, C.D. (1994). Plans can matter! The role of land use plans and state planning mandates in limiting the development of hazardous areas. *Public Administration Review*, 54(3), 229–238.
Christensen, T. W. (2009). *Crisis leadership: A study of leadership practice*. Doctoral dissertation, Capella University.
Craye, M., Funtowicz, S., & Van Der Sluijs, J. P. (2009). A reflexive approach to dealing with uncertainties in environmental health risk science and policy. *International Journal of Risk Assessment and Management*, 5, 216–236.
Djalante, R., Holley, C., & Thomalla, F. (2011). Adaptive governance and managing resilience to natural hazards. *International Journal of Disaster Risk Science*, 2(4), 1–14.
Fang, C., Li, M., & Wu, F. (2009). Analysis on urbanization and the spatial differentiation and flow of the migrant workers in Wenchuan earthquake-stricken regions. *Journal of Mountain Science*, 126(4), 482–489.
Federal Emergency Management Agency and National Emergency Management Association. (2001). *State capability assessment for readiness*. Washington, DC: The Agency Press.
Fisher, S., & Surminski, S. (2012). *The roles of public and private actors in the governance of adaptation: The case of agricultural insurance in India*. Centre for Climate Change Economics and Policy working paper no. 102. http://eprints.lse.ac.uk/46400/1/WP89.pdf. Accessed 14 Nov 2013.
Gao, X. (2008). Achievements and development in construction of emergency management system with Chinese characteristics. *Chinese Public Administration*, 11, 18–24.
Gopalakrishnan, C., & Odaka, N. (2007). Designing new institutions for implementing integrated disaster risk management: Key elements and future directions. *Disasters*, 31(4), 353–372.
Hou, J., & Yang, Y. (2008). Xiangcaixianguan's effect on township financial expenditure. *Journal of Public Management*, 5, 61–67.
Ikeda, S. (2006). An integrated risk analysis framework for emerging disaster risks: Toward a better risk management of flood disaster in urban communities. In S. Ikeda, T. Fukuzono & T. Sato (Eds.), *A better integrated management of disaster risks: Toward resilient society to emerging disaster risks in mega-cities* (pp. 1–21). Tokyo: Terra Scientific Publishing Company.
Intergovernmental Panel on Climate Change. (2012). Glossary of terms. In C.B. Field, V. Barros, T.F. Stocker, D. Qin, D.J. Dokken, K.L. Ebi, M.D. Mastrandrea, K.J. Mach, G.-K. Plattner, S.K. Allen, M. Tignor & P.M. Midgley (Eds.), *Managing the risks of extreme events and disasters to advance climate change adaptation*. A special report of working groups I and II of the Intergovernmental Panel on Climate Change (IPCC) (pp. 555–564). Cambridge/New York: Cambridge University Press.
Janssen, M. (2010). Advances in multi-agency disaster management: Key elements in disaster research. *Information Systems Frontiers*, 12(1), 1–7.
Jin, C. (2007). *Beichuan statistics year book 2007*. Bureau of Statistics of Beichuan, Beichuan.
Johnson, G.F. (2005). Taking stock: The normative foundations of positivist and non-positivist policy analysis and ethical implications of the emergent risk society. *Journal of Comparative Policy Analysis*, 7, 137–153.
Jones, S., Aryal, K., & Collins, A. (2013). Local level governance of risk and resilience in Nepal. *Disasters*, 37(3), 442–467.
Kim, J. K., Sharman, R., Rao, H. R., & Upadhyaya, S. (2007). Efficiency of critical incident management systems: Instrument development and validation. *Decision Support Systems*, 44(1), 235–250.

Lall, S., Wang, H.G. (2012). *China urbanization review: Balancing urban transformation and spatial inclusion*. An eye on East Asia and Pacific, no. 6. Washington, DC: World Bank. http://documents.worldbank.org/curated/en/2012/01/16211663/china-urbanization-review-balancing-urban-transformation-spatial-inclusion#. Accessed 12 Apr 2014.

Li, X. (2012). Several important issues on China's urbanization. In: *Urban China initiative annual forum*, 9 September 2012, Beijing.

Manyena, S. B. (2006). The concept of resilience revisited. *Disasters*, 30(4), 433–450.

Melé, D. (2003). Organizational humanizing cultures: Do they generate social capital? *Journal of Business Ethics*, 45(1–2), 3–14.

Mili, L. (2003). Mitigating the vulnerability of critical infrastructure in developing countries. In A. Kreimer, M. Arnold, & A. Carlin (Eds.), *Building safer cities: The future of disaster risk* (pp. 273–288). Washington, DC:World Bank.

Naess, L. O., Bang, G., Eriksen, S., & Vevatne, J. (2005). Institutional adaptation to climate change: Flood responses at the municipal level in Norway. *Global Environmental Change*, 15(2), 125–138.

Nagasaka, T. (2006). New mode of risk governance enhanced by an e-community platform. In S. Ikeda, T. Fukuzono, & T. Sato (Eds.), *A better integrated management of disaster risks: Toward resilient society to emerging disaster risks in mega-cities* (pp. 89–107). Tokyo: Terra Scientific Publishing Company.

National Development and Reform Commission. (2014). *National new-type urbanization plan* (2014–2020). http://www.gov.cn/xinwen/2014-03/16/content_2639841.htm. Assessed 12 Apr 2014

Organisation for Economic Co-operation and Development. (2012). *Disaster risk assessment and risk financing: A G20/OECD methodological framework*. http://www.oecd.org/gov/risk/G20disasterriskmanagement.pdf. Accessed 14 Oct 2013.

Overseas Development Institute. (2012). *Disaster risk management in post-2015 development goals: Potential targets and indicators*. http://www.odi.org.uk/sites/odi.org.uk/files/odi-assets/publications-opinionfiles/8354.pdf. Accessed 10 Dec 2013.

Pan, J., & Wei, H. (2010). *Annual report of urban development of China*. China: Social Sciences Academic Press.

Paton, D., & Johnston, D. (2001). Disasters and communities: Vulnerability, resilience, and preparedness. *Disaster Prevention and Management* 10(4), 270–277.

Pelling, M., & Dill, K, (2010). Disaster politics: Tipping points for change in the adaptation of sociopolitical regimes. *Progress in Human Geography*, 34(1), 21–37.

Pelling, M., & High, C. (2005a). *Social learning and adaptation to climate change*. Benfield Hazard Research Centre Disaster Studies Working Paper, 11, pp. 1–19.

Pelling, M., & High, C. (2005b). Understanding adaptation: what can social capital offer assessments of adaptive capacity? *Global Environmental Change*, 15(4), 308–319.

Sabatier, P. A. (1998). The advocacy coalition framework: Revisions and relevance for Europe. *Journal of European Public Policy*, 5(1), 98–130.

Shi, P., et al. (2011). *Atlas of natural disaster risk of China*. China: Science Press.

Shieh, S., & Deng, G. (2011). An emerging civil society: The impact of the 2008 Sichuan Earthquake on grass-roots associations in China. *The China Journal*, 65, 181–194.

Sichuan Provincial Government. (2011). *The 12th five-year urbanization plan of Sichuan Province*. http://www.sc.gov.cn/10462/10883/11066/2012/1/6/10195285.shtml. Accessed 14 Oct 2013.

The State Council of P.R.C (2007). *Regulations to strengthen disaster risk management of local governments*. http://www.gov.cn/gongbao/content/2007/content_744155.htm. Accessed 14 Oct 2013.

The State Council of P.R.C. (2008a). *Support for reconstruction of post-disaster areas of Wenchuan Earthquake*. http://www.gov.cn/zwgk/2008-06/30/content_1031467.htm. Accessed 14 Oct 2013.

The State Council of P.R.C. (2008b). *Reconstruction plan of Wenchuan Earthquake*. http://www.gov.cn/zwgk/2008-09/23/content_1103686.htm. Accessed 14 Oct 2013.

The State Council of P.R.C. (2013a). *Support for reconstruction of post-disaster areas of Lushan Earthquake.* http://www.gov.cn/zwgk/2013-07/19/content_2451314.htm. Accessed 14 Oct 2013.

The State Council of P.R.C. (2013b). *Reconstruction plan of Lushan Earthquake.* http://www.gov.cn/zwgk/2013-07/15/content_2445989.htm. Accessed 14 Oct 2013.

Tierney, K. (2012). Disaster governance: Social, political, and economic dimensions. *The Annual Review of Environment and Resources*, 37, 341–363.

Tompkins, E. L., Lemos, M. C., & Boyd, E. (2008). A less disastrous disaster: Managing response to climate-driven hazards in the Cayman Islands and NE Brazil. *Global Environmental Change*, 18(4), 736–745.

United Nations International Strategy for Disaster Reduction (2004). Terminology: Basic terms of disaster risk reduction. http://www.unisdr.org/2004/wcdr-dialogue/terminology.htm. Accessed 14 Nov 2013.

Vatsa, S. K. (2004). Risk, vulnerability, and asset-based approach to disaster risk management. *International Journal of Sociology and Social Policy*, 24(10), 1–48.

Wamsler, C. (2006). Mainstreaming risk reduction in urban planning and housing: A challenge for international aid organisations. *Disasters*, 30(2), 151–177.

Waugh, W.L., & Streib, G. (2006). Collaboration and leadership for effective emergency management. *Public Administration Review*, 66, 131–140.

World Bank. (2013a). *World development report 2014: Risk and opportunity managing risk for development.* http://siteresources.worldbank.org/EXTNWDR2013/Resources/8258024-1352909193861/8936935-1356011448215/8986901-1380046989056/WDR-2014_Complete_Report.pdf. Accessed 21 Nov 2013.

World Bank. (2013b). *Building resilience: Integrating climate and disaster risk into development.* http://www.worldbank.org/content/dam/Worldbank/document/SDN/Full_Report_Building_Resilience_Integrating_Climate_Disaster_Risk_Development.pdf. Accessed 21 Nov 2013.

World Bank. (2013c). *Strong, safe, and resilient: A strategic policy guide for disaster risk management in East Asia and the Pacific.* http://www-wds.worldbank.org/external/default/WDSContentServer/WDSP/IB/2013/03/08/000333037_20130308112907/Rendered/PDF/758470PUB0EPI0001300PUBDATE02028013.pdf. Accessed 21 Nov 2013.

Ya'an Statistics Bureau. (2013). *The Ya'an statistics year book 2013.* Bureau of Statistics of Ya'an, Ya'an.

Yao, L. (2009). Township reform trend: Analysis from perspective of fiscal and administrative power. *Agricultural Economics*, 2, 8–13.

Yao, H., & Xu, Z. (2013). *Report on economic development in western region of China 2013.* China: China Renmin University Press Co., LTD.

Zhang, H. (2010). *Emergency management evaluation.* China: China Human Resource & Social Security Publishing Group Co., Ltd.

Zhang, Q. (2014). *Disaster governance and rising power: A historical review of China's disaster risk management from 1949.* School of Public Policy and Social Development, Beijing Normal University working paper 2014–01.

Chapter 12
Post-Disaster Reconstruction Models: The Governance of Urban Disasters in China, Iran and Myanmar

Helen James

Abstract A preponderance of the world's population is expected to live in urban environments by 2050 (ADB, Asia 2050: realizing the Asian century. ADB, Manila, 2011). Cities as sites of vulnerability for disasters originating from both natural and man-made causes are attracting considerable research on understanding the cross-cultural and governance dynamics associated with large-scale mortalities (Paton and Jang, Disaster resilience: exploring all-hazards and cross-cultural perspectives. In: Miller D, Rivera J (eds) Community disaster recovery ad resiliency: exploring global opportunities and challenges (pp 81–100). Taylor and Francis, Oxford, 2011; Paton and Violanti, Working in high risk environments: developing sustained resilience. Charles C. Thomas, Springfield, Illinois, 2012; Pelling, Vulnerability of cities: natural disasters and social resilience. Earthscan, London, 2003, Adaptation to climate change: from resilience to transformation. Routledge, London/New York, 2011). Disaster risk reduction policies of various Asian regional governments are currently exploring pragmatic approaches to recover and reconstruct lives, families and livelihoods of those affected. Examination of adaptation to trauma arising from large-scale losses in various cross-cultural contexts and different governance regimes presents the possibility of deriving new insights into practical disaster reconstruction models and policies. As such, this paper highlights the fundamental contributions of specific sociocultural and governance frameworks in disaster reconstruction policy. In so doing, the chapter investigates various urban disaster sites—namely, Bam and Tabriz (Iran), Pyapon, Bogale and Labutta (Myanmar) and Beichuan and Yingxue (People's Republic of China, PRC)—where large-scale mortalities arising from earthquakes and a cyclone warrant attention by those researching on disaster resilience, recovery and reconstruction. Based on the fieldwork in the aforementioned disaster sites, this chapter suggests that while the physical reconstruction of a livable habitat is important, the sociocultural factors in enabling

Field work was done in the PRC and Iran in October 2012, while in Myanmar it took place in June and November 2012 and November 2013.

H. James (✉)
Department of Anthropology, School of Culture History and Language, College of Asia Pacific, The Australian National University, Canberra, ACT, Australia
e-mail: Helen.james@anu.edu.au

M.A. Miller, M. Douglass (eds.), *Disaster Governance in Urbanising Asia*,
DOI 10.1007/978-981-287-649-2_12

disaster-impacted communities to reconstruct peoples' daily lives are of greater importance in the long-term recovery. Through the lens of civil society, difficulties in adapting to new realities around an engaged future are highlighted.

Keywords Disaster reconstruction • Governance • Civil society • Transformation • Cross-cultural contexts

12.1 Introduction

With a preponderance of the world's population expected to live in urban environments by 2050 (ADB 2011), cities as sites of vulnerability to disasters originating from both "natural" and man-made causes are attracting considerable research on understanding the cross-cultural and governance dynamics associated with large-scale mortalities (Pelling 2003, 2011; IFRCRCS 2009; Paton and Jang 2011; Paton and Violanti 2012). While a "natural" hazard does not become an environmentally triggered disaster until it connects with people, governance is clearly a primary factor in determining whether the hazard results in large-scale mortalities and extensive losses of physical, human and social assets. Globally, environment is theorised to be potentially one of the key drivers of human movement (UNDP 2010) as climate change results in more frequent and more intense environmentally centred events. Across Asia, large segments of population are moving from rural areas to cities in search of employment, education and a better life (Athukorala et al. 2000; Xin et al. 2010). According to ADB (2011), by 2030 it is anticipated that 1.1 billion more people in Asia will be living in cities; by 2015, Asia is anticipated to have 12 megacities with populations over 10 million. China alone is projected to have 115 cities with populations over 750,000 people. Many of these large cities are situated in coastal regions that are exposed to severe environmental events such as typhoons, storm surges, landslides, floods and cyclones. Other disaster events, such as earthquakes, tsunamis and volcanic eruptions, may not be precipitated by environmental triggers, but originate in geological forces which can result in catastrophic destruction in both coastal and rural areas. As rural to urban migration across Asia produces increasingly large cities, it also places increasing numbers of people in the path of environmentally or geologically originated hazard events.

The joint combination of the growth of these mega-urban conurbations and their geographic locations in coastal regions, which are prone to typhoons (hurricanes), storm surges, floods, mudslides and cyclones, highlights the potential complex disaster context. In such a context, the quality and efficiency of disaster governance would have a direct impact on the number of mortalities and affected people, as well as the efficacy of relief, recovery and reconstruction efforts. As a result of such disaster events, when transport, communications and logistical issues (e.g. delivery of relief, shelter and health services) are affected, even the most experienced emergency management and recovery operations can be thrown into disarray. For instance, in the case of the Great East Japan undersea earthquake and the 2011 tsunami (the Tohoku disaster), the triple-effect compound disaster of the initial events was overlaid by the nuclear reactors meltdown at Fukushima, which emphasised the weaknesses of Japanese disaster emergency governance. Although Japan is much experienced in preparing for and recovering from tsunamis and earthquakes, during the 2011 tsunami, the sea wall and well-prepared disaster evacuation shelters were inadequate to withstand the strength and height of the tsunami (Birmingham and McNeill 2012; Willacy 2013). Even worse, during the nuclear reactors meltdown, the spotlight on weaknesses was much highlighted through the survivors' expressions of dissatisfaction and lack of trust, not only in emergency management but also on the capacity of local- and national-level governance to take timely actions to address their recovery needs (Birmingham and McNeill 2012). Given the interconnected complexities of urban lifestyles and large numbers of populations involved and their varying demographic characteristics[1], the likely compound nature of future environmental and geological disasters when they strike large urban conurbations suggests a fundamental revision of both national-level governance mechanisms and emergency response and recovery governance.

The awareness of the implications for societal and governmental stability of large-scale human, social and physical losses that are potentially arising from these types of disaster events has led disaster risk reduction policies of various Asian regional governments to look for pragmatic approaches to assist those who are affected. Examination of adaptation to trauma arising from such large-scale losses in various cross-cultural contexts and different governance regimes presents the possibility of deriving new insights into pragmatic disaster reconstruction models. By investigating various urban disaster sites at Bam and Tabriz (Iran); Pyapon, Bogale and Labutta (Myanmar); and Beichuan and Yingxue (People's Republic of China) where large-scale mortalities arising from earthquakes and a cyclone warrant attention by those researching disaster resilience, recovery and reconstruction, this paper highlights the fundamental contributions of specific sociocultural and governance frameworks in disaster reconstruction policy.

The abovementioned urban sites were chosen not only because of the disaster events that occurred in these locations in the past decade (2003–2013) and the subsequent opportunities they provided to explore the long-term impact on vulnerable populations, but most importantly, because they present the possibility of investigating how disaster recovery and reconstruction frameworks have been utilised in completely different sociocultural and governance contexts. For instance, Bam (2003) and Tabriz (2012) disaster emergency management framework operated in the Islamic sociocultural context of an authoritarian regime, whereas Beichuan and Yingxue (2008), located in Sichuan province, Southwest (SW) China, operated in the "top-down" authoritarian context of the communist state machinery which was wary of civil unrest, and Pyapon, Bogale and Labutta (2008) in Myanmar operated

[1] For instance, in the Tohoku case, the events struck an area with primary ageing characteristics, thus reducing the survivors' willingness to relocate and increasing their need for governmental assistance to recover.

in the Buddhist sociocultural context, which at the time of the disaster was an authoritarian military state that was about to transit to democratic governance forms. In the case of Myanmar, the disaster events (a major cyclone) had a significant impact in urban coastal areas where the majority of mortalities occurred, although it also impacted large rural areas as the storm surge flooded across the low-lying agricultural areas of the Ayeyarwaddy delta. In Iran and China, the disaster events (earthquakes) impacted mostly the well-established inland areas, including a rural area outside Tabriz in northern Iran (East Azerbaijan) where there had never been an earthquake; and although seismic activity had been evident in the SW China Wenchuan earthquake for some time before 12 May 2008, in the case of Bam, Iran, there had not been an earthquake for some 2,000 years. Hence, like the people of the Burmese delta, the inhabitants of Bam had no experience of such a catastrophic event or disaster emergency response and management. The differences in the locations of these sites (inland/coastal) and their sociocultural and governance frameworks provide an opportunity to investigate the implications of the longer-term consequences of particular approaches to disaster recovery and reconstruction and to theorise a possible universal application for post-disaster reconstruction in the cross-cultural context.

Based on the previously mentioned fieldwork, this paper suggests that while the physical reconstruction of a livable habitat is important, of even greater importance in long-term recovery are the sociocultural and governance factors in enabling disaster-impacted communities to reconstruct their daily lives and recover from the trauma of the event. In carrying out this fieldwork, I have drawn on the standard methodologies of qualitative research, employed direct observation on site and conducted both structured and semi-structured interviews. These interviews include 40 non-governmental organisations (NGOs) and international NGOs (INGOs); academics from four universities; 16 high-level government officials from the three countries including the deputy governor of Kerman province, Iran, wherein Bam is located, and the mayor of Beichuan, Sichuan province, SW China; 20 private citizens; and the chief executive of the privately funded Bam rehabilitation centre. Furthermore, three focus group discussions[2] comprising of 40–50 persons from the three villages in the Burmese delta and smaller groups (of women) in three villages outside Tabriz, northern Iran (Jigha, Ghizghapan and Shahsawar), were also conducted. In addition, surveys and secondary sources were also used. Through the lens of civil society, I have sought to capture survivors' difficulties in adapting to new realities and transforming their lives around an engaged future.

12.2 Perspectives on the Urban Governance of Disasters

Much attention in disaster governance has been focused on relief, mitigation and preparedness measures for rural communities, especially in the developing world. However, population dynamics and the realities of risk governance mandate that

[2] Each FGD had about 40–50 people.

considerably more attention needs to be directed to the urban governance of disasters particularly in heavily populated megacities with the potential for catastrophic losses. In China, for instance, among my informants at the Southwestern University of Finance and Economics, Chengdu, there was deep concern lest a similar earthquake to that of the 12 May 2008 event (the Wenchuan earthquake) strikes the densely populated provincial capital of Chengdu (population around 15 million). Memories of the catastrophic Tangshan earthquake of 1976 when over 250,000 people died are still very raw among Chinese people. Likewise, whenever taller buildings dot the skyline of China's cities, there is considerable unease about the potential impact of another large-scale disaster event, both in political economy and sociocultural perspective. The megacity of Shanghai, for instance, has been divided into squares by the emergency management local authorities, and in each section a local official responsible for mobilising people and relief measures in his/her section is in charge in the event of a disaster. The potential far-reaching compound implications of environmentally triggered and geologically originated disaster events in megacity urban context in Asia are now rising to prominence among those responsible for disaster risk reduction and mitigation measures.

Since the advent of the UN-sponsored Hyogo Framework for Action (HFA) in 2005 following the 1995 Kobe earthquake in which over 6,000 people lost their lives, nations have been asked to prioritise enhancing resilience[3] of disaster-impacted people in their disaster risk reduction plans. In practical terms, this often means that investments of state and personal resources should be more targeted at preparedness, relief and mitigation policies in an attempt to reduce the financial and human costs of disasters. Rarely is there a concerted long-term policy dimension to these plans seen from the perspective of enabling disaster-impacted communities to effect full recovery of their livelihoods and lifeways. Even more rarely is there a concerted focus on the social development of these communities as part of enhancing their resilience and adaptive capacities in the event of a disaster. Moreover, state-led reconstruction policies frequently focus on restoration of the necessary physical environment—houses, roads, hospitals, bridges, schools—with little attention to restoration of the social assets, networks and cultural landscape which collectively elaborate the texture of human interactions and which provide meaning to individual lives. Transformation as part of policy agenda to achieve innovative adaptive capacities which may yield a more robust community and an individual in future is still at initial stages of the policy lexicon (Pelling 2011).

In the post-Hyogo 2015 framework, the transformation perspective is being encapsulated as "build back better" rather than seeking to return to a pre-disaster situation, which in reality is not possible. However, even this prescription places emphasis on the reconstruction of physical assets and overlooks the sociocultural

[3] UNISDR (2005) defines resilience as: "The capacity of a system, community or society potentially exposed to hazards to adapt, by resisting or changing in order to reach and maintain an acceptable level of functioning and structure. This is determined by the degree to which the social system is capable of organising itself to increase this capacity for learning from past disasters for better future protection and to improve risk reduction measures" (p. 4).

assets that not only impact populations' value but provide the essential basis for their resilience and psychosocial recovery. The growing emphasis is on using disaster as a catalyst for change and, according to the Recommendations for Recovery and Reconstruction in the Post-2015 Global Framework for DRR suggested by the International Recovery Forum (IRF Kobe, Japan, 22 January 2013), decentralisation of recovery design to recognise and support long-term local solutions. These are to be achieved through strengthening national/local government links and institutionalising mechanisms for capturing collective memory through community involvement, in effect giving priority to the "bottom-up" approaches of participatory governance. The goal to "build back better" through inter-sectoral coordination and shared responsibility by all stakeholders (community, national and local governments, public and private institutions) is intended to facilitate sustainable linkages between recovery and development. It is this integrated perspective in the new approaches to post-disaster reconstruction, particularly in the case of compound disasters evincing multiple dimensions as in the case of disasters which impact large urbanised populations, which shows most promise of delivering an effective long-term recovery.

Implicit in this new integrated approach is recognition of the significance of social and cultural assets in enhancing resilience and providing a base for effective long-term disaster recovery. This perspective is attracting increasing attention in disaster research (De La Paz 2013; Hayashi 2010; Miichi 2014; Paton and Jang 2011; Zayas et al. 2010; Zayas 2014). Since the research of Shaw and Goda (2004) on the Kobe earthquake of 1995 and Nakagawa and Shaw (2004) on the Kobe and Gujarat earthquakes of 2001, there has been increasing awareness of the importance that survivors attach to these social assets and networks of interaction. In their research, Shaw and Goda quantified the greater importance that survivors attach to restoration of their social environment than to government-initiated rebuilding programmes, although these too are appreciated. Lo (2010) has also given similar evidence from the 921 Chi-Chi earthquake (1999) in Taiwan. Adding to this, the extensive research by Paton and Jang (2011) on the significance of cultural and social assets among the Hakka in Taiwan as a key element in their recovery from Cyclone Morakot (2009) is a landmark in disaster resilience research.

In similar vein, Ronan and Johnston (2005) also examined this perspective in their pioneering volume on New Zealand's Mt. Ruapehu earthquake. With a focus on communities, schools, youth and families, Ronan and Johnston proposed that these entities should become the core of disaster preparedness, resilience policy and practice. They recognise the centrality of social assets and networks in the major contribution communities, schools, youth and families make to disaster preparedness activities. As Cook (2005) has argued, the social relations inherent in kinship and family networks are built around trust, which she calls the "glue" that holds communities and societies together, thus providing the possibility of social cohesion despite traumatic losses. By exploring public and private sector approaches to restoring social assets and networks in the longer-term reconstruction of the lifeways of disaster-impacted populations in Southwest China (Sichuan), the Myanmar Delta (Myanmar) and Bam and Tabriz (Iran), this paper suggests that the formulation—communities, schools, youth and families—may also become the core sites for more effective long-term post-disaster reconstruction policy in both the urban and rural contexts. As indicated above, the selection of these sites has been guided by the cross-cultural and varied governance contexts they offer to disaster research, as well as the possibilities of implementing the findings across different disaster, governance and cultural contexts.

12.2.1 Brief Overview of the Three Disaster Sites: Southwest China, Iran and Myanmar

In Sichuan province, Southwest China, on 12 May 2008, a major earthquake of 8.0 struck ten heavily populated provinces, causing somewhere between 80,000 (official figures) and 400,000 (Taiwan researcher's figures) mortalities and leaving over four million people homeless. It destroyed every school in the disaster zone causing enormous grief to large numbers of parents and relatives. Over 5,000 children were among those killed (Yang 2013). There were immediate cries of corruption levelled at the building contractors who were perceived to have used substandard building materials in the construction of the schools, thus contributing to their collapse. Afraid of civil unrest, the Chinese government took a "top-down" approach consistent with its authoritarian governance structures. It sent military troops to the search and rescue operation guided by former Premier Wen Jiabao and commenced the construction of new highways and tunnels through the mountains to facilitate access to the disaster-impacted area and support the recovery operations. It also directed ten of the most prosperous provinces to take financial responsibility for rebuilding each of the affected disaster-impacted counties.[4] It was then determined that the epicentre, Yingxue, would be rebuilt in situ, while Beichuan, a destroyed town where over 30,000 people lost their lives, would be left as a memorial to the dead and a completely new city rebuilt some kilometres away in a site considered to be safer. The cost of this rebuilt city was in the order of $2 billion which was borne by Shandong Province. Thus a "top-down" approach was taken for reconstruction of the physical and social assets with little involvement of the affected community or social organisations.

The Chinese approach to governance and reconstruction after the Wenchuan earthquake may be contrasted with that in the Burmese delta when the category 3 Cyclone Nargis struck on 2–3 May 2008. It left a trail of destruction across more than 50 townships, 37 of the most severely impacted townships being in the heavily populated Yangon and Ayeyarwaddy divisions. It thus affected both urban and rural communities. The cyclone resulted in at least 84,537 people dead, 53,836 missing and 19,359 injured (Tripartite Core Group 2008).[5] It was the worst natural disaster

[4] Personal communication with mayor of Beichuan, October 2012

[5] The Tripartite Core Group (2008, pp. 1–4) provides these figures as of July 2008 based on the Village Tract Assessment conducted one month after the Cyclone. Out of an estimated population in the affected areas of 7.35 million, around 2.4 million people were severely affected by the Cyclone. An informant from a local NGO told me in November 2012 that the Bogale River was "choked with bodies"; many washed up weeks later in Sri Lanka, India and on the eastern beaches of Myanmar. The exact death toll may never be known

in Burmese history. When it made landfall some 250 km southwest of Yangon, it carried wind speeds of up to 200 km/h and very heavy rain. In the Ayeyarwaddy delta where the damage was most severe, a storm surge of 3.6 m (12 ft) accompanied the strong winds. As the population had not experienced such a severe event previously, it had no disaster risk reduction template by which to guide their reactions to the imminent danger.[6] Immediately after the Cyclone struck, around 800,000 people were estimated to be displaced, and around 260,000 were thought to be in camps or refuges around the delta including in monasteries.

The unsuspecting population was not accustomed to cyclones taking this path; usually they tracked across the Bay of Bengal up to the Bangladesh coast. But deforestation in the headwaters of the Ayeyarwaddy River which had been bringing tons of sediment downstream for years had been extending the delta seawards by 3 km per year for some years. The resulting shallowing of the waters off the southern tip of the delta may have contributed to the unusual turning path of the Cyclone. From the fieldwork in this area in 2012 and 2013, I found that the integrated approach to relief, recovery and reconstruction which was set in train by the Tripartite Corps Group (TCG) in 2008[7] meant that private entities, local NGOs and some INGOs were working in partnership with government and community groups to mainstream disaster risk reduction (DRR) and development. As a result, the delta has been divided between various NGOs/INGOs in an integrated private/government approach to mainstreaming DRR and development, which has seen the tragedy becoming an opportunity for the emergence of new social capital (CPCS 2009). In particular, I focus on some of the most severely impacted coastal towns—Bogale, Labutta, and Pyapon—where mortalities were greatest. Indeed, according to the TCG Village Tract Survey of July 2008, Bogale suffered 35,000 confirmed deaths while 2,500 were either missing or presumed to be dead. According to the same survey, Labutta suffered 33,000 confirmed deaths, while 48,000 were either missing or presumed to be dead. Likewise, the fishing town of Pyapon had 5,000 confirmed deaths and 2,500 of missing or presumed to be dead cases (see Table 12.1). In terms of mortalities, these three towns suffered the most losses. I also compare three villages, Aung Kone, near Amar, a Karen village which had no mortalities; Byi Chaung, a Burmese village at the foot of the delta which lost one third of its population of 300; and Tamarpin in the centre delta, in a rural area inland from Pyapon, which also had no mortalities.

In fieldwork in Iran in October 2012, I visited Bam and Kerman in eastern Iran and Tabriz in the north, East Azerbaijan. Bam, which was struck by a massive earthquake measuring 6.6 on the Richter scale in 2003 that resulted in 30,000 mortalities,

[6] Myanmar is hazard prone. Some 11 cyclones have impacted Myanmar in the past 60 years, but only two made landfall in the Delta, the country's major rice growing area. Floods, urban fires, major windstorms (14 since 1910) and six earthquakes have struck the country. Further, about 11 tsunamis over the past 250 years have impacted the Bay of Bengal and Andaman Sea regions. The 2004 Indian Ocean tsunami killed 60 people in Myanmar and left some 2,500 homeless in Myanmar's coastal areas

[7] TCG consisted of ASEAN, the Myanmar Government and the UN agencies.

Table 12.1 Number of casualties: Dead and missing persons by division and township (TCG 2008)

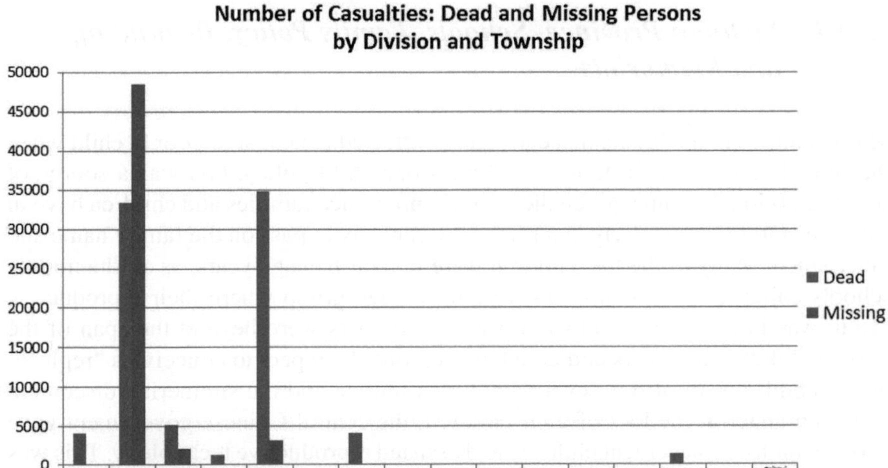

	Napudaw	Labutta	Mawgyun	Phyarpone	Bogale	Kyeik Latt	Dadeye	Thanlym	Kyauktan	Kayan	Thone Kwa	Twantay	Kawhmu	Kungyangon	Dala	Seikkyi Kenaungto
	Aryeyardy Division							Yangon Divioson								
Dead	4178	33344	5250	1258	34744	12	4111	3	13	1	6	25	130	1446	14	2
Missing	10	48464	2127	10	3198	0	19	7	0	0	0	0	0	0	1	0

has still not fully recovered, and the intense grief still felt by survivors is palpable. Many collapsed buildings are still not rebuilt; and health issues are an ongoing concern.[8] In August 2012, outside Tabriz (just 37 km from the town), another major earthquake struck 185 villages in a rural area, which resulted in 700 mortalities. Here, local NGOs are coordinating the relief and reconstruction efforts and providing psychosocial services to survivors. In both cases, a dichotomous private/government approach is being taken to the recovery and reconstruction effort, in that each sector has pursued its own agenda without planned interaction with the other. However, it differs considerably from that in Myanmar. This is because in Myanmar, it is not an integrated approach, but pursued in parallel lines with the government-derived post-disaster reconstruction efforts, which may have inadvertently contributed to social problems in the recovery phase. By examining the approach to reconstruction of social assets in these three different cultural and governance contexts from the perspective of family policy, rebuilding in situ or relocation and schools and other memorials, it is possible to derive a composite picture of the efficacy of disaster policy and its impact on the long-term recovery of the affected people.

[8] All health facilities in Bam were destroyed and half of the medical staff lost their lives in the initial earthquake.

12.3 Reconstruction of Social Assets

12.3.1 Sichuan Province, Schools, Family Policy, Rebuilding and Memorials

Many families in the Wenchuan earthquake-affected zone lost their only child when the schools collapsed. In terms of China's one-child policy, this was a source of severe emotional trauma, given the central importance families and children have in Chinese culture, particularly in terms of the capacity to pass on the family name and inheritance. Many of the lost children were in their teenage years, as all the middle schools collapsed. Their mothers were in the age group where their reproductive health was in question. In other words, the mothers were beyond the span of the normal childbearing years and could not reasonably expect to conceive a "replacement" child. Aware of this severe emotional trauma and the simmering discontent and even anger at the loss of their children, the central Chinese government gave permission for replacement children and assisted reproductive technology. This was seen as practical public policy to assuage the grief and try to abate the survivors' anger. The government thus took into account the sociocultural issues in the reconstruction phase, not just the reconstruction of physical infrastructure, although this too was extensive and, as indicated above, included a multilane highway and tunnels through the mountains to facilitate access and relief in case of a future event.

Government assistance with reproductive health technology therefore has become a key policy in the approach to reconstruction of sociocultural assets. The mayor of Beichuan, the devastated town which lost over 30,000 people, is also the local family planning centre's chief. He is centrally involved in this policy of social engineering which is seen as a critical element in psychosocial recovery. However, only a little over 76 % of eligible couples have availed themselves of this policy, while about 16 % decided that they would not, and others are undecided (Zhou 2013). Reasons for couples deciding not to participate in the government programme range from being "too old", marriage not being very robust, to financial concerns involved in raising a new child.

I visited both Yingxue and Beichuan in October 2012. Both testify to a different approach to the physical reconstruction where Yingxue has been rebuilt in situ, while Beichuan has been left as a memorial to the dead while the new city has been rebuilt some distance away. In Yingxue, the collapsed middle school with its fallen clock stopped at the time the earthquake struck (2.20 pm) has been turned into a public memorial park, with yellow flowers of remembrance lining the steps from the street level to the clock. A steady stream of mourners comes to pay their respects. Across the school, a new earthquake museum has been built as part of the government's approach to recognising the need to mourn but also for survivors to have a tangible focus for their mourning. The beautifully presented new city of Yingxue itself is part of the memorial with attractive rotunda along the river's edge inviting visitors to sit and contemplate.

The horror of the earthquake is fully felt in Beichuan as the entire town has become a memorial to the dead. On the collapsed mountain side where the school once stood, another haunting memorial has been established, again with yellow flowers, tinkling Taoist-type music playing continuously and a constant stream of mourners. Across the city centre, where the central business district used to be, another formal public memorial has been erected, close to the new river. When the earthquake struck, the earth moved both horizontally and vertically; the CBD literally fell through the earth, taking the lives of the 10,000 people who worked there, while the river rose and now covers the former CBD site by 10 m.[9] In addition to the public memorials, there are also numerous private memorials. One which stays with me has been erected by a grieving husband "to my darling wife and daughter"; amidst the debris of their former home, a red "Hello Kitty" doll still lies lost without her former owner. Around the town, the everyday reminders of former lives—a perfume container, a jewellery box, a laptop, a chair—still lie as they were cast on that fateful day, part of the ongoing memorials to lost loved ones.

12.3.2 *Iran, Bam and Tabriz: Schools, Family Policy, Rebuilding and Memorials*

Like China, family policy and family reconstruction is deemed as important in the Islamic culture of Iran. When the 2003 earthquake struck the historic city of Bam, it not only took the lives of an estimated 30,000 people but also destroyed the 2,000-year-old citadel outside the city—a cultural heritage site which formerly attracted many visitors. Since no earthquake had occurred in Bam for 2,000 years, the aforementioned event was quite unexpected. The government provided incentives to assist family re-formation which took the shape of funds in cases of remarriage and to assist rebuilding of houses. In view of these financial incentives, many non-Bam people went to Bam to marry survivors, as they wanted the funds and land as well as assets to be gained by remarriage. Some 80–90 % of widows remarried in a culture where there are strong cultural and legal imperatives for a female to have a male protector, a husband, a brother, a father or a son. Where the remarriage occurred with another Bam survivor, the marriage seemed to have lasted; however, where the marriage was with a person from another part of Iran, in many cases the new family formation failed.[10] However, many have obviously thrived; research on well-constructed data here by Hosseini-Chavoshi and Abbasi-Shavazi (2013) has revealed that the family reconstruction policy resulted in a spike in fertility following the disaster as survivors sought to reconstruct their lifestyles and social assets.

At Bam, rebuilding has been in situ the prevalent mode of recovery after disaster. An aspect of the reconstruction policy is evident in the partially built houses which

---

[9] Personal communication with mayor of Beichuan and colleagues from the Southwestern University of Finance and Economics, Chengdu, Sichuan Province, October 2012

[10] Personal communication with Deputy Governor of Kerman province, Bam, October 2012

disfigure the city of Bam as it still struggles to arise from the destruction. Bam is not thriving as it was once. With the funding from the government, many people have taken the precept, "build back better" to be "build back bigger", that is, to build a bigger house. However, the funding has often been insufficient for the type of dwelling to be constructed, and many remain unfinished.

On the other hand, the involvement of the private sector in reconstruction policy can be gauged by the state-of-the-art rehabilitation centre outside Bam, funded by private German and Iranian business sponsors, which takes care of the many disabled, orphans and elderly who have lost all their relatives and single female-headed households. At this centre income generation programmes provide training for widows and female-headed households; elderly survivors and orphans have live-in accommodation; the disabled are provided with physiotherapy and employment programmes. In the case of Bam, the government and private sectors seem to be dichotomous and operating in parallel tracks instead of implementing an integrated approach to reconstruction policy. The reconstruction of the destroyed Bam citadel being undertaken with government funding seems an attempt to provide a focal point for a public memorial to assist survivors recover from their traumatic memories. It may also be part of a strategy to attract tourists to return to the area and support the local economy.

Health and psychosocial issues in Bam continue to be major public policy problems. In a study of 916 survivors by Montazeri et al. (2005), 41 % reported that they had lost three to five members of their family in the earthquake. During fieldwork in Bam in October 2012, the deputy governor of Kerman province said that he lost five of his brothers. His sorrow was etched into his face. Bam is not thriving and for many survivors the trauma seems to be still present. The Montazeri study found that 58 % of those surveyed suffered from severe mental health distress, "three times higher than the reported psychological distress among the general population" (Montazeri et al. 2005, p. 1). It is unknown how many suicide cases this has led to because there is no official data, given that suicide is a deadly sin in Islam. However, there has been an "inexplicable" rise in the incidents of fatal motor vehicle accidents in the years since the earthquake. Montazeri et al. (2005) also found that "female gender, lower education, unemployment, and loss of family members were …associated with more severe psychological morbidity among survivors" (p. 5).

Despite the government's attention to immediate concerns for shelter, water purity and sanitation, post-health issues have received comparatively less attention. The study by Firoozabadi et al. (2011) of 256 survivors referred to Kerman hospitals after the Bam earthquake showed that urogenital system trauma occurred in 74.5 % of cases examined—a reflection of survivors being caught under piles of brick, cement and other debris (p. 51). Commenting on the usefulness of mortality data as indicators of post-disaster health care and policy, Zolala (2010) observed that with respect to the data from the Bam earthquake:

> [The] Vital Registry is the only organization that recorded the deaths caused by the earthquake. Information from interviews and viewing some of the data produced by the Vital Registry found that these data are available in cross tables based on gender, age groups, and causes of death. The data are not accessible from the original forms in order to maintain

> confidentiality, unless there is judicial permission. The data are entered electronically using software designed by the Vital Registry, which is different from the software being used in the Health Ministry. (Zolala 2010: 541)

Thus her research throws doubt on the validity of much of the data on mortalities collected after the Bam earthquake. She adds:

> At the local level, the Bam Health Network had collapsed after the earthquake, and there was no place to hand in the data collected from the urban and rural health centres. These data included the routinely collected mortality data from the rural and urban health centres. Also, there was no way to enter data for a few months, those involved with the data collection and data processing said there was no place to do the work and no working computers. The interviews also revealed that after the earthquake the authorities were distracted from routine data collection, and focused on specific data collection, such as infectious disease. The large scale of the tragedy created by the earthquake made the routine data collection less of a priority. (Zolala 2010, pp. 541–547)

Possibly the most thoroughgoing study of the Bam earthquake survivors is that undertaken by Ardalan et al. (2011), which focuses on elderly survivors 2 months and 5 years after the event through the use of a cluster survey. In policy terms they employ the lens of ageing societies to examine the impact on the vulnerability of the elderly when the infrastructure which supports their ability to function independently in their communities is destroyed (a feature also of post-disaster reconstruction after the 2011 Japan Tohoku earthquake and tsunami). Their research focused on 210 participants in the urban parts of Bam and Baravat and their rural surroundings. The participants (51 % women and 49 % men) were aged 60–90 years at the time of the earthquake or 65–95 years at the time of the survey. The majority were illiterate. Thirty had suffered severe injury, and 32 had lost one of their household members from the earthquake. In terms of health issues, 52 have had a chronic disease at the time of the earthquake, while 176 had a chronic disease 5 years later, at the time of the survey. The conclusions of Ardalan et al. (2011) are that the earthquake adversely affected the functional capacity of the elderly survivors. They were disadvantaged in obtaining relief assistance, both material and services, a finding similar to that of Thailand's post-2004 tsunami and of some parts of the Burmese delta (e.g, Dedaye, Bogale) after Cyclone Nargis. "Relief and medical services providers did not take into account the health status and functional capacity of the elderly [in the Bam area]" when delivering services (Ardalan et al. 2011, p. 107). Living with others was one of the risk factors for the elderly survivors, while those living alone were found to be more functionally independent. In terms of policy parameters, they recommend that echoing the findings of HelpAge International, disaster response organisations need to identify and prioritise the needs of the elderly in disaster risk mitigation and post-disaster reconstruction strategies (Ardalan et al. 2011).[11]

The case of the earthquake sites in East Azerbaijan provides a contrast to Bam in some ways. I visited three villages—Jigha, Shahsawar and Ghizghapan—that are

[11] Ardalan et al. (2011) note that at the time of the study, two elder-care centres were operative in Bam, serving around 100 people.

part of a reconstruction project centred on 30 affected villages. Reconstruction is conducted by an Iranian NGO named Iranian Psychological Services—a privately funded NGO run by CEO Dr Amir Askari. In Jigha, responsibility for reconstruction lies with the government, and there are obvious difficulties as many vulnerable people, especially the elderly and disabled, strive to rebuild their mud-brick homes themselves. However, although there were two fatalities among the school children (who died when their homes collapsed), the school itself did not collapse and was again functioning when we visited in October 2012. Shahsawar, supported by civil society and charity organisations and merchants from Tabriz, is doing much better. Women's groups here came out to talk with us and seemed very happy to see us. The school here also did not collapse as it was constructed of earthquake-resistant interlocking bricks. The third village, Ghizghapan, was being entirely rebuilt by the private sector, including installation of solar power in what was obviously an adaptive, developmental perspective. In these latter two villages, recovery efforts were more advanced than in the first village—a fact that the local people were delighted to inform us.

Comparing the post-disaster status of the urban and rural populations at the sites studied, in the cases of the Sichuan and Wenchuan earthquake and the Iranian Bam and Tabriz earthquakes, it seemed that the rural populations were more self-reliant and able to "bounce back" better than the urban populations, although considerably more research on this aspect needs to be undertaken. In both the urban populations studied, their responses were severely conditioned by the unexpected nature of the disaster event, the lack of preparation time, the sense of impotence and frustration at not being able to assist loved ones. In the case of the rural population in the three villages outside Tabriz, although they too had not experienced an earthquake in the area previously and had lost fewer people to the disaster, their comparatively greater poverty (they lived in mud-brick houses) did not hinder them in mobilising their collective social capital within their villages to commence the rebuilding themselves. They did not wait for others to arrive and assist. However, being close to the wealthy city of Tabriz was also advantageous for them because the merchants were heavily engaged through Islamic charities with what could be called "bridging" social capital (Lin et al. 2001). They mobilised to assist villagers with psychosocial counselling, rebuilding and provision of shelter, clothes and medicines in the immediate aftermath of the event. The view of those with whom we spoke was clearly that the NGOs/charity organisations were more effective in assisting the survivors to recover than the government.

12.3.3 Burmese Delta: Schools, Family Policy, Rebuilding and Memorials

In the Burmese delta, the scale of the destruction meant that no one sector could be expected to have the resources by itself to bring about the reconstruction required, even of the physical assets. Here, some 60 % of public and monastic schools and most of the health facilities were destroyed (Tripartite Core Group 2008). The tragedy presented an opportunity for integrated development programmes drawing on the resources of private, international and the government sectors. A series of periodic surveys by the TCG since 2008 has assembled much useful information on the state of the recovery processes which document that food security is still an issue. However, it also indicates that most housing has been rebuilt; many of the 450,000 housing units destroyed by the cyclone have been rebuilt in flimsy materials and to smaller dimensions than those which previously housed their owners. Nonetheless, livelihoods have been only partially restored, and there are ongoing psychosocial issues. Many fear another similar event. Being divided into project zones for which various NGOs take developmental responsibility has meant that the Delta towns and villages are gaining access to increased resources for capacity-building and development projects, as bonding (horizontal) and bridging (vertical) social capital are applied to the reconstruction programme.

Rebuilding in situ has been the most prominent approach to the recovery effort in the towns and villages of the Ayeyarwaddy delta. Rebuilding community and economic life, over the past few years, has progressed vigorously among the towns and villages located along the Pyapon-Bogale road which runs the length of the eastern side of the delta from Yangon down to Labutta at the foot of the delta. This rebuilding has been a major recovery effort which evokes the "build back better" formulae in the development approach to disaster recovery. While the majority of the people are farmers, fishermen[12] and labourers, with a few businessmen, traders or workers in the service industries, there is a high proportion of landless labourers. According to the pre-Nargis UNDP (2007) Integrated Household Living Conditions Survey (IHLC), in Ayeyarwaddy division—though not the poorest part of the country—32 % of the landless in the delta worked in agriculture while the remainder worked in the fisheries sector or in salt production, trade and transportation. Some 44 % of the landless lived below the government poverty line (UNDP 2007; Tripartite Core Group 2008).

While the 2007 UNDP IHLC Survey provides a snapshot of the area pre-Nargis, the Post-Nargis Joint Assessment (PONJA) conducted 1 month after the cyclone provides baseline data of the immediate impact of the storm.[13] It utilises two sets of data: primary data on the human impact from the Village Tract Assessment (VTA) conducted in early June 2008 and secondary data from Myanmar government ministries, UN agencies and interviews during field visits, which formed the basis for the PONJA Damage and Loss Assessment (DaLA) of infrastructure damage. The

[12] Fishing is a major occupation, especially in the coastal towns of Pyapon, Bogale and Labutta

[13] Following the 25 May 2008 ASEAN-UN International Pledging Conference, the agreed Tripartite Core Group (TCG) coordinated relief efforts by the Government of Myanmar, the UN and ASEAN. On 31 May, the TCG agreed to conduct the Post-Nargis Joint Assessment (PONJA) to assess the extent of the destruction and the requirements for humanitarian assistance as well as medium- and long-term recovery. About 250 staff from the GOM, ASEAN, ADB and WB, UN agencies and NGOs conducted the VTA and the DaLA. Staff from 18 government ministries were involved in field visits and data interpretation; some 70 people condensed this data into the Post-Nargis Joint Assessment Report (July 2008).

VTA survey of households, key informants (teachers, village leaders) and focus groups involved 250 enumerators in 291 villages across 30 of the most impacted townships undertaken in 10 days in early June 2008. In the 15 days preceding the VTA survey, over 65 % of households reported health problems ranging from colds (39 %), fever (37 %) and diarrhoea (34 %) to mental health problems (23 % of households) and injuries (8 %). As a result of the damage, health staff reported a decline in health service provision in immunisation (from 83 to 66 %) and communicable diseases (from 43 to 34 %) and in maternal and child health services especially for pregnant women (from 81 to 71 %). There was a 21 % increase in health staff saying that their health facility was not adequately supplied with essential medicines.

Based on the VTA survey, government data and UN agencies, some 50–60% of public schools, including monastic schools as well as libraries and educational resources, were destroyed or damaged by the cyclone. Many schools were used as evacuation shelters. Since Nargis, the government, private sector, NGOs and international donors have contributed to the rebuilding, including trauma counselling for children and the provision of early childhood centres as safe places for children. Reopening of schools was a high priority as part of the strategy to provide an early return to normal living structures for children and parents. Pre-Nargis figures on enrolment (UNICEF, Multiple Indicator Cluster Survey 2000) suggest that around 20% of children from the poorest quintile did not enrol in primary school compared to less than 5% of their wealthier counterparts. By the end of primary school (at about age 11), about 60% of students from the richest quintile were usually transferred to middle school while only 10% of students in the poorest quintile were eligible for the same. As the severe impact of Nargis on livelihoods and incomes became apparent, the risk of children not going to school or not continuing their education increased as many were needed to assist in replanting the sea-soaked fields or in helping to re-establish the family livelihoods. In the village of Byi Chaung, in the far south of the Delta, where no middle school is readily accessible, it is rare for any child to progress to middle school,[14] and the majority stops their education at standard four. This situation is also confirmed by the Post-Nargis Periodic Survey IV undertaken in 2010.

Many schooling children also mention that they still find it difficult to concentrate and fear another storm. Many have not returned to school. Although no deaths occurred, in the central delta village of Tamarpin (population about 500), I found (in 2013) children walking along a rough path for about 45 min to the next village to attend primary school. This is because Tamarpin is an "unregistered" village, hence, no school too. Only a few students from Tamarpin attend middle school and none attend high school. No one has ever attended university. Its poverty is a stark contrast to the situation in the Karen village of Aung Kone, which has a well-built school, a vibrant, supportive parents group and from which a number of members (15) have graduated from university.

[14] Focus group discussion, November 2012

Food security was a real concern in the immediate aftermath of the storm. Those who survived took out their off-season stored provisions (where these had not been damaged by the storm) and shared with others. In the five worst affected townships of Labutta, Bogale, Pyapon, Dedaye and Kyaiklat, more than half the households reported that they had lost all food stocks. On the day of the VTA survey in early June 2008, 55 % of households had 1 day of food left; however, they said they could source for food from local markets. PONJA reported that by 30 June 2008, 676,000 people had been provided relief food items; but 924,000 vulnerable people in Ayeyarwaddy and Yangon divisions were in urgent need of basic food supplies until the next harvest in November that year. In this context, where it was considered essential to give priority to re-establishing food security, within a short time after the storm, the government and NGOs encouraged survivors to leave their refuges in monasteries and return to work. The population changed its diet: consumption of fish and eggs dropped by 25 %, while consumption of vegetables and fruits decreased by 9 %. Infant feeding practices were disrupted, posing a major threat to the nutrition of children under 5 years. Survivors were at risk of malnutrition and micronutrient deficiencies.

One of the important sources of protein for the agricultural communities in the delta (which usually provides around 45 % of GDP) is the farm animals. In the wake of the storm, concerted attention was given—with assistance of NGOs, INGOs and government agencies—to restock ducks, geese, buffaloes, cattle, goats, pigs, chickens and other farm animals so that eventually it could contribute to both the offsetting of the food insecurity and set in motion the critical infrastructure for long-term recovery of farm livelihoods. The fisheries sector, both marine and inland, where damaged ponds, hatcheries and jetties, boats, nets and equipment needed to be replaced was also affected by the loss of ice production plants and cold storage facilities, fish processing, transport and marketing infrastructure. In the salt production sector, concentrated in the lower delta, every salt worker and all their families lost their lives, some 20,000 people in total (TGC, Post Nargis Joint Assessment 2008, pp. 91, 93). The interacting effects of both these human and industry losses have been extensive, far beyond the capacity of any one sector to cope with on its own. Hence in developing comprehensive policies for the long-term recovery phase of agriculture, fisheries and commercial/industry ventures, an integrated approach drawing on the community, NGOs, INGOs and government sectors as well as resources has been essential. This has resulted in public-private partnerships by which new forms of bridging and linking social capital have emerged.

Recovery of livelihoods has necessarily required income generation schemes for stimulation of off-farm employment to contribute to that of the agricultural sector. Although in the wake of the storm, casual labour was withdrawn and employment opportunities were destroyed, it returned to normal during the recovery phase. The rice mills, dried fish/shrimp and fish paste production, factories in the industrial parks, small- and medium-sized enterprises and microenterprises were all urged to resume production and provide employment opportunities to assist in the restoration of livelihoods. However, few had access to small grants or credits to get their businesses started again; PONJA (2009, p. 24) reported that only 27 % of households had access to this type of financial assistance. Over 95 % of households who were able to access this type of financial assistance did so from family and friends—one of the indicators of the strong bonding social capital present among the towns and villages in the delta. In fieldwork in the delta in June and November 2012 and November 2013, I observed that these integrated approaches appear to be effective, as the normal tenor of rural and urban life appeared, on the surface, to have resumed.

An important part of this recovery outcome has been the restoration of housing. Since some 800,000 housing units were estimated to have been destroyed by the storm (450,000 totally destroyed; 350,000 damaged), housing reconstruction has benefited from the integrated approach. According to the VTA survey, 57 % of houses in the path of the storm across both Ayeyarwaddy and Yangon division were totally destroyed, while 25 % were partially damaged and 16 % were less damaged. Only 2 % had no damage. To address this enormous need for immediate shelter, the government, NGOs, INGOs and individual citizens provided tents, tarpaulins and temporary homes to over 195,000 cyclone-impacted households, who represented over 30 % of the families requiring such assistance.

Many villagers across the delta helped each other to rebuild and did not wait for assistance from the private, NGO or government sectors (CPCS 2010). The main styles of housing in the affected areas are traditional (of thatch and wood) and modern (brick, cement). In Bogale, one of the most affected towns, only three houses were left standing after the cyclone, one of which belonged to the community librarian who sheltered up to 30 people in his house. Yet, today, Bogale has been rebuilt; an observer unaware of the tragedy which occurred here would find it difficult to believe that such a disaster had swept through the town.

Likewise, in many villages deep in the delta, where outside help did not arrive for some time, villagers began rebuilding themselves, as bonding-level social capital swung into action. These villages include the Karen village of Aung Kone near Amar which did not suffer any mortalities, but where every house was destroyed by the winds and storm surge, and the Burmese village of Byi Chaung which lost one third of its total population (300), of which 70 were children. The local community leader in Byi Chaung informed me that her house[15] was rebuilt by neighbours and that all assisted each other. The PONJA report comments on the "tremendous collaborative effort" among the devastated communities which had resulted in an estimated 80% of houses being rebuilt by the time of the PONJA assessment. By mid-June 2008, the total displaced population of an estimated 800,000 people, many of whom had sought shelter with friends, family, neighbours or at monasteries and other religious places, had mostly returned to their communities of origin to rebuild in situ. Only 14 % of the impacted communities still had temporary settlements. Most of the relief camps set up to provide immediate relief in the first days after the storm had been dismantled (PONJA 2008). In discussions here in 2012, I found that there had been almost no permanent out-migration from delta communities as a result of the cyclone. This is a factor of kinship and family networks, social

[15] Later, in a similar disaster event, her house has also been designated as an evacuation point.

networks, cultural ties and land title deeds to the land the farming and fishing families work and which provide their livelihoods. At the far foot of the delta where there was most destruction, one island of fishing families was entirely washed away; on the other, the government initially sought to move survivors to a new location on the mainland. However the fishing families moved back to their island home and rebuilt a little further inland, away from the coast. They felt that the location offered by the government had negative impact on their capacity to rebuild their fishing livelihoods; they also preferred the location they used to live and where they had their sociocultural ties.[16]

A snapshot of demographic change in the ten worst affected sites can be drawn from the VTA. In accordance, 26 % of females and 12 % of males in the 18–60 age group have died. The mortalities could be expected to have significant impacts on the roles and relationships between the genders in these villages and may in future have an effect on local, intra-village migration patterns. There could be many remarriages, early marriages and a possible increase in intervillage marriages as survivors seek partners further afield from their original villages. In addition, out-migration from affected areas also seems possible in the future. The VTA (2008) states "... poor and uncertain conditions make it less likely that men will bring in new wives from outside the delta region. This will likely affect the social fabric of villages and kinship systems" (p. 161). The Post-Nargis Periodic Review IV (TCG 2010) conducted a survey of 1,400 households in the delta. Among these households, they found that 335 women had become pregnant in the 2 years since Cyclone Nargis and 82 % of them had delivered a child. At the time of the survey, some 12 % were awaiting a birth and another 7 % had aborted or had a still birth. The TCG report (2010) states that "...3 % of the 274 children [in this survey group] born in the 2 years since Cyclone Nargis died compared to 4 % of births recorded in the previous survey" (p. 17).

While there has been little migration out of the delta, there has been some movement within the delta either to other villages in search of a spouse, better house or employment. The Post-Nargis Periodic Survey IV observes that since the cyclone there has been a "slow increase" in households living in dwellings other than their pre-cyclone houses, although it is not clear whether this is a result of poverty or the wish to make a new start. Since the first survey in 2008, the TCG has recorded a slow increase from 12 to 13 % and then 15 and 21 % of households who recorded changes in place of dwelling. Much of this mobility was a direct result of the severe destruction—for example, at Labutta and Bogale. An indication that mobility may have arisen as a result of households wishing to make a new start comes from the greater proportion of male-headed households who moved to a new dwelling: 22 % compared to 12 % of female-headed households. Households whose head was younger than 60 years (n = 1,157) were more likely to have moved: 23 % compared to 12 % of households where the head was over 60 years old (n = 243). This final survey by the TCG in July 2010 found that "only a minority" (p. 85) of households who no longer lived in their same house had moved from their primary community.

[16] Personal communication with NGO leader, November 2012

Of those whose dwelling had been completely destroyed by the cyclone, only 6 % had moved to a new community; while only 3 % of those whose dwelling was partially damaged had moved. This suggests that the vast majority of survivors returned to their normal place of residence to rebuild, establish new families and restore their livelihoods, with only minimal intra- or intervillage mobility.

In the Karen area at Aung Kone near Amar, the domestic NGO Metta Development Foundation has been rebuilding schools based on the Orissa model—that is, substantial, high-set two-storey brick and cement schools which have dual purpose: both for education and for disaster risk reduction. In terms of disaster risk reduction, in the event of a similar disaster, the schools are intended to be used as evacuation centres/places of refuge. Across the other side of the delta, in Byi Chaung, the newly built cyclone-resistant school is also intended to have a similar dual function. However, it is one storey, built of interlocking bricks. The community again is part of Metta Development Foundation's projects. These schools are forming the focal points for developmental activities in the community and functioning as community service centres (Metta Development Foundation 2010).

Social capital, however, is not uniformly strong across the delta. According to the testimony of the community librarian in Bogale, the people in Bogale did not expect the storm to be so severe. They had heard around lunchtime that the storm had swept through Labutta, a town right at the foot of the delta which was almost entirely destroyed, but people in Bogale had not thought of making specific preparations or taking mitigation action—indications perhaps of both weak social capital and lack of leadership in the community. Had social capital in Bogale been stronger, evacuation measures might have been implemented in a timely fashion which could have saved lives. Likewise, had community leadership been appropriately exercised, as in the case of the Vietnamese community based in New Orleans at the time of Hurricane Katrina (Airriess et al. 2008), not only might lives and assets have been saved, but also the basis for a robust recovery might have been implemented in a more effective fashion. Vulnerability of place conjoined with social vulnerability at Bogale contributed to the significant number of mortalities and widespread destruction of assets (Cutter et al. 2003).

The librarian's account is consistent with the findings of a survey by HelpAge International (2009) which compared the situation of older persons in Bogale, Kyaiklat and Dedaye and found that variations in social capital resonant in the services provided for vulnerable older people had significant consequences for their recovery and ongoing psychosocial well-being. Of the three seriously impacted sites, social capital in Bogale was found to be the weakest, giving rise to ongoing fears of another cyclone, fears for food and health security and livelihood security among the older persons included in the study.

Recovery in the delta and across Ayeyarwaddy and Yangon Divisions in general has been the outcome of this integrated model of disaster response. It is called an integrated model because the rebuilding, relief and recovery initially began at community-level, by using their own resources, drawing on traditional mores and customs which stem from the indigenous values systems of the Myanmar sociocultural structures. Throughout the Post-Nargis Periodic Survey Reports and in the developmental activities of the domestic NGOs I spoke with, there is an obvious concerted focus on social development and the reconstruction of the lifestyles of the impacted communities; some respondents openly call for additional social activities such as the "community kitchen" approach to psychosocial recovery fostered by Metta Development Foundation in which all come to share food and stories as part of the healing process. This was implemented for 4 months from July 2008. In the Karen village of Aung Kone (population of 1,400), 133 babies had been born between 2008 and 2012. In the focus group discussion of some 50 women present, there was much hilarity and a strong sense of well-being. Although the village was close to the coast (one could smell the sea air), the mangroves surrounding the village had helped to deflect the severity of the storm surge.

In Byi Chaung before Nargis, since 2008 there had been 23 weddings and 85 babies born between 2009 and 2012. This village had undergone some psychosocial counselling but had not conducted the community kitchen. Social capital was clearly strong, but there were still issues of food security, health and education. This was a mixed fishing/agriculture village—with a high percentage of casual labour—that was only beginning to return to the area in 2012.

The urban areas in the delta as represented by Bogale, Labutta and Pyapon, although physically rebuilt, did not seem as resilient as the villages in the rural, agricultural areas of the delta, although, as with the sites in Iran and China, more research on this aspect is necessary. The Burmese delta, both in rural and urban areas, has historically survived on the traditional "self-reliance" culture wherein neighbours help each other; and these qualities have clearly been advantageous in recovering from Cyclone Nargis (Paratharayil 2010). However, as HelpAge International found many variations occur particularly in the urban areas surveyed, despite having been physically rebuilt. Considerable ongoing work in terms of economic and social development in both rural and urban areas is being undertaken by the NGOs including attention to the psychosocial needs of the survivors. Differences in governance, culture and leadership, as well as differences in the nature of the disaster event, impact on the efficacy and design of post-disaster reconstruction policies.

12.4 Conclusions

In the Burmese delta towns and villages, greater emphasis has been placed on rebuilding lives and livelihoods from a developmental perspective. I have not seen any publicly funded memorials similar to those in China or Iran. However, I have seen privately constructed, large-scale new stupas along the Pyapon-Bogale road which obviously housed the remains of loved ones and serve as mass cremation sites. These large new stupas often sit in an empty field where once a cluster of villages might have stood. Reconstruction of the social assets of survivors is taking shape around families, schools, communities and youth, the latter often in the form of Buddhist induction ceremonies into the *sangha* for a short time which provides the opportunity for the village to celebrate. As Samuels (2010) has argued, in the Burmese delta, as in Aceh, survivors show the importance they attach to social relations in the reconstruction of communities and livelihoods.

In the three disaster sites, there is a clear approach to seeking to integrate disaster and development policy while mainstreaming disaster risk reduction. Both private and government sectors are seeking to assist rebuilding of social assets through memorials, family policy and social development policy. The longer-term reconstruction of lifestyles is centred on the mores of collective societies drawing on the traditional strong bonding level of social capital.[17] This is being enhanced in certain cases through access to extra assets inherent in the bridging and linking levels of social capital. The roles of the private and public sectors should not be dichotomous, but interactive.

There is a good deal of adaptation and transformation in progress from the use of solar panels in East Azerbaijan to that of interlocking bricks for schools in the Burmese delta. Nonetheless, in overall, the urban disaster sites appear to be doing less well than those in rural areas. Reconstruction of social assets is occurring, but is still long term and incipient. It is clear that community ownership of the healing and recovery processes is seen as essential while at the same time drawing on the extra capacity-building resources available from the NGO and government sectors. As communities slowly reconstruct their social assets, there is the impression that while sorrow is still present, there is also a gradual looking towards its passing.

The distinction between the urban and rural settings in terms of disaster governance may be an illusion. While it is easy to theorise and visualise the impact of a major hydrological or geological event on a densely populated coastal conurbation, the above discussion has highlighted the impact of a compound disaster on inland urban areas, rural areas as well as coastal fishing towns. In an event, when a cyclone like Nargis sweeps through an area like the Burmese delta, or an earthquake strikes agricultural villages like those outside Tabriz, the long-lasting sociocultural and economic effects on devastated rural communities, destroyed crops and livelihoods are just as complex as in the urban settings. Devastated Bogale and Beichuan and Bam in fact are dependent on their surrounding rural environs for food in the immediate aftermath of the disaster event. The health issues affecting the psychosocial well-being of survivors, maternal and child health outcomes, immediate sanitation, shelter, purification of water and requirement to revive interrupted livelihoods are no less urgent and complex in both settings. It could be said that the urban and rural settings are interdependent, both in normal times and in times of disaster impact, relief, recovery and reconstruction as the resources from one setting are often deployed to assist the other. As survivors in cities and towns look beyond the immediate recovery phase to the later reconstruction phase, sociocultural features of their lives often come to the fore as wellsprings on which they draw to provide inner strength to resume the tasks of daily living.

---

[17] See Paton et al. (2008) for an insightful evaluation of the "collective efficacy" of collective societies in facilitating community recovery from a disaster.

Such sociocultural features often have their origins in traditional networks of religion, art, dancing, song, feasting and celebration of the major milestones of human life, marriage and birth. As those managing complex urban disasters seek new and more effective approaches to post-disaster reconstruction, the opportunity to draw on the sociocultural features which characterise all human societies should not be neglected.

References

Airriess, C., Li, W., Leong, K. J., Chen, A. C., & Keith, V. M. (2008). Church-based social capital, networks and geographical scale: Katrina evacuation, relocation, and recovery in a New Orleans Vietnamese American community. *Geoforum, 39*(3), 1333–1346.

Ardalan, A., Mazaheri, M., Mowafi, H., Van Rooyen, M., Teimoori, F., & Abbasi, R. (2011). Impact of the 26 December 2003 Bam earthquake on activities of daily living and instrumental activities of daily living of older. *Prehospital and Disaster Medicine, 26*(2), 99–108.

Asian Development Bank (ADB). (2011). *Asia 2050: Realizing the Asian century*. Manila: ADB.

Athukorala, P., Manning, C., & Wickaramasekara, P. (Eds.). (2000). *Growth, employment and migration in Southeast Asia: Structural change in the greater Mekong countries*. Cheltenham: Edward Elgar.

Birmingham, L., & McNeill, D. (2012). *Strong in the rain: Surviving Japan's earthquake, tsunami and Fukushima nuclear disaster*. New York: Palgrave Macmillan.

Centre for Peace and Conflict Studies (CPCS). (2009). *Listening to voices from inside: Myanmar civil society's response to Cyclone Nargis*. Phnom Penh: CPCS.

Centre for Peace and Conflict Studies (CPCS). (2010). *Listening to voices from inside: Ethnic people speak*. Phnom Penh: CPCS.

Cook, K. (2005). Networks, norms and trust: The social psychology of social capital. *Social Psychology Quarterly, 68*(1), 4–14.

Cutter, S. L., Boruff, B. J., & Shirley, W. L. (2003). Social vulnerability to environmental hazards. *Social Science Quarterly, 84*(2), 242–261.

De La Paz, C. (2013, September 18–20). *The discourse of disasters in Philippine festivals: Culture, local governance and the construction of historical memory.* Paper presented at the international conference on the demography of disasters: Implications for future policy on development and resilience. Canberra: Australian National University.

Firoozabadi, M., Abedinzadeh, M., & Moslemi, M. (2011). Genitourinary system trauma after 2003 Bam earthquake in Kerman, Iran. *Therapeutics and Clinical Risk Management, 7*, 49–52.

Hayashi, I. (2010). Building a disaster resilient culture: Activities of volunteer associations in Kushimoto-cho Wakayama Prefecture, Japan. In C. Zayas, V. Hernandez, & A. Fajardo (Eds.), *Overcoming disasters: Lessons from post-disaster interventions in Japan and Southeast Asia* (pp. 15–26). Quezon City: Center for International Studies, University of the Philippines.

HelpAgeInternational. (2009). *The situation of older people in cyclone affected Myanmar*. Chiang Mai: HelpAgeInternational.

Hosseini-Chavoshi, M., & Abbasi-Shavazi, J. (2013, September 18–20). *Demographic consequences of the 2003 Bam Earthquake in Iran.* Paper presented at the international conference on the demography of disasters: Implications for future policy on development and resilience. Canberra: Australian National University.

International Federation of Red Cross and Red Crescent Societies. (2009). *World disasters report: Early warning, early action*. Geneva: IFRCRCS.

Lin, N., Cook, K., & Burt, R. S. (2001). *Social capital*. New York: Aldine de Gruyter.

Lo, J. C. (2010). The impact of the Chi-Chi Earthquake on demographic changes: An event history analysis. In S. Kurosu, T. Bengtsson, & C. Campbell (Eds.), *Demographic responses to economic and environmental crises* (pp. 193–203). Kashiwa/Chiba: Proceedings of IUSSP seminar, Reitaku University.

Metta Development Foundation. (2010). *Post-Cyclone Nargis trauma healing and recovery initiatives*. Yangon: Metta.

Miichi, K. (2014, forthcoming). Saving folk performing arts for the future: Challenges of Unotori Kagura after the East Japan Great Earthquake in 2011: Chapter 11. In H. James & D. Paton (Eds.), *The consequences of disasters: Demographic, planning and policy implications*. Springfield, IL: C. H. Thomas.

Montazeri, A., Baradaran, H., Omidvari, S., Azin, S. A., Ebadi, M., Garmaroudi, G., et al. (2005). Psychological distress among Bam earthquake survivors in Iran: A population-based study. *BMC Public Health, 5*(4), 1–6.

Nakagawa, Y., & Shaw, R. (2004). Social capital: The missing link in disaster recovery. *International Journal of Mass Emergencies and Disasters, 22*(1), 5–34.

Paratharayil, M. (2010). Basic versus focused psychosocial interventions for community well-being: Lessons following the Nargis cyclone interventions in Burma/Myanmar. *Intervention, 8*(2), 148–157.

Paton, D., & Jang, L. (2011). Disaster resilience: Exploring all-hazards and cross-cultural perspectives. In D. Miller & J. Rivera (Eds.), *Community disaster recovery ad resiliency: Exploring global opportunities and challenges* (pp. 81–100). Oxford: Taylor and Francis.

Paton, D., & Violanti, J. (Eds.). (2012). *Working in high risk environments: Developing sustained resilience*. Springfield, IL: Charles C. Thomas.

Paton, D., Gregg, C. E., Houghton, B. F., Lachman, R., Lachman, J., Johnston, D. M., et al. (2008). The impact of the 2004 tsunami on coastal Thai communities: Assessing adaptive capacity. *Disasters, 32*(1), 106–119.

Pelling, M. (2003). *Vulnerability of cities: Natural disasters and social resilience*. London: Earthscan.

Pelling, M. (2011). *Adaptation to climate change: From resilience to transformation*. London/New York: Routledge.

Ronan, K., & Johnston, D. (2005). *Promoting community resilience in disasters: The role for schools, youth and families*. New York: Springer.

Samuels, A. (2010). Remaking neighborhoods in Banda Aceh: Post-tsunami reconstruction of everyday life. In M. Clarke, I. Fanany, & S. Kenny (Eds.), *Post-disaster reconstruction: Lessons from Aceh* (pp. 211–223). London/Washington, DC: Earthscan.

Shaw, R., & Goda, K. (2004). From disaster to sustainable civil society: The Kobe experience. *Disasters, 28*(1), 16–40.

Tripartite Core Group (TCG). (2008). *Post-Nargis joint assessment. Village tract assessment, June 2008*. Yangon: TCG/UN.

Tripartite Core Group (TCG). (2008–2010). *Post-Nargis periodic reviews I-IV*. Yangon: TCG/UN.

UNDP. (2007). *Integrated household living conditions survey (IHLC)*. Yangon: UNDP and Ministry of Immigration and Population.

UNDP. (2010). *Human development report 2010*. New York: UNDP.

UNICEF. (2000). *Myanmar Multiple Indicator Cluster Survey*. Yangon: UNICEF.

United Nations International Strategy for Disaster Reduction (UNISDR). (2005). *Hyogo framework for action, 2005–2015: Building the resilience of nations and communities to disasters*. Kobe: UNISDR.

Willacy, M. (2013). *Fukushima: Japan's tsunami and the inside story of the nuclear meltdowns*. Sydney: Pan Macmillan Australia.

Xin, M., Manning, C., & Nur Effendi, T. (Eds.). (2010). *The great migration: Rural–urban Migration in China and Indonesia*. Cheltenham: Edward Elgar.

Yang, C. (2013, September 18–20). *Population livelihoods in the process of rebuilding after the 5/12 Wenchuan earthquake, South West China.* Paper presented at the international conference on the demography of disasters: Implications for future policy on development and resilience. Canberra: Australian National University.

Zayas, C., (2014, forthcoming) 'Land is life and life is land: Development, resilience, family and knowledge for the Ayta of Mt Pinatubo' chapter 6. In H. James, & D. Paton (Eds.), *The consequences of disasters: Demographic, planning and policy implications.* Springfield, IL: C H Thomas Publishers.

Zayas, C., Hernandez, V., & Fajardo, A. (Eds.). (2010). *Overcoming disasters: Lessons from post-disaster interventions in Japan and Southeast Asia.* Quezon City: Center for International Studies, University of the Philippines.

Zhou, K. (2013, September 18–20). *Population reconstruction in the post-disaster context: Lost-child families' willingness on fertility and associated relevant factors.* Paper presented at the international conference on the demography of disasters: Implications for future policy on development and resilience. Canberra: Australian National University.

Zolala, F. (2010). Data collection after massive natural disasters (focusing on Bam earthquake, Iran). *Disaster Prevention and Management, 19*(5), 541–547.

Chapter 13
Nuclear Risk Governance in Japan and the Fukushima Triple Disaster: Lessons Unlearned

Pablo Figueroa

Abstract Nuclear risk governance in Japan came to the fore of public concern following the triple disasters of March 11, 2011. For the first time since atomic energy was incepted, citizens became aware that despite pervasive narratives of technological safety, nuclear power poses a tangible risk on people's lives. In this context, 3.11 reveals the urban and compound nature of contemporary disasters. Regardless of having taken place in a nonurban region, the effects of the earthquake, tsunami and cascading accidents at the Fukushima plant were severely compounded by inadequate decision-making in the capital. Lack of transparency, lack of governance and poor risk communication emerged as salient issues affecting disaster prevention and response. This chapter interrogates the historical and political context that paved the road to the worst nuclear accident in recent times. It then discusses institutional weaknesses in safety assessments and describes conflicts that characterized the expansion of nuclear power. Finally, it argues that in order not to repeat the same mistakes that led to the Fukushima catastrophe, important lessons in nuclear risk governance have yet to be learned. These include the need for greater transparency, strengthening the independence of regulatory bodies, the importance of taking into account people's perceptions of nuclear energy, facilitating citizen participation in the construction of risk discourses and acknowledging the vital role of cities (especially Tokyo) in disaster governance.

Keywords Nuclear risk governance • Transparency • Fukushima • Japan • Compound disaster

---

P. Figueroa (✉)
Center for International Education, Waseda University, Tokyo, Japan
e-mail: pablo.figueroa@aoni.waseda.jp

M.A. Miller, M. Douglass (eds.), *Disaster Governance in Urbanising Asia*,
DOI 10.1007/978-981-287-649-2_13

13.1 Introduction

The discussion of the significance and implications of governance has gained enormous popularity in recent years, not only in academia but in government organizations and civil society as well. At a national level, governance refers to the structures and processes for collective decision-making involving governmental and nongovernmental actors (Nye and Donahue 2000, cited in Renn 2008). It has been argued that "risk governance" is the application of the core principles of governance to the context of risk decision-making (Renn 2008). Risk governance involves, but also extends beyond, risk assessment, risk management and risk communication. Moreover, risk governance is concerned with how risk information is collected, analysed and communicated and how decision-making is carried out through rules, processes and mechanisms that characterize complex webs of actors (Renn 2008, 8–9). In this chapter, "nuclear risk governance" will refer to the abovementioned processes in the context of national nuclear energy policy, with a focus on decision-making, transparency and citizen participation.

Nuclear risk governance in Japan came to the fore of public concern following the Great East Japan Earthquake and the triple disasters of March 11, 2011 (hereafter 3.11). For the first time since atomic energy was incepted more than 50 years ago, citizens became aware that despite pervasive narratives of technological safety, nuclear power poses a tangible risk on people's daily lives. Transparency of nuclear oversight was deeply questioned during the months following the catastrophe. An independent investigation commission appointed by the Diet concluded that governance in safety had been lacking due to collusion between the government, the regulator and the nuclear industry. In this context, 3.11 reveals the intrinsic urban and compound nature of contemporary disasters (Douglass 2013; Aldrich 2012b). Regardless of having taken place in a nonurban region, the effects of the earthquake, tsunami and cascading accidents at the Fukushima plant were severely compounded by inadequate decision-making in the capital. Long-lasting economic and societal implications resonated well beyond the disaster zones reaching core metropolitan areas. Production networks were disrupted due to damages to infrastructure, displacement of the local populations and profound uncertainty surrounding nuclear contamination. Moreover, the meltdowns accentuated existing schisms between pronuclear and anti-nuclear sectors of society; massive protests (the largest in postwar Japan) were staged in Tokyo and elsewhere. A calamity that carried transnational consequences for nuclear energy policy, the March 11 events threw to the ground the simplistic idea that disasters occur in isolation.

Although triggered by an earthquake and tsunami, the Fukushima nuclear accident was not a natural catastrophe but rather a man-made disaster with foreseeable causes (Noggerath et al. 2011, 45; Kingston 2012a, 2). Observers identified technical issues, governance and regulatory weaknesses and systemic failure of nuclear safety assessments (Gundersen 2012, 37). Poor risk management and an institutionalized complacency about risks were central to increasing the probabilities of a catastrophic accident and an uncertain crisis response (Kingston 2012a, 1). Therefore, the Fukushima disaster was not just predetermined by the earthquake and tsunami and the technological deficiencies to cope with the accident, but also by sociopolitical factors that worsened the "technological" component of the disaster. The dangerous proposition of placing multiple reactors at one single site (a common siting pattern throughout Japan) proved catastrophic at Fukushima, and this inadequacy can be explained by examining government policy and financial support for nuclear siting (Juraku 2013, 41–43).

The electric utility Tokyo Electric Power Company (TEPCO) played a decisive role in the rapid worsening of the crisis. To justify their inability to cope with the disaster, the company claimed at first that the tsunami and the earthquake were "beyond imagination" (*sōtegai*): they could have not been anticipated. However, a strong body of historical evidence contradicts this assertion. Despite having had successive opportunities throughout the years for revising safety standards, TEPCO and the regulators failed to pay attention to repeated warnings coming from multiple sources and did not introduce necessary safety measures.

The Japanese government was also a main force responsible for the sociopolitical shaping of the disaster. Although state narratives posed that Japan's nuclear plants were absolutely safe, 3.11 showed that official assumptions did not take into account the probability of serious accidents. The Economy, Trade and Industry Minister at the time of the accident, Mr. Banri Kaieda, affirmed that there was a "myth of safety" surrounding nuclear power plants and that regulators always thought that measures against emergency situations were unnecessary because such situations would never occur (NHK Special Documentary 2011). However, justifying the lack of preparedness by a safety myth conveniently dilutes political responsibility. To gain valuable lessons from Fukushima, it is crucial to examine critically the secretive political culture that has permeated contemporary Japan (Bowen 2003).

Some of the most virulent criticism towards the institutions entrusted with public safety came from the Fukushima Nuclear Accident Independent Investigation Commission. The National Diet of Japan appointed in 2011 the first independent delegation in the history of the country's constitutional government to investigate the direct and indirect causes of the Fukushima disaster. The commission's report concluded that the accident was "profoundly manmade" and that:

> The TEPCO Fukushima Nuclear Power Plant accident was the result of collusion between the government, the regulators and TEPCO, and the lack of governance by said parties. They effectively betrayed the nation's right to be safe from nuclear accidents. Therefore, we conclude that the accident was clearly "manmade." We believe that the root causes were the organizational and regulatory systems that supported faulty rationales for decisions and actions, rather than issues relating to the competency of any specific individual. (The National Diet of Japan 2012, 16)

Nuclear risk governance was therefore a key issue affecting the Fukushima disaster: lack of governance by the operators, the regulators and the government paved the road to a nuclear catastrophe that is still unfolding. In the following sections, I shall interrogate the political context that moulded Japan's nuclear programme while paying attention to the urban dimensions of the disaster, e.g. the connections between government, the nuclear industry, the safety agencies, citizen groups and the general public.

This article draws upon anthropological fieldwork I conducted in Fukushima in 2011 and in Tokyo between 2011 and 2014. It is informed by in-depth, first-hand interviews with Fukushima residents and a variety of actors involved with the disasters (leaders of NGOs carrying relief operations, long-time anti-nuclear activists, radiation-concerned citizens, volunteers, photojournalists, filmmakers, academics) and the analysis of official documents, media reports and digital sources such as blogs, websites and social media. My intention is to frame nuclear risk governance from an interdisciplinary perspective that combines sociocultural anthropology with disaster governance, risk management and disaster prevention. In doing so, I will aim to shed light into issues of nuclear risk governance and the urban dimension of nuclear catastrophes.

13.2 Historical and Political Context

Japan started its nuclear programme in 1954, when the first budget for nuclear research and development was proposed to the Diet. Post-Second World War II politicians envisioned the peaceful application of nuclear technology to the generation of electricity; since then, the Japanese government has had a strong, steady commitment to nuclear energy policies (Tabusa 1992; Kondoh 2007; Hara 2013). The stable generation of electricity to cope with increasing demands was considered a matter of national security and economic growth: the government designed, very early on, an energy scheme heavily dependent on nuclear power. One of the key figures recognized as the drive behind Japan's nuclear programme was Yasuhiro Nakasone, a member of the House of Representatives who would later become prime minister. Drafted by Nakasone and passed on Dec. 19, 1955, the Atomic Energy Agency Law called for "independent, democratic and transparent activities in nuclear research" (Carpenter 2012). This democratic law, however, appears to have been broken from the very beginning. During construction planning of the first commercial nuclear power plant in Tōkai village, the Science and Technology Agency (*Kagaku gijutsu-chō*) estimated that the cost of an accident at the Tōkai Nuclear Power Plant would equal twice the national budget at the time. The agency, however, concealed the report denying its existence until 1999 and continued to expand nuclear power fully aware of the risks involved in the enterprise (Onitsuka 2012).

A plethora of nuclear agencies concerned with the development of nuclear fuel, nuclear power and its commercialization were quickly created in the following years: the Japan Atomic Industrial Forum (JAIF), the Japan Atomic Energy Research Institute (JAERI), the Atomic Fuel Corporation, the Japan Atomic Power Company (JAPCO), the Japan Nuclear Fuel Corporation and the Power Reactor and Nuclear Fuel Development Corporation (PNC). Japanese commercial power plants went online in the mid-1960s and their construction boomed until the mid-1990s (Hara

2013, 23; Juraku 2013, 43), when a prolonged economic recession slowed down the break-neck pace sustained until then. Nevertheless, by 2010, about 30% of Japan's electricity was supplied by 54 reactors sited at 17 plants, with an ambitious goal of further boosting nuclear power to 50% by 2030 with the construction of 14 new reactors.[1]

Juraku points out that the concentrated siting of reactors at single plant was a key strategy utilized by advocates of nuclear power to achieve rapid expansion by eschewing local opposition; fundamental issues that may have slowed progress were played down or ignored altogether (Juraku 2013, 52). This dangerous overlook in terms of nuclear governance[2] helped shape the "myth" that critical problems would never arise in nuclear power plants in Japan. Onitsuka examines the fact that Japanese power plants hold multiple reactors concentrated in specific regions, usually impoverished rural areas. Once the depopulating, decaying towns got used to receiving generous amounts of cash coming from the government and the nuclear industry, they fell into a vicious relationship of nuclear dependency. This is explained by the nexuses between local governments and the state: local governments and their officials have the final say in the decision-making process; residents of the communities hosting power plants are excluded from nuclear energy policy-making. Local governments, in turn, have little autonomy because they rely heavily on the subsidies and grants they receive from the central government. State grants represent approximately 70% of local incomes (Onitsuka 2012), and these come with strings attached. The Tohoku prefectures of Miyagi, Fukushima and Iwate had been deteriorating financially since the 1990s and were already in a precarious state when hit by the 2011 earthquake and tsunami (Samuels 2013). Towns hosting power plants have traditionally occupied a space of vulnerability in the nuclear scheme: with little self-reliance, they are subsumed under the logic of the state, which determines conditions of production, distribution and consumption of energy. Ironically, although local towns are central to the functioning of the system at a national level, they nevertheless remain peripheral nodes of the nuclear network. The core-periphery system leads to "disaster incubation", in which hazardous infrastructure is usually placed in peripheral areas where oversight is weak (see Douglas earlier in this volume). In the case of nuclear power in Japan, it has been a common practice for the utilities to carry out safety tests by themselves. The results were then sent to the Nuclear and Industrial Safety Agency (NISA) and the Nuclear Safety Commission of Japan (NSC) for final validation (Kingston 2012b).

Such potentially dangerous practices based on weak oversight have been the norm in Japan and can be traced back to the nuclear agencies' mindset. The proliferation of reactors concentrated in specific regions, along with lax regulations, and

[1] In 2010, the cabinet of Japan passed the Basic Energy Plan created by the Ministry of Economy, Trade and Industry (METI). The plan, criticized as unrealistic by some, proposed the construction of at least 14 new nuclear reactors.

[2] The typical array of nuclear power plants in Japan severely undermines public safety: the high concentration of reactors one next to another at one single site augments risk exponentially if things go wrong in an accident, as it was the case with Fukushima.

the practice of sweeping problems under the rug was made possible due to the cosy ties between the government, the regulators and the nuclear industry. In the beginning, two main offices were in charge of promoting nuclear development. These were the Science and Technology Agency (STA) and the Ministry of International Trade and Industry (MITI, now transformed into Ministry of Economy, Trade and Industry [METI]). Although their functions differed, the two agencies' interests frequently overlapped, and competition among them was not unusual.

In 1995, there was a nuclear accident at the Monju Nuclear Power Plant[3] in Fukui Prefecture, and radiation was released into the atmosphere. The incident's handling by the operator and the regulator generated public doubts about proper risk management; the agency and the operator tried to cover up the mishap and downplayed the environmental damages. This manoeuvre was exposed,[4] badly hurting the credibility of the STA. The STA was dissolved a few years later under government structural reform (Hara 2013).

Three agencies were then created, absorbing the power the STA once had: the METI, the Cabinet Office and the Ministry of Education, Culture, Sports, Science and Technology (MEXT). Among these, the agency that emerged as a key player in nuclear energy policy was the METI, which had a strong commitment to the promotion of nuclear power. METI then created the Agency for Natural Resources and Energy (ANRE) and NISA, the main nuclear regulator. The fact that the nuclear watchdog was placed under the umbrella of a ministry that had an outspoken mission to expand nuclear power did not pass unnoticed to the public; citizens perceived an inherent conflict of interest because the regulator worked hand in hand with a staunch promoter of nuclear energy. This awkward situation had been in the public eye for many years before the Fukushima accident happened. Despite official claims of nuclear safety, the thin, often blurred line separating regulator and regulated was already a matter of concern for informed people in mainstream society.

It is worth noting that in Japan, information of the nuclear risk has tended to have the purpose of manipulating the perceptions of the public (Figueroa 2013). In the government stance, public opinion is malleable and it should be modified to fit into the administration's objectives (Aldrich 2012b). This perspective has reinforced the mindset and practices of the so-called Iron Triangle, which is composed by politicians, state bureaucracy and the business sector. Together, they are responsible for

[3] For an in-depth discussion of the accident and a history of the Monju reactor, see http://www.cnic.jp/english/newsletter/nit134/nit134articles/monju.html.

[4] In 1996, an official appointed to the internal investigation of the sodium coolant leak killed himself. The official's relatives subsequently sued the operator for damages, claiming he took his life after being pressured to lie to the public. The Supreme Court rejected the suit and ruled in Japan Atomic Energy Agency's (JAEA) favour. Fukui residents, where the Monju Fast Breeder Reactor sits, meanwhile sued the state, claiming the reactor was unsafe. The Supreme Court in 2005 overruled a lower court ruling and dismissed the suit (Japan Times May 13, 2013). But the Monju controversies did not end there. In November 2012, it was revealed that JAEA had failed to conduct regular inspections on almost 10,000 out of a total 39,000 pieces of equipment at Monju, including safety-critical equipment. This led to the resignation of the head of the JAEA president Atsuyuki Suzuki (World Nuclear News May 20, 2013).

the creation of a state narrative that emphasizes the safety and need of nuclear power. With little attention being paid to transparency and participative decision-making, nuclear risk governance has been missing from the picture. Moreover, due to assessment weaknesses there has been no realistic consideration of the possibility of a serious nuclear disaster. Worst-case scenarios have been downplayed or discarded altogether.

13.3 Assessment Weaknesses

Proper risk assessment (without exaggerating or minimizing the hazards and dangers) is critical for responsible utilization of nuclear power because potential risks may extend well beyond the vicinities of power plants. In worst-case scenarios, a catastrophic accident may affect the public health of hundreds of thousands of people in local towns and cities during an extended period of time. At a regional level, it may spoil the environment and contaminate the food chain, wildlife, livestock and other resources that are central to the functioning of the local and regional economies, as it was the case in Fukushima.[5]

At the same time, poor risk assessment enhances the high disaster risk faced by city dwellers: when flooding, earthquakes and other catastrophes strike urban agglomerations, large numbers of people are usually unable to mobilize resources by themselves and can be caught in bottlenecks, with precarious access to food, water and shelter. This enhanced vulnerability of the urban populations is brought about by the transition from rural to urban societies and the hyper-density of human settlements in mega-urban regions (MURs) (Douglass 2013).

Critics of nuclear power have pointed out that the nuclear industry projects a false sense of security by providing estimations that are unreliable. Since the initial construction of nuclear power plants in the 1960s, operators and regulators have utilized a method known as probabilistic risk assessment (PRA). This method has proved poor in terms of its accuracy for preventing severe nuclear accidents; in recent years, the PRA has been contested on theoretical and empirical levels. According to independent reactor analysts, although the nuclear industry and the safety regulatory agencies claim that the possibility of catastrophic accidents is so small that it is not even worth considering, the actual frequency of severe incidents is much higher than the figures expected by PRA-based calculations. For instance, PRA studies do not take into account institutional malfunctioning at the regulatory level (Gundersen 2012). In addition, studies have shown that a chain-of-event conception of accidents cannot account for the indirect, non-linear and feedback relationships that characterize many accidents in complex systems (Ramana 2011).

[5] For instance, farmers and fishermen see their occupation not only as a means of survival but also as a way of life: their identities and sense of land are tied to their intimate knowledge of farming and fishing practices.

The Fukushima Nuclear Accident Independent Investigation Commission emphatically criticized arbitrary interpretation and selection of a probability theory. Their report states that:

> TEPCO tried to justify the belief that there was a low probability of tsunami, and used the results of a biased calculation process as grounds to ignore the need for countermeasures. TEPCO also argued that basing any safety assessment against tsunami on a probabilistic approach would be using a methodology of technical uncertainties, and used that argument to postpone considering countermeasures for tsunami. (The National Diet of Japan 2012)

The abovementioned suggests that in addition to utilizing an unreliable method for assessing risk, the nuclear industry has manipulated scientific data with the aim of highlighting technical uncertainties. Thus, uncertainty was the justification for biased calculations and for avoiding the need to address safety measures. This clearly goes against current global trends of nuclear risk governance and urban risk governance that call for co-participative strategies taking into account multiple perceptions and discourses of risk, e.g. stakeholders, laypeople, business sectors, etc.

13.4 Opposition

Public opposition to nuclear energy rose dramatically after the Fukushima disaster, especially in urban areas. However, perceptions of people living in towns hosting power plants are divided: while some worry about the overall nuclear safety, others see their survival tied to the income brought by the nuclear industry through incentives and grants. Identical findings have been reported elsewhere (Aldrich 2012b). Nevertheless, past and present research indicates that the aggressive pursuit of nuclear energy in Japan has often met resistance at local and national levels, and this is the reason why the central government has devoted many resources to induce people to embrace nuclear power. Policy instruments historically utilized by the state include: financial incentives, pronuclear education included in children's school syllabuses, advertorials in newspapers and jobs in local industries (Aldrich 2012b, 136).

It has been often the case that local communities oppose the construction plans of power plants (Tabusa 1992; Lesbirel 1998). Anti-nuclear groups filed administrative and civil lawsuits against the construction of the Ikata nuclear power plant in Ehime prefecture, the Tokai No. 2 nuclear power plant in Ibaragi, the Fukushima No. 2 nuclear power plant in Fukushima, the Kashiwazaki-Kariwa nuclear power plant in Niigata and the Onagawa nuclear power plant in Miyagi. Residents disputed siting projects on the grounds that safety related to the construction and the operation of power plants was inadequate. Despite citizens' concerns about having nuclear plants in their vicinities, all suits were ruled in favour of the nuclear power companies (Tabusa 1992, 273–297). Lesbirel described different intensities of opposition (from little to high) to the siting of the following plants: Mihama, Taketomi, Saijo, Tokai, Hamaoka, Mihama, Date, Matsushima, Kashiwazaki, Gobo, Tomari, Ashihama and Namie (Lesbirel 1998, 5). Some of these siting disputes carried on for 10 or more years.

Moreover, various studies suggest a stark contrast between official narratives of nuclear safety and the citizens' standpoint (Kingston 2004, 17; Kondoh 2009, 64). Public polls conducted by *Asahi* newspaper indicate a continuous decrease in the support for nuclear power after the 1986 Chernobyl accident and the 1999 Tokaimura accident (Kondoh 2009, 64; Kotler and Hillman 2000, 23). During the years from 1986 through 1992 anti-nuclear grassroots activities gained momentum in Japan's metropolitan areas in the form of concerned women entering the stage. This movement of "concerned mothers" that took shape in response to the Chernobyl disaster was composed of highly educated housewives living in urban areas, and, although they were unable to expand their agenda to reach national politics, they achieved ample social recognition (Hasegawa 1995, 30). In 2011, the government's mishandling of issues of radiation fueled the anger of many concerned mothers of Fukushima children. For example, one informant said:

> I'm a chief of a parent-teacher association (PTA). After the incident, the authorities pressed me hard to not speak about radiation or to say that nuclear power is dangerous. In my view, because of high levels of radiation, I had expected that children would be evacuated from here. But they said that the prefectural government would decide whether it was dangerous or not, that it was none my business, and I was ordered not to make any public statements. At first they were careful about not letting the children play outside, but after two months they said it was fine to do so. I am concerned and dissatisfied with this attitude. (A mother in Fukushima)

Local residents who were discontent with the authorities created the Fukushima Network for Protecting Children from Radiation, an association that denounced "reckless radiation limits" and argued for the need to "protect hundreds of thousands of Japanese children from a lifetime of possible increased cancer risk" (Green Action Japan 2011). Concerned mothers continue to emphasize their discontent towards the government's handling of the radiation problem at regional and national levels.

In addition, recent surveys of public opinion indicate that support for nuclear energy plunged in 2011 after the Fukushima disaster, with as much as 74% of respondents agreeing with the replacement of nuclear energy with alternative sources (NHK Poll November 4, 2011; Mainichi Daily News Poll August 22, 2011[6]). Considered together, all the above suggests that a majority of people, especially those living in urban areas, are against the further expansion of nuclear energy and agree with the implementation of safer alternative sources.

13.5 Protests

The Fukushima disaster awakened the anti-nuclear movement across Japan. In the summer of 2012, weekly protests were staged in front of the prime minister's office in Tokyo. These rallies were organized by the Metropolitan Coalition Against

[6]The *Mainichi Shimbun* poll results are taken from the following website: http://www.jice.or.jp/sinsai/sinsai_detail.php?id=1497. Accessed 20 June 2014.

Nukes, and their immediate aim was to shut down Oi nuclear power plant in Fukui Prefecture.[7] At first, the protests were not significant in size but the demonstrations grew steadily reaching (according to the organizers) 200,000 demonstrators at their peak (Ogawa 2013). In an unprecedented move, then prime minister Noda agreed to talk with a delegation of the abovementioned coalition. The meeting, however, did not bring the opposing positions closer. According to the *Asahi Shimbun* newspaper: "Yoshihiko Noda conceded nothing and angry protesters stormed out of the room after a 30-min face-to-face confab between the prime minister and representatives of the Metropolitan Coalition Against Nukes" (*Asahi Shimbun* August 23, 2012b). The rallies spread to other parts of Japan including KEPCO's headquarters in Osaka.

Despite prevailing mistrust and massive anti-nuclear demonstrations, the nuclear industry decided to ignore the citizens' views and pushed the government for a hasty restart of idled reactors. The government's decision to connect to the grid again reactors 3 and 4 in Oi nuclear power plant[8] was felt by the general public as a deep betrayal in favour of the interests of the "nuclear village".[9]

People voiced their concerns but the government privileged the interests of the nuclear industry by exercising the same politics prior to 3.11. In line with a carefully crafted state narrative of nuclear power, laypersons' views were dismissed as irrational and uninformed. Public rage against nuclear energy in the wake of the Fukushima disaster has not resulted in an open reconsideration of Japan's long-term energy production strategies. In this sense, although it generated widespread opposition and mistrust that continued to this day, the Fukushima disaster has not been a game-changing event (Kingston 2012b).

13.6 Lack of Transparency

Historically, lack of transparency in the decision-making process has been a salient issue affecting nuclear power in Japan. The construction of the Fukushima plants is a case in point. When the project was first considered, TEPCO, the prefectural office and the local mayors negotiated the purchase of the required lands from the local residents in complete secrecy. People in the communities of Futaba town and Okuma town were deliberately kept in the dark to avert public resistance. The concealment went so far as to establish a special committee to prevent the project's

---

[7] This power plant was the only one in Japan to operate two nuclear reactors in the wake of Fukushima. There has been controversy over the safety of the plant because it was thought that the reactors were built on an active seismic fault line. However, a recent panel of experts agreed on September 2, 2013, that the fault underneath Oi nuclear power plant is not active.

[8] These reactors were switched off again in September 2013 due to regular checkups. Given the current public opposition, there is no restart in sight.

[9] Simply put, the term "nuclear village" refers to advocates of nuclear power in a wide range of sectors such as utilities, government, bureaucracy, media, banks and academia.

information from reaching the public; incognito engineers visited potential construction sites with female companions pretending they were regular tourists. Local residents were informed they would be hosting a power plant only 2 years after the decision had been made by their towns' representatives (Onitsuka 2012).

TEPCO's lack of transparency stands out on its own. The electric utility has a history of cover-ups, widespread falsification of safety checks and submission of fake technical data (Aldrich 2012a, 5; Gundersen 2012, 39; Kingston 2004, 19). Known incidents include: (a) the falsification of inspection records in 2002 to hide cracks in reactors at 13 of its 17 nuclear plants. It should be underscored that the nuclear regulator did not carry out the safety checks but entrusted the electric utility to do so instead; (b) the falsification of test data of reactor containers of Fukushima Daiichi during the 1990s; (c) the falsification of records of coolant water temperatures during the mid-1980s; (d) the concealment of the true extent of damage at the Kashiwazaki-Kariwa nuclear plant in Niigata after an earthquake in 2007 caused a fire and leakage of hundreds of gallons of radioactive water; and (e) the failure to properly inspect cooling systems and spent fuel pools at Fukushima Daiichi before the accident. What is more, it was discovered that for over 20 years TEPCO had been secretly giving donations of about two billion yen per year (the equivalent of $26 million) to local governments near nuclear facilities. The payments were designated as "funds to deal with local communities" (*Asahi Shimbun*, September 15, 2011). According to critics, these lax regulations created an environment in which TEPCO officials could withhold important information. For decades, TEPCO continued to falsify, therefore prioritizing the company's profits over public safety (Gundersen 2012).

Other nuclear power companies and regulatory agencies have shown lack of transparency as well. For instance, in 2000 Kansai Electric Power Co. (KEPCO) penalized its vice president and five other executives over the falsification of data on plutonium-uranium mixed oxide (MOX[10]) fuel shipped to Japan by a British supplier the previous year. This became a public scandal: previously, the utility had consistently denied there were any problems with the fuel.[11] In yet another episode illustrating connivance between the regulator and the nuclear industry, it transpired that an ex-chief of a regional nuclear safety regulator in Hokkaido asked Hokkaido Electric Power Co. to gather opinions in favour of the use of mixed oxide fuel MOX in the Tomari nuclear power plant. This request was made because government surveys showed people were clearly against the use of MOX. It is worth noting the financial arrangement behind the request: the central government was to give prefectures accepting MOX ¥1 billion (approx. US$ 10 million) before the use of the nuclear fuel started and ¥ 5 billion over 5 years afterwards. In addition, ANRE and the nuclear watchdog NISA asked five electric utilities throughout Japan to instruct employees and other people to express opinions favouring nuclear power generation in public meetings, symposiums and other events (Japan Times November 28, 2011b).

[10] A nuclear fuel consisting of uranium and plutonium

[11] For several months, KEPCO executives told environmental groups, MITI officials and the Fukui Prefectural Assembly that the fuel was safe and had been properly checked.

In the wake of Fukushima, the government's failure to keep any records of the 23 meetings held during 2011 to deal with the crisis further exemplifies the sort of opacity that pervades political culture in Japan (BBC Online January 27, 2012). An editorial of the *Asahi Shimbun* called the lack of minutes of the meetings and the response to the disaster a "monumental level of government ineptitude" (*Asahi Shimbun* January 26, 2012a). The newspaper implied that the omission was most likely deliberate and that it had the aim of protecting officials by preventing their blunders from coming to light.

13.7 Lack of Governance

Lack of governance is perhaps the most serious failure among nuclear safety regulators. As of March 2011, NISA was a department of the METI, which was fully committed to the promotion of nuclear power. This conflict of interest generated distrust among the public. The ministry promoting nuclear energy had under its wing the nuclear safety regulator; frequent personnel exchange and revolving doors between NISA and METI called into doubt the transparency of nuclear power-promoting officials working within a nuclear safety regulatory body.

As a result of lack of governance among regulators and operators, previous signs of the vulnerability of the Fukushima Daiichi nuclear plant were ignored (Noggerath et al. 2011; The National Diet of Japan 2012). Regulators' experiments suggested that tsunami waves could easily topple the Fukushima Daiichi seawall, which was designed to withstand just 5.7 m compared to the 15-m-high waves that flooded the facility. Seismologist Katsuhiko Ishibashi and palaeontologist Koji Minoura had warned lawmakers that power plants in Japan were not prepared to endure earthquake and tsunami worst-case scenarios. However, despite having had abundant opportunities for revising safety measures, the incumbent parties did not reassess risk for nuclear power plants (Hasegawa 2012; Figueroa 2013). In 2012, the chairperson of the NSC Haruki Madarame admitted in a testimony to the Diet that the country's regulations were flawed, outdated and did not meet global standards. He stated that Japanese safety requirements were too loose and many officials have looked the other way trying to avoid changes: "We ended up wasting our time looking for excuses that these measures are not needed in Japan" Madarame said (cited in Kingston 2012a).

On March 12, Madarame accompanied then prime minister Naoto Kan to the Fukushima No. 1 plant to check the situation onsite, just hours before a series of explosions left the nuclear complex severely crippled. It was later reported that Kan lost trust in Madarame after the latter told him that explosions at the rectors were unlikely.

These structural flaws in Japan's regulatory administration created a set of bodies that prioritized their own institutional wellbeing over public safety. Such attitudes were the opposite of the mindset required for good nuclear governance and oversight.

After 3.11, discussions about the urgent need to implement governance started to be held in meetings at the City Hall in Fukushima City. The last recorded video (to this date) of the regular meeting plenary session of March 12, 2014,[12] shows how speakers underscore the lack of governance and call for transparency, open access to information that affects citizens' lives and other forms of co-participatory action, including the establishment of strategic alliances with NGOs, NPOs and other non-governmental actors.

13.8 Poor Risk Communication

Risk communication has been defined as "actions, words, and other interactions that incorporate and respect the perceptions of the information recipients, intended to help people make more informed decisions about threats to their health and safety" (Ropeik 2008, 59). Among the main functions of risk communication are the creation of confidence in institutions responsible for the assessment and management of risk and the involvement in risk-related decisions and conflict resolution (Renn 2008, 203). Although at the core of current conceptions of risk communication is an emphasis on the interactive process of exchange of information and opinions among individuals, groups and institutions, this process does not always take place in a twofold manner but often is, in the tradition of the sender-receiver model, a one-sided action. Furthermore, it has been argued that risk communication is in most cases a fundamentally asymmetrical relationship between unequal parties (Hayenhjelm 2006). This view can be applied to Japan where nuclear risk communication has taken the form of a one-way, source-to-target process. Rather than encouraging an open dialogue, information regarding nuclear energy has served the purpose of inducing public acceptance of nuclear power.

Prior to Fukushima, discussions about the need of improving risk communication were held in 2005 and 2006 at regulatory meetings. However, the function of risk communication was generally reduced to persuasion: "The use of risk information is also a task of the NSC (...) and better *risk communication for public understanding*" (NSC 2006, emphasis added).

Risk communication failures during the Fukushima disaster were multiple and profound. They include: the way the government, the regulators and TEPCO dealt with communicating to people the venting of the ailing reactors, the unclear evacuation orders to the local residents and downplaying for months the seriousness of the crisis (Figueroa 2013). The above resulted in ongoing confusion and deep mistrust towards the government and the nuclear industry. A survey conducted 5 months after the catastrophe showed that 60% of citizens perceived the central government as the least reliable source of information after a disaster (Samuels 2013).

In terms of risk communication, the government's biggest fault was its reluctance to share with the public the discussion of possible worst-case scenarios

[12] http://www.city.fukushima.fukushima.jp/site/gikai/13.html (Accessed on March 25, 2014)

(Sandman 2004, 2011). On March 25, 2011, then prime minister Naoto Kan received a report produced by the Japan Atomic Energy Commission that contemplated the possibility of having to evacuate all people living within 255 km of the Fukushima plant—an area including greater Tokyo (population 35 million) and the cities of Sendai and Fukushima (Japan Times September 19, 2011a). The report was buried and treated as if it never existed under fears that it would cause panic among citizens (Quintana 2012). Yet, risk communication experts agree that panic is rare (Fischhoff 2011; Sandman 2011). Experience suggests that people tend to feel reassured during a serious crisis if they are told honestly and responsibly about a possible worst-case scenario, and not the opposite.

The Fukushima disaster did not alter the nuclear industry's unilateral views: even in the wake of the accident, power companies rallied for a nuclear future despite strong public opposition. In November 2011, TEPCO's ex-president Mr. Toshio Nishizawa told the *Financial Times* that the utility saw the triple meltdowns as only a temporary setback: "The sooner we get back to normal, the better. Whatever happens, our core business will not change" (Financial Times 2011/11/15). TEPCO's ex-president's words show there is a contradiction between the mindset of the company's executives and the public image TEPCO wants to project. The company's website states that "TEPCO strongly wishes to be a nuclear power plant operator which has the world's highest level of safety awareness, engineering capabilities and <u>risk communication ability</u> with society" (TEPCO website, emphasis added).

In addition, a recent statement by the president of the Japan Atomic Energy Forum Mr. Takuya Hattori claims that energy supply without nuclear power after Fukushima has been possible due to the operation of thermal plants, which he sees as "old" and "deteriorated". He describes the enterprise as very demanding—"like walking on a tightrope"—and argues for the restart of nuclear power plants in order to provide "good-quality electricity, *safe,* affordable and stable". Mr. Hattori argues:

> As part of the effort, it is important to steadily implement work to assure conformity with the new safety standards and to restart NPPs whose safety has been confirmed. In this respect, we ask the national government, the regulatory authority and nuclear operators strive to improve transparency in those processes and <u>give thorough explanations of the results to the people</u> in order to <u>obtain understanding</u> from society. Being part of the nuclear industry, we, the Japan Atomic Energy Forum, Inc., will endeavor to <u>make people understand the necessity</u> of nuclear energy and to share information on the industry's safety-improvement activities. (Japan Atomic Energy Forum website 2013/09/17, emphasis added)

All the above suggests that in the nuclear industry and the regulators' views, the role of the public is reduced to *understanding*. Thus, people are required to "understand" and "accept" nuclear power without an open discussion of the risks involved. This attitude has permeated risk communication in Japan for decades along with the denial of nuclear risk. Such denial became a serious hindrance for risk communication of the Fukushima disaster: risks that are not previously acknowledged cannot be communicated effectively during a nuclear crisis.[13]

---

[13] As discussed above, there had been numerous red flags regarding the vulnerability of Fukushima Daiichi, especially in regard to an unsubstantial seawall and the high risk of being struck by a

13.9 Lessons Unlearned

In order not to repeat the same mistakes that led to the Fukushima disaster, it is imperative to consider how decisions were made. First and foremost, given the cosy relationship between the regulators and the nuclear industry that existed prior to the accident, the autonomy of nuclear regulatory organizations must be strengthened. This was partly achieved by the creation in 2012 of the Nuclear Regulatory Agency (NRA) (*Genshiryoku Kisei Iinkai*), which is an extra-ministerial department of the Ministry of Environment (previously, the Nuclear and Industrial Safety Agency [NISA] had been placed under the Ministry of Economy, Trade and Industry [METI]). However, although the NRA is presided by a chairman and five commissioners who are more independent compared to the previous regulatory agency, its secretariat—which is in charge of implementing all practical measures of inspection, regulation and safety management—is staffed with almost the same members that previously composed the body of NISA. It remains to be seen how the new regulatory organization NRA performs its duty of public safety and whether or not the secretariat will be able to change its mindset.

A second (unlearned) lesson in nuclear risk governance is that people's perceptions of the nuclear issue *do* matter for proper risk management, risk assessment and risk communication. Governance is meant to stress the importance of having stakeholders and the public participate in risk participation processes (Renn 2008, 273). This includes the empowering of all actors to participate constructively in the discourse and co-designing the frame for the risk issue with multiple groups. Although people's opinions were momentarily taken into account in 2011 by ex-prime minister Naoto Kan,[14] the 2012 landslide victory of the conservative Liberal Democratic Party (LDP)[15] put current prime minister Shinzo Abe in control of the national energy policies. Mr. Abe has expressed many times that he is in favour of restarting the reactors if their safety can be assured. This change in government was widely seen as a sharp shift to the right. Indeed, the LDP has erased from the political landscape any advances towards a dialogue with the public that may have been achieved in the previous administration. Despite public opposition, the nuclear industry is pushing hard to get back to business quickly, as evidenced by five nuclear operators that hastily filed for checks under the post-Fukushima new safety rules. Early approval, however, seems unlikely given the widespread mistrust towards the nuclear industry. What is more, even if the safety checks are validated by the regulator, the consent of the local communities hosting the plant is needed. Many local residents throughout the country are against the restart of reactors at this point.

massive earthquake. Despite having had recurrent opportunities for reviewing safety measures, the institutions in charge of public safety did not reassess earthquake and tsunami risk for nuclear power plants.

[14] Kan announced plans for zero nuclear power in Japan by 2030, which some described as a last ditch attempt to remain in office.

[15] With the exception of a short period between 2009 and 2012, the LDP has been in power since 1955.

A third (unlearned) lesson has to do with transparency. The crucial need for greater transparency in nuclear risk assessment, risk management and risk communication has been widely argued at local, national and global levels. As discussed above, Japan's political culture has been secretive and opaque. Many expected that Fukushima would provide a "window for change" and that information concerning public safety would be made more readily accessible for public consultation and debate. In this regard, although there is a new tendency towards disclosing information through the online posting of videos and transcripts of official meetings, some of the sessions are designated as "private" (*Hikōkai*) and are not released for public viewing. What happens during those meetings remains a secret. Furthermore, the government's recent passing of a state secret protection law concerning foreign and defence policies has caused uproar among citizens, academics and journalists. The bill was approved despite widespread opposition, and it punishes the leak of state secrets. The penalties for violators are heavy: 10 years in prison for civil servants who leak classified information and 5 years for citizens convicted of encouraging leaks. The law covers defence, diplomacy, counterterrorism and counterintelligence. But it also empowers bureaucrats throughout government to lock away documents for up to 60 years and even destroy those documents before the marked timeline for release. The media is especially concerned because they believe this law goes against the journalistic freedom and people's right to know. Moreover, critics say nuclear policy (and nuclear oversight) is particularly vulnerable to abuse under the category of "state secret" because all nuclear-related information could be withheld if considered a matter of national security, e.g. to prevent terrorism or to protect the so-called professional secrecy (*shokugyō-jō no himitsu*) or intellectual property that allows manufacturers to keep technical data undisclosed. In any case, it is difficult not to see how the passing of this bill may severely undermine transparency in nuclear energy policy and decision-making.

A fourth (unlearned) lesson refers to citizen participation. In a rare display of anger, the Japanese public showed their opposition to nuclear power in the wake of Fukushima. But the anti-nuclear movement was not able to capitalize on this extraordinary set of circumstances to renegotiate their role in the broader political arena. Although visually striking, the massive demonstrations did not apply enough political pressure to produce significant changes in energy policy. Growing as a political force would have implied clearer action strategies, negotiating with existing political parties and the creation of transnational networks. At the moment, public interest in the nuclear issue appears to have decreased, and the rallies have weakened. Leaders of anti-nuclear citizen groups reported that people feel disoriented because they do not know how to pass across their message. With the LDP's shift to the right and the strengthening of pronuclear policies, anti-nuclear groups face the challenge of reorienting their goals in a less than promising context.

A fifth lesson (perhaps learned by the people and organizations more directly affected by the 2011 triple disasters) is the need for greater resilience during disaster times. With a death toll of over 16,000 and more than 150,000 displaced, large numbers of Tohoku residents were critically affected by the tsunami and the nuclear catastrophe. In the wake of crisis, local networks and small-scale organizations proved enormously dynamic in offering first aid, rescue and relief to the victims. On the other hand, the prefectural and national governments were widely criticized for their ineffective and cumbersome response. It is to be expected that valuable lessons for resilience such as multi-organizational collaboration, the increase of preparedness activities, the improvement of regional warning systems and a shift from a centralized hierarchical control mechanism to a more balanced contributive governance processes (Kapucu et al. 2010) will be learned throughout Tohoku by local governments, communities, NGOs and other decentralized disaster response organizations. To pass on this know-how to other regions of Japan and the world constitutes an important institutional challenge and a matter of political responsibility.

However, the lack of a unified approach towards tackling the problem of social mistrust of the nuclear industry, along with the deep uncertainty surrounding the reconstruction of the devastated areas in Tohoku, remediation and cleanup of the environment, and the safe operation of nuclear power plants in a highly seismic country may result in undermined resilience throughout Japan should another nuclear disaster strike. It is imperative to have a more profound theoretical, political and practical consideration of the abovementioned issues in order to promote nuclear risk governance, supervision and resilience to disaster.

13.10 Conclusions

In this chapter I have aimed to analyse nuclear risk governance in Japan surrounding the Fukushima disaster while paying attention to the urban dimension of the catastrophe. After interrogating the historical and political context that paved the road to the worst nuclear accident in recent times, I discussed institutional weaknesses in safety assessments and described conflicts that characterized the expansion of nuclear power. Lack of transparency, lack of governance and poor risk communication emerged as salient issues affecting disaster prevention and response. My contention is that in order not to repeat the same mistakes that led to the Fukushima disaster, important lessons in risk governance have yet to be learned.

Several conclusions can be drawn from the preceding pages. The first one is that the Fukushima nuclear accident was not a natural catastrophe but rather a man-made disaster with foreseeable causes. Previous red flags about technical and design issues were systematically ignored. Collusion between the government, the regulator and the nuclear industry played a central role in the making of the accident. State politics of nuclear power were based on rapid expansion by eschewing local opposition; citizens' views were dismissed as irrational and uninformed. The aggressive growth of Japan's nuclear programme was achieved through a set of tactics that included turning a blind eye to potential hazards; secrecy in the construction, operation and regulation of nuclear energy; and a carefully crafted state myth of technological safety. Combined, all these aspects exponentially increased the probability of a nuclear disaster in an earthquake-prone country.

A detailed examination of the processes and mechanisms that have characterized nuclear risk governance in Japan reveals that public opposition to nuclear energy has been a constant and powerful force throughout time. With varying degrees of intensity, citizens have disputed narratives of safety in the use of nuclear energy. Poor risk communication of the Fukushima disaster compounded psychological damages in the population by generating deep mistrust of the government; risks were denied and the severity of the crisis downplayed. In the nuclear village's mind-set, the role of the public is to accept atomic energy, without a serious and open discussion of the risks involved. This attitude was repeatedly exemplified in the government's withholding of vital information regarding radiation levels and the reluctance to openly discuss worst-case scenarios during the accident.

Furthermore, lack of transparency and lack of governance in decision-making processes affecting nuclear energy stand out on their own. Secrecy in the construction plans, cover-ups, falsification of safety tests and concealment of failures of power plants have been orchestrated by the electric utilities in connivance with the regulators and, ultimately, the government. These structural flaws in nuclear oversight go against the core principles of regulatory culture and good governance.

Among the lessons that need to be learned are the importance of taking into account people's perceptions of risk and facilitating citizen participation. Governance is meant to stress the importance of engaging the public and a variety of stakeholders in risk participation processes. In this context, co-designing the frame for risk and empowering all actors to actively participate in the discourse is crucial. This element has been missing in the management, assessment and communication of nuclear risk, before and after 3.11.

If resilience and disaster prevention are to be taken seriously, the role of cities (especially Tokyo) in risk governance needs to be more clearly acknowledged and addressed. It is only through the sustained effort in strengthening nuclear risk governance that a similar catastrophe can be prevented.

References

Aldrich, D. (2012a). *Post-crisis Japanese nuclear policy: From top-down directives to bottom-up activism.* http://www.eastwestcenter.org/sites/default/files/private/api103.pdf. Accessed 21 Feb 2012.

Aldrich, D. (2012b). *Networks of power: Institutions and local residents in post-Tohoku Japan. Natural disaster and nuclear crisis in Japan.* New York: Routledge.

Asahi Shimbun. (2011, September 15). http://ajw.asahi.com/article/0311disaster/recovery/AJ2011091510558. Accessed 20 June 2014.

Asahi Shimbun. (2012a, January 26). http://ajw.asahi.com/article/views/editorial/AJ201201260056. Accessed 20 June 2014.

Asahi Shimbun. (2012b, August 23). http://ajw.asahi.com/article/0311disaster/AJ201208230074. Accessed 20 June 2014.

BBC Online. (2012, January 27). http://www.bbc.co.uk/news/world-asia-16754891. Accessed 20 June 2014.

Bowen, R. (2003). *Japan's dysfunctional democracy: The Liberal Democratic Party and structural corruption.* New York: M.E. Sharpe.

Carpenter, S. (2012). *Japan's nuclear crisis: The routes to responsibility*. New York: Palgrave Macmillan.

Douglass, M. (2013). *The urban transition of environmental disaster governance in Asia*. ARI working paper no. 210. Singapore: Asia Research Institute, National University of Singapore.

Figueroa, P. (2013). Risk communication surrounding the Fukushima nuclear disaster: An anthropological approach. *Asia Europe Journal*, 11, 53–64.

Financial Times. (2011, November 15). http://www.ft.com/intl/cms/s/0/a37b37ca-0f7e-11e1-88cc-00144feabdc0.html#axzz35A8bBrgQ. Accessed 20 June 2014.

Fischhoff, B (2011). *The emotions of nuclear experts*. http://www.thebulletin.org/web-edition/features/the-emotions-of-nuclear-experts. Accessed 20 Feb 2012.

Green Action Japan. (2011). http://fukushima.greenaction-japan.com/2011/05/13/petition/. Accessed 20 June 2014.

Gundersen, A. (2012). The echo chamber: Regulatory capture and the Fukushima Daiichi disaster. *Greenpeace International*. http://www.greenpeace.org. Accessed 20 June 2014.

Hara, T. (2013). Social shaping of nuclear safety: Before and after the disaster. In R. Hindmarsh (Ed.), *Nuclear disaster at Fukushima Daiichi: Social, political and environmental issues* (pp. 41–56). London: Routledge.

Hasegawa, K. (1995). A comparative study of social movements in the post-nuclear energy era in Japan and the United States. *International Journal of Japanese Sociology*, 4, 21–36.

Hasegawa, K. (2012). *Toward a post-nuclear society: Examining the 3/11 disaster and nuclear risks*. Paper presented at the symposium *Towards Long-Term Sustainability: In Response to the 3/11 Earthquake and the Fukushima Nuclear Disaster*. The Center for Japanese Studies, University of California, Berkeley, April 20–21, 2012.

Hayenhjelm, M. (2006). Asymmetries in risk communication. *Risk Management*, 8(1), 1–15. http://www.jstor.org/discover/10.2307/3867940?uid=3738992&uid=2&uid=4&sid=21104144282567. Accessed 20 Oct 2011.

Japan Atomic Energy Forum website. http://www.jaif.or.jp/english/news_images/.../ENGNEWS02_1379483625P.pdf. Accessed 20 June 2014.

Japan Times. (2011a, September 19). http://www.japantimes.co.jp/news/2011/09/19/national/tokyo-faced-evacuation-scenario-kan/. Accessed 20 June 2014.

Japan Times. (2011b, November 28). http://www.japantimes.co.jp/opinion/2011/11/28/editorials/rigging-opinions-on-nuclear-power/. Accessed 20 June 2014.

Japan Times. (2013, May 13). http://www.japantimes.co.jp/news/2013/05/15/reference/monju-generating-only-misfortune/. Accessed 20 June 2014.

Juraku, K. (2013). Social structure and nuclear power siting problems revealed. In R. Hindmarsh (Ed.), *Nuclear disaster at Fukushima Daiichi: Social, political and environmental issues* (pp. 41–56). London: Routledge.

Kapucu, N., Arslan, T., & Coolins, M. (2010). Examining intergovernmental and interorganizational response to catastrophic disasters: Toward a network-centered approach. *Administration and Society*, 42(2), 222–247.

Kingston, J. (2004). *Japan's quiet transformation: Social change and civil society in the twenty-first century*. New York: RoutledgeCurzon.

Kingston, J. (2012a). Mismanaging risk and the Fukushima nuclear crisis. *The Asia-Pacific Journal: Japan Focus*, 10(12, No. 4), 1–12.

Kingston, J. (2012b). Japan's nuclear village. *The Asia-Pacific Journal: Japan Focus*, 10 (37, No. 1), 1–11.

Kondoh, K. (2007). The challenge of climate change and energy policies for building a sustainable society in Japan. *Organization and Environment,* 22(1), 52–74, Retrieved July 28, 2011, from http://oae.sagepub.com/content/22/1/52.full.pdf+html. Accessed 28 July 2011.

Kondoh, K. (2009). The challenge of climate change and energy policies for building a sustainable society in Japan. *Organization & Environment*, 22(1), 52–74.

Kotler, M. L., & Hillman, I. T. (2000). *Japanese nuclear energy policy and public opinion*. http://bakerinstitute.org/research/japanese-nuclear-energy-policy-and-public-opinion/. Accessed 20 June 2014.

Lesbirel, S. (1998). *NIMBY politics in Japan: Energy siting and the management of environmental conflict*. Ithaca: Cornell University Press.
NHK Poll. (2011). http://www.nhk.or.jp/bunken/yoron/social/page_02.html. Accessed 20 June 2014.
NHK Special Documentary. (2011). *Japan's nuclear crisis*. http://www.nippon-sekai.com/main/articles/fukushima-daiichi-nuclear-power-plant-crisis/nhk-special-japans-nuclear-crisis-part-1/. Accessed 2 Nov 2011.
Noggerath, J., Geller, R., & Gusiakov, K. (2011). Fukushima: The myth of safety, the reality of geoscience. *Bulletin of the Atomic Scientists*, 67(5), 37–46. http://bos.sagepub.com/content/67/5/37.full.pdf+html. Accessed 13 Oct 2011.
Nuclear Safety Commission, Japan (NSC). (2006). *Hakusho*. http://www.nsr.go.jp/archive/nsc/NSCenglish/.../hakusho2006_part2-conclu.pdf. Accessed 20 June 2014.
Ogawa, A. (2013). Demanding a safer tomorrow: Japan's anti-nuclear rallies in the summer of 2012. *Anthropology Today*, 29(1), 21–24.
Onitsuka, H. (2012). Hooked on nuclear power: Japanese state-local relations and the vicious cycle of nuclear dependence. *The Asia-Pacific Journal* 10 (3 No 1), January 16.
Quintana, M. (2012). *Fukushima crisis concealed: Japanese government kept worst-case scenario under wraps*. http://japanfocus.org/events/view/129. Accessed 8 Feb 2012.
Ramana, M. V. (2011). *Beyond our imagination: Fukushima and the problem of assessing risk*. http://thebulletin.org/web-edition/features/beyond-ourimagination-fukushima-and-the-problemof-assessing-risk. Accessed 20 June 2014.
Renn, O. (2008). *Risk governance: Coping with uncertainty in a complex world*. London: Earthscan.
Ropeik, D. (2008). Risk communication: More than facts and feelings. *IAEA Bulletin*, 50(1). http://www.iaea.org/Publications/Magazines/Bulletin/Bull501/Risk_Communication.html. Accessed 22 Feb 2014.
Samuels, R. (2013). *3.11. Disaster and change in Japan*. Ithaca: Cornell University Press.
Sandman, P. (2004). *Worst case scenarios*. http://www.psandman.com/col/birdflu.htm. Accessed 20 Feb 2012.
Sandman, P. (2011). *Interview on the futurist*. http://www.wfs.org/content/futurist-interviews-crisis-communications-expert-peter-sandman-fukushima-daiichi-nuclear-mel. Accessed 20 Feb 2012.
Tabusa, K. (1992). *Nuclear politics: Exploring the nexus between citizens' movements and public policy in Japan*. PhD dissertation, Columbia University, New York.
TEPCO website. http://www.tepco.co.jp/en/nu_reform/index-e.html. Accessed 20 June 2014.
The National Diet of Japan (2012). *The official report of the Fukushima nuclear accident independent investigation commission*. Tokyo: The National Diet of Japan.
World Nuclear News. (2013, May 20). http://www.world-nuclear-news.org/c-jaea_head_resigns_following_monju_lapses-2005134.html. Accessed 20 June 2014.

Index

A
Accountability, 27, 28, 179, 191, 198
Aceh. *See also* Free Aceh Movement (Gerakan Aceh Merdeka (GAM)); Law on Governing Aceh (LoGA)
  earthquake, 50
  independence, 92
  peace process, 88
  tsunami, 85–104
Adaptation social, 3
Agency for Natural Resources and Energy (ANRE), 268, 273
Agglomeration effects, 15–19, 23, 24
Aging, 2, 21, 147, 229, 231, 248–250, 276
Agno river
  Integrated Irrigation Project (ARIIP), 158
  Irrigation System Extension Project (ARISEP), 158
Agriculture, 15, 56, 131, 138, 151–153, 156, 160, 184, 199, 201, 215, 251, 253, 257
Ambayaoan-Dipalo Irrigation System (ADRIS), 158
Ambon, 46, 52, 57
Andhra Pradesh, 130, 131, 136
Anthropocene age, 2
Anthropology, 128, 150, 266
Aquino, President, 186, 190
ASEAN economic community, 207
Aseismic, 52
Ash deposits, 51
Asia Development Bank (ADB), 14, 15, 23–26, 35, 37, 177, 178, 238, 251
Asian Disaster Reduction Center (ADRC), 50
Atomic energy
  agency, japan, 266–268, 277
  agency law, 266
  commission, Japan, 267, 276
  forum, Japan, 266, 276
  JAERI, 266
Atomic Fuel Corporation, 266
Atomic Industrial Forum, Japan (JAIF), 266
Atomic Power Company, Japan (JAPCO), 266
Australian Government, 179, 181, 187–188
Australian Mekong Resource Centre, Ayeyarwaddy Delta, 240, 244, 251

B
Baan Mankong, 30, 198
Babad Ing Sangkala, 54
Badan Rehabilitasi dan Rekonstruksi (BRR), 91, 93, 94
Baguio city, 150
Bali, 48, 49, 51
Bam earthquake, 237, 240, 242, 244, 245, 247–250, 258
Banda Aceh, 53, 57, 89, 91, 92, 94, 104
Bandung, 32, 56
Bangkok, 9, 17, 25, 30, 46, 47, 195–207
Bangkok disaster, 198–199
Bangus farming, 159–160, 172, 174
Barangay
  Disaster Risk Reduction and Management Council (BDRRMC), 148–151, 153, 181, 183, 187, 188
  Disaster Risk Reduction and Management Plan (BDRRMP), 189

M.A. Miller, M. Douglass (eds.), *Disaster Governance in Urbanising Asia*, DOI 10.1007/978-981-287-649-2

Batavia, 53, 55, 56
Bayaoas River, 154
Beichuan, 214, 218–220, 239, 240, 243, 246, 247, 258
Beijing, 17, 19, 25, 31
Bengal, 7, 46, 111, 113–116, 119, 135, 136, 146, 244
Bhopal Gas Leak, 128
Bhuj earthquake, 141
Bogale, 239, 243, 244, 249, 251, 253–258
Borobudur, 52, 54, 55
Buddhism temples, 54
Bued-Angalacan River, 154
Bureaucracy, 56, 116, 191, 199, 200, 268, 272
Burma. *See* Myanmar

C
Cak Irsyad's community, 75, 78
Calmay River, 151, 154, 158, 170
Canals, 20, 67
Capacities, 2, 7–10, 15, 16, 18, 21, 24, 27, 28, 33, 34, 38, 135, 139, 148, 149, 161, 162, 173, 181, 192, 213, 214, 219, 227, 229–232, 241
Capitalism, 47, 197, 199
Cascading accidents, 264
Catastrophes, 128, 190, 191, 222, 266, 269
Central Java, 51–54
Centre for Peace and Conflict Studies (CPCS), 244, 254
Centre for Research on the Epidemiology of Disasters (CRED), 146
Chao Phraya River Basin, 205
Chengdu, 17, 218, 241, 247
Chennai, 17, 136, 137
Chiang Mai, 56
Chi-Chi earthquake, 242
China urbanisation, 4
Chongqing, 17, 212
Cilacap, 55, 57
Cities
  coastal, 6, 19, 45, 47, 56, 57, 91, 96, 131, 135, 136, 146, 151, 159, 258
  compact, 31
  depopulating, 22
  eco, 31
  global, 15
  green, 31
  hill, 114, 116, 120, 121
  linear, 31
  port, 24
  private, 22, 240
  resilient, 27, 112, 120
  secondary, 7, 8
  smart, 31
City Agriculture Office (CAO), 153, 160
City Disaster Risk Reduction and Management Council, 153
City Region, 6, 14, 15, 19, 21, 27, 31–33, 37, 38
Civil Defence Act of 1979, 200
Civil society, 10, 18, 27, 35, 38, 87, 89, 90, 111, 117, 122, 123, 164, 177–192, 226, 232, 240, 250, 264
Civil society organisations, 87, 177–192
Climate
  change, 1, 6, 10, 11, 14, 15, 18–20, 24, 34, 51, 110, 114, 151, 154, 160, 162, 164, 172, 174, 182, 183, 187, 188, 196, 198, 200, 206, 214, 238
  change Act, 151, 183
  research, 57
Cluster Coordinated Development Program (CCDP), 32
Coastal areas, 7, 19–21, 31, 89, 93, 95, 96, 98, 128, 132, 240, 244
Coastlines, 27, 89, 94, 99
Collaborative governance, 179, 191
Community
  disaster recovery, 30, 91, 104, 128, 136, 230, 239, 240, 242, 251
  Disaster risk reduction and management (CBDRRM), 189
  kitchen approach, 257
  rural, 240, 243, 258
Compensation, 6, 63, 66–68, 70–73, 76, 78–80, 132, 227
Compound Disasters, 10, 11, 15, 23–26, 33, 36, 115, 239, 242, 258
Conflict
  armed, 86, 89, 91, 99
  social, 7
  violent, 86–90, 92, 93, 95, 97, 101, 102, 104
Conflict resolution, 35, 275
Conservation, environmental, 8
Contestation, 3, 10, 17, 28
Contingency plans (CPs), 189, 214, 216, 217, 220–222, 228
Corruption, 29, 90, 91, 97, 202, 204, 205, 243
Cotabato City, 146
Crisis
  leadership, 229
  management, 93, 133, 229, 233
Crisis Management Initiative (CMI), 93
Crop Failures, 51
Cross-cultural contexts, 239, 240

Cuddalore, 137, 139
Cultural capital, 10
Culture material, 171
Cyclical pattern, 49
Cyclone Warning Centre, 136
Cyclonic storm, 114, 118, 136

D

Dagupan City Coastal Fisheries Resources Management Ordinance of 2003, 159, 160
Damage and Loss Assessment (DaLA), 251
Dams, 14, 19, 27, 34, 35, 37, 38, 153, 154, 200, 201, 213
Darjeeling, 111, 113–119, 121–123
Darjeeling District of West Bengal, 111, 113–119, 121–123
Dawel River, 154
Deaths, 46, 51, 52, 57, 131, 132, 146, 198, 244, 248
Decentralisation, 94, 242
Deforestation, 14, 24, 31, 115, 200, 244
Degradation, environmental, 30, 117, 200
Delhi, 17, 25, 56, 117, 130, 134, 139
Deltas, 6, 7, 153, 154
Democratic accommodation, 9
Democratisation, 9, 195–207
Demography, 47
Depopulation rural, 18
Development
  community-driven, 30
  sectors, 28
  sustainable, 28, 183, 212, 214, 228, 233
Disabilities, 159, 162, 189
Disaster
  environmental, 2–8, 10, 13–16, 18–21, 23–28, 30, 35–38, 63–65, 86–90, 94, 96, 97, 104
  escalating, 63
  history, 156–157
  Incubation, 23–26
  incubation, 23–26, 267
  literature, 110, 123
  mitigation, 63, 66, 129
  nuclear, 269, 279
  planning, 29
  prevention, 29, 200, 231, 266, 279, 280
  reconstruction, 239
  relief, 18, 29, 226
  responses, 2
  risk management, 29, 110, 111, 116, 117, 119, 122, 123, 213–219, 221, 230, 231, 233
  victims, 16, 80
  vulnerability, 64, 178, 179, 191, 192
  zone, 70, 243, 264
Disaster Administration Theory, 4
Disaster Capitalism, 197
Disaster Governance
  capacities, 8, 28
  functions, 7
  model, 10
  participatory, 15
  programs, 8, 11
  regimes, 3, 10
Disaster Management (DRDM), 130, 131, 133, 137–139
Disaster Management Act (DMA), 129, 133, 140
Disaster Management Expert-centered, 139
Disaster Prevention and Mitigation Act 2007, 200
Disaster Risk management (DRM), 213–217, 219, 221–223, 225–232
Disaster Risk Reduction (DRR), 182–184, 188, 189, 242, 244
Disaster Risk Reduction Network Philippines (DRRNet Philippines), 183, 188–192
Disempowerment, 64–66, 76, 191
Displacement Monitoring Centre, 14
Diverse Urbanising Societies, 5–11
Dongguan, 17
Drain, 153
Drilling, 63, 65, 73, 74
Dykes, 48

E

Early warning, 148, 163–166, 170, 172, 188, 189, 214, 216, 221, 223, 228, 229, 232
Earth, 26, 52, 114, 247
Earthquake, 10, 204. *See also* Tsunami
  historic, 52
  resistance, 61–81, 226, 270, 272
  tsunami-generating, 49, 57
Earth Systems, 2, 3
East Asia, 20, 29
Ecology
  impact effects, 26–27
  local, 26
Economy
  dependent, 22
  depressed, 22
  global, 204
  local, 33, 96, 248
Education and training, 30, 214, 221, 226, 228, 232

EM-DAT database, 128
Emergency
management agency, 220
management system, 216, 220
mass, 134
response, 134, 135, 188, 214, 216, 221, 223–224, 228, 239, 240
Empowerment, 61–81, 187
Energy policies, 266, 277
Entrepreneurial vendors, 81
Environment(al)
change, 26
conservation, 8
degradation, 30, 117, 200
determinist, 3
disasters, 2–8, 10, 14–16, 18, 19, 21, 24–27, 30, 35, 37, 38, 63, 65, 86–88, 104
hazard, 23, 118, 149, 173
justice, 28, 30, 38
learning, 7, 111, 118–120
risks, 1, 112, 123, 149
transitions, 14
Eruptions, 46, 48, 50–52, 62, 63, 65–68, 71, 76, 146
Ethnicity, 64, 117, 157
Evacuation, 47, 57, 65, 133, 137, 150, 164, 166–168, 185, 252, 254, 256, 275

F
Factories, 70, 72, 196, 201–203, 206, 253
Families, 21, 30, 62, 76, 88, 92, 98, 103, 131, 132, 137, 146, 152, 153, 158, 177–192, 227, 242, 246, 253–257
Family policy, 245–258
Federal Parliamentary Constitutional Republic of India, 128
Fire Defense Act of 1999, 168, 200
Fishermen, 22, 132, 135, 136, 151, 251, 269
Flood
control, 20, 165, 184, 1987
management, 20, 184, 188, 196, 197, 203, 205
preparedness, 30
repeated, 30
risk management, 184
Flood Relief Operations Centre (FROC), 201, 202
Food security, 20, 251, 253, 257
Free Aceh Movement (Gerakan Aceh Merdeka (GAM)), 86
French India, 130
Fukushima
daiichi power plant nuclear accident, 10
disaster, 265, 270–272, 275–277, 279, 280
Nuclear Accident Independent Investigation Commission, 265, 270
plant, 264, 272, 276
triple disaster, 263–280

G
Gender, 64, 157, 248, 255
Geographical scales, 6, 29, 38
Geological disasters, 50, 222, 223, 239
Gerakan Aceh Merdeka (GAM). *See* Free Aceh Movement
Global assessment report, 178
Global climate change, 1, 6, 11, 14, 18–20, 24, 33, 51, 114, 196, 198, 200
Global supply chains, 25, 196
Global urban population, 2, 112
Global Water Partnership (GWP), 35
Governance
across-border, 38
capacity, 7, 117, 123
decentralized, 279
effective, 2, 5, 123, 202
environmental, 28
public, 22
regimes, 2–5, 9, 10, 110, 239
responsive, 8
unified, 31
universalized, 4
urban, 2, 9, 10, 28, 61, 110, 117, 150, 195–207, 240–245
Government agencies, 9, 90, 91, 186, 188, 192, 215, 231, 253
Great East Japan earthquake, 22, 23, 187, 264
Great Hanshin-Awaji Earthquake, 26
Groundwater, 24, 31, 32, 70, 198
Guangzhou, 17, 24
Gujarat earthquake, 128, 242

H
Hangzhou, 17
Hanoi, 31, 32
Hazards, 1, 13, 19, 23, 27, 30, 33, 63, 87, 110, 114–120, 123, 124, 128, 129, 131, 133, 137, 146, 148–150, 157, 159, 160, 162, 171, 173, 174, 179, 181–183, 185, 187, 198, 214, 215, 222, 223, 229, 238, 241, 249, 267, 269, 279

Health, 10, 15, 33, 56, 76, 110, 133, 162, 198, 212, 238, 245, 246, 248, 249, 251, 252, 256–258, 269, 275
HelpAge International disaster response organisations, 249
Heritage, 247
Himalayas, 114
Hokkaido Electric Power Co., 273
Hong Kong, 17, 18, 24
Housing
  development schemes, 191
  social, 188
Housing and Urban Development Coordinating Council (HUDCC), 146
Human flourishing, 66
Hydrometeorological hazards, 179, 182, 185
Hyogo Framework for Action (HFA), 241

I
Iceland, 23
India, 4, 7, 8, 17, 20, 24, 111–115, 118, 124, 127–132, 134–136, 139–142, 243
Indian National Centre for Ocean Information Services (INCOIS), 136
Indian Ocean Tsunami Response, 132
Indian Vulnerability Atlas, 128
Indonesia revolution, 50
Industrialisation, 1
Informal settlements, 2, 8, 20, 192
Informal settler families, 146, 153, 158, 177–192
Infrastructure
  improvements, 3
  physical, 2, 3, 8, 246
Institute of Philippine Culture (IPC), 181
Institutions, 6, 27, 28, 30, 36–38, 65, 80, 110, 116–118, 120, 122, 123, 129, 130, 134, 142, 149, 157, 178, 183, 188, 197, 199, 201–207, 214–217, 220–222, 228, 232, 242, 265, 275, 277
Insurance companies, 10, 66
Integrated Household Living Conditions Survey (IHLC), 251
Integrated Water Resource Management (IWRM), 35, 36
Intergovernmental Panel on Climate Change (IPCC), 114, 214
International Centre for Environmental Management (ICEM), 34
International Decade for Natural Disaster Reduction (IDNDR), 129
International Fund for Agricultural Development (IFAD), 96
International Recovery Forum, 242
Iran, 4, 10, 237–259
Isabela City, 146

J
Jakarta, 17, 19, 25, 32, 33, 45–47, 55, 56, 70, 74, 79, 89, 91–94
Janatha Vimukthi Peramuna (JVP), 100, 101
Japan
  japan nuclear fuel corporation, 266
  japan nuclear risk governance, 263–280
Justice, social, 8, 17

K
Kansai Electric Power Co. (KEPCO), 272, 273
Kashmir earthquake, 128
Kedarnath flood, 128
Kerala, 130–132
Ketsana Tropical Storm, 178–183, 185, 187
Knowledge systems, 1, 8
Kobe, 26
  earthquake, 241
  Great Hanshin-Awaji Earthquake, 26
Kolkata, 17, 24, 25, 115, 116
Korban Lapindo Menggugat (KLM), 73, 77, 79
Krakatau Eruption, 50
Kuala Lumpur, 17, 56

L
Land
  development schemes, 22
  habitable, 26, 159
  hazard-prone, 13
  subsidence, 24, 31, 159
  unstable, 15
Landslide, 2, 7, 19, 33, 67, 95, 111, 114–119, 121, 122, 128, 146, 184, 200, 213, 222, 223, 238, 277
Lapindo
  lapindo brantas, 63, 65, 76
  mudflow disaster, 6, 61–81
Latur earthquake, 128
Law on Governing Aceh (LoGA), 94
Levee. *See* Dykes
Liberal Democratic Party (LDP), 277, 278
Liberation Tigers of Tamil Eelam (LTTE), 86, 95–97, 100–102
Life spaces, 38, 63–65, 71, 79–81
Literacy, 134, 162

Livelihoods adaptive, 6
Local knowledge
framework for understanding, 161–163
identification of, 148
integration of, 148, 149, 173
Logistics management, 214, 221, 224–225
Lombok volcano eruption, 51
London, 25
Lower Agno River Irrigation System (LARIS), 158
Lushan earthquake, 9, 211–233

M
Mahinda Chintana policy, 104
Management, Disaster. *See* Disaster governance
Mandaluyong, 186, 189
Manila, 4, 8, 9, 17, 19, 24, 25, 30, 46, 49, 55, 56, 66
disaster, 145–174, 177–192
Disaster Risk Reduction and Management Council (MMDRRMC), 183, 184
Earthquake Impact Reduction Study (MMEIRS), 182
Integrated Flood Risk Management Master Plan, 184
Post-Ketsana Recovery and Reconstruction Program, 181
Markets, 21, 25, 53, 67, 96, 190, 253
Megacities asian, 7, 47
Mega-Urban regions (MURs), 6, 7, 15–17, 19–24, 31–33, 269
Mekong River Commission (MRC), 35, 36
Mekong River Mekong River Basin, 35, 36
Merapi volcano, 53
Metropolitan authorities, 31
Metta development foundation, 256, 257
Migration rural to urban, 179, 238
Military coup, 197, 201, 204
Millennium Development Goals Fund (MDGF), 182
Ministry of Economy, Trade and Industry (METI), 25, 267, 268, 274, 277
Minorities, 102
Modelling software, 134, 135
Monju nuclear power plan, 268
Monsoon season, 114
Montazeri study, 248
Mount Pinatubo eruption, 46, 48
Mt. Ruapehu earthquake, 242
Mudflow lapindo, 6, 7, 65–66, 74, 81
Mud volcano, 62, 63, 65–72, 76, 77, 79, 81
Multiscalar planning, 29–37
Mumbai, 17, 24, 25
Myanmar, 4, 10, 22, 35, 237–259

N
Nanjing, 17
Nargis cyclone, 46, 243, 249
Nature, 3, 5, 8, 10, 14, 19, 27, 38, 81, 87, 103, 104, 116, 121, 132, 148, 149, 159, 170, 172–174, 183, 192, 197–199, 215, 239, 250, 257, 264
Neighborhoods, 223
Neoliberalism, 204
Nepal, 116
Network
distribution systems, 25
effects, 23–26
production, 26, 264
spatial effects, 24
New Delhi, 130, 139, 142
New York, 25
Non-Governmental Organisations (NGOs), 74, 79, 87, 91, 93, 111, 132, 133, 135, 136, 138, 139, 141, 197, 204, 240, 243–245, 250–258, 266, 275, 279
Non-state actors, 9, 86, 101, 123, 128, 129, 132, 134–137
North Maluku, 52
North Sumatra, 50, 91
Nuclear
crisis, 276
power plant, 24, 37, 265–270, 272–274, 276, 277, 279
Regulatory Agency (NRA), 277
Risk Governance, 23, 116, 240, 263–280
Safety Commission of Japan (NSC), 267, 274, 275
Nuclear and Industrial Safety Agency (NISA), 267, 268, 273, 274, 277

O
Ondoy Tropical Storm, 150
One-child policy, 246
Orissa supercyclone, 128, 141
Outstrip environmental learning, 118

P
Padang, 46, 49, 50, 55, 57
Pantal River, 154, 157
Participation
community, 184
disaster governance, 148

Pasig-Marikina River Basin, 181
Patogcaoen River, 154
Pearl River Delta, 19, 32, 46
Permanent residential registration system, 212
Phailin cyclone, 141
Philippine Atmospheric Geophysical and Astronomical Services Administrations (PAGASA), 164, 165, 168, 180, 182
Philippine Disaster Risk Reduction and Management Act, 151, 183
Philippine Institute of Volcanology and Seismology (PHIVOLCS), 168, 182
Philippines Republic Act (RA), 151, 164, 183
Planning
  environmental, 111
  natural disaster, 3
  participatory, 29, 38
  relational, 38
  riparian, 37
  theory, 4
  urban, 146, 187, 228, 232
Politics
  global, 5
  local, 5
  regional, 5
Population
  rural, 7, 131, 250, 177231
  urban, 2, 4, 6–8, 10, 19, 20, 26, 27, 37, 112, 124, 131, 146, 162, 172, 178, 218, 250, 269
Porong, 61–81
Ports, 6, 24, 26, 46, 47, 52, 53, 56, 57, 136
Post-Disaster
  experiences, 150
  models, 237–259
  planning, 227
  recovery, 104, 128, 230
Post-Tsunami Operational Management Structure (P-TOMS), 100, 101, 103
Power Reactor and Nuclear Fuel Development Corporation (PNC), 266
Preparedness, 4, 8, 24–26, 28, 30, 33, 47, 48, 117, 122, 128–130, 132, 133, 135–142, 148–150, 153, 157, 164, 170, 172, 173, 181, 186–190, 214, 219, 222, 226, 229–232, 240–242, 265, 279
Prevention, disaster, 29, 200, 231, 266, 279, 280
Private sector, 27, 38, 117, 123, 139, 191, 196, 205, 206, 215, 225, 226, 232, 242, 248, 250, 252
Project Implementation Agency (PIA), 133, 135, 137
P-TOMS. *See* Post-Tsunami Operational Management Structure (P-TOMS)
Public-private partnership (PPP), 225, 226, 228
Public services, 15, 21, 157, 212, 218, 226, 227, 229, 231
Public spaces, 22, 72, 74, 160
Puducherry Decision Support System (PDSS), 134

Q

Quezon city, 180, 186, 189

R

Radiation, 266, 268, 271, 280
Reconstruction, 10, 38, 52, 90–94, 97–99, 101–103, 132, 181, 199, 213, 224, 227–230, 237–259, 279
Recovery
  dynamics, 87
  physical, 37
  planning, 104, 214, 219, 221, 227, 228, 232
  post, 128
Red cross, 132, 162, 166, 224, 228
Reforms, 8, 9, 18, 33, 184, 229, 268
Refugees, 56, 79, 99
Regulatory agencies, 269, 273
Rehabilitation, 91, 100, 128, 133, 138, 148, 184, 198, 240, 248
Relief, 18, 29, 86, 90–92, 97–101, 103, 132, 133, 138, 166, 167, 180, 185, 201, 221, 222, 224, 226, 229, 238, 240, 241, 244–246, 249, 251, 253, 254, 256, 258, 266, 279
Relief and Rehabilitation Commissioner (RRC), 133
Relocation, 2, 30, 57, 78, 187, 190, 191, 223, 224, 226, 227, 230, 245
Rescue period, 223, 224
Research Institute, Japan (JAERI), 266
Resilience, 6–10, 18, 19, 21, 26, 27, 29, 30, 33, 37, 38, 61–81, 110–113, 120, 129, 133, 138, 147, 161, 162, 164, 173, 179, 182–192, 198, 199, 206, 214, 215, 229, 239, 241, 242, 257, 278–280
  social, 237
  strategies, 6
Resilient cities, 27, 112, 120
Resources
  energy, 14
  human, 132, 162, 217, 223, 229, 231
  international, 132
  local, 132
  mobilizing, 29
Richter scale, 49, 150, 157, 213, 221, 244

Riparian Regions, 6, 14, 19–21, 33–38
Risk
  analyses, 214, 221, 223, 228
  assessment, 23, 63, 64, 179, 189, 190, 215, 230, 264, 269, 277, 278
  governance, 23, 116, 240, 263–280
Rivers
  Agno River, 153, 154, 158
  deltas, 6, 7, 153, 154
  flooding, 145–174
Ruapehu Earthquake, 242
Rural, urban linkages, 24

S
Safety management, 277
Sanggar alfaz, 62, 75–79
Sea levels, 11, 18, 24, 47, 147, 153, 160
Secondary cities, 7
Seismotectonics, 49, 128
Seoul, 17, 20, 24–26, 31
Settlements informal, 2, 8, 20
Shanghai, 17, 19, 24, 25, 31, 241
Shenyang, 17
Shenzhen, 17, 25
Sichuan
  earthquake, 213
  provincial government, 213, 226, 232
Sidoarjo mud volcano disaster, 78
Sikkim earthquake, 120
Singapore, 17, 26, 51, 61, 88
Slope
  failures, 117, 119
  instability, 115
Slums, 2, 9, 20, 22, 37, 131, 185
Social
  capital, 188, 191, 198, 231, 244, 250, 251, 253, 254, 256–258
  change, 64, 70, 142
  identity, 7, 81
  inclusion, 139–141
  mobilisation, 73, 74
  services, 146, 153, 245
South Korea, 17, 19, 20
Spatial polarisation, 6, 199
Sri Lanka, 4, 6, 7, 85–104, 243
St. Andrew's Day earthquake, 55
State
  actors, 8, 138
  institutions, 129
  intervention, 27
Storms, 114, 134, 201
Subduction
  plate, 49
  zone, 45–57
Suburbs low-income, 97
Sumbawa, 51
Supply Chains, 22, 25, 26, 196
Sustainability, 31, 32

T
Tabriz earthquake, 214, 218, 220, 221, 237, 239, 240, 242–245, 247–250, 258
Tambora volcanic eruption, 50
Tamil Nadu Government, 138
Tamil Rehabilitation Organisation (TRO), 100
Taring Padi Group, 62
Tectonic
  events, 57
  frontiers, 57
  interaction, 47
  plate grinds, 49
Temples, 54, 134
Thailand, 4, 9, 17, 24, 25, 30, 53, 195–207, 249
Thailand 2011 floods, 195–207
Thai Rak Thai administration, 200, 201
Thane Cyclone, 8, 128, 131, 136–139, 141, 142
Tianjin, 17, 31
Tohoku earthquake, 46, 52, 249
Tokai Nuclear Power Plant, 266, 273
Tokyo
  Tokyo Electric Power Company (TEPCO), 265, 270, 272, 273, 275, 276
  tokyo metropolitan region, 31
Township authorities, 220, 223, 224, 227, 230
Transborder
  agreements, 35
  collaboration, 36
Transition rural to urban, 269
Transparency, 10, 37, 98, 215, 264, 269, 272–276, 278–280
Transportation, 15, 16, 23, 31, 35, 120, 121, 166, 212, 219, 222, 225, 228, 251
Trauma, 10, 53, 119, 120, 239, 240, 242, 246, 248, 252
Tripartite Corps Group (TCG), 244, 245, 251, 255
Tsunami
  reconstruction process, 97
  recovery plan, 214, 219, 221, 227, 228, 232
  response, 86, 88, 95, 101, 103, 132
  victims, 93, 99
Typhoon Ketsana, 179–182
Typhoons, 19, 46, 146, 149–151, 160, 168, 170–174, 179–182, 238
Tyranny of distance, 8, 111, 115–117

U
Unemployment, 96, 140, 206, 248
Union Territory of Pondicherry, 131
United Nations (UN), 2, 28, 112, 129, 146, 157, 177, 178, 185, 205, 214
  United Nations Global Assessment Report on Disaster Risk Reduction, 178
  United Nations International Strategy for Disaster Reduction (UNISDR), 2, 13, 14, 18, 19, 21, 27, 28, 30, 157, 162, 178, 214, 241
Uppsala Conflict Data Program, 957
Urban
  agglomerations, 6, 7, 20, 23, 37, 46, 269
  Cluster Coordinated Development Program (CCDP), 32
  Development House Act (UDHA), 186
  form, 32
  governance, 2, 9, 10, 28, 33, 110, 117–118, 195–207, 240–245
  Poor Consortium (UPC), 79
  resilience, 120
  settlement, 5, 7, 8, 21, 112, 146, 147, 149, 153, 173, 174, 215
Urbanisation, 2, 4–7, 9, 10, 47, 56, 104, 147, 173, 177–179, 184, 199
  rapid, 9, 47, 179, 199
Uttarakhand State Disaster Management Authority, 130

V
Victims, 15, 16, 66, 70–74, 77, 79–81, 93, 95, 99, 132, 191, 202, 279
Village Knowledge Centre (VKC), 135, 136
Villages, 7, 8, 18, 55, 62, 67, 68, 70, 72, 74, 76, 92, 103, 113, 118, 131–133, 137, 151–153, 159, 160, 164, 240, 245, 249–252, 254, 255, 257, 258
Villupuram, 137
Volcanic Explosivity Index (VEI), 48, 50, 51
Volcanoes, 51
Vulnerability
  assessment, 22
  compound, 15
  database, 35
  effects, 20–23

W
Warning System Early, 129, 148, 164, 165, 184, 188, 189, 205, 214, 221, 223, 232, 279
War Victims, 99
Water
  crisis, 41
  diversion, 35
  drinking, 121, 180
  governance, 197, 199, 200, 205, 206
  programme, 184
Waterways, 8, 31, 32, 146, 177–192
Weather events, 2, 11, 14, 19, 24, 31, 33, 181
Wenchuan earthquake, 9, 211–233, 240, 241, 243, 246, 250
West Bengal, 7, 111, 113–116, 119
West Java, 32, 50, 51, 56
West Sumatra, 57
World Bank, 28, 89, 91, 93, 95, 96, 104, 110, 112, 113, 133, 178, 181, 195, 196, 214, 215
World Commission on Dams (WCD), 34
Wuhan, 17

X
Xi'an, 17

Y
Yogyakarta, 50, 54, 56, 62, 65

Y
[illegible]

MIX
Papier aus verantwortungsvollen Quellen
Paper from responsible sources
FSC® C105338

If you have any concerns about our products,
you can contact us on
ProductSafety@springernature.com

In case Publisher is established outside the EU,
the EU authorized representative is:
Springer Nature Customer Service Center GmbH
Europaplatz 3, 69115 Heidelberg, Germany

Printed by Libri Plureos GmbH
in Hamburg, Germany